纺织服装高等教育"十三五"部委级规划教材

功能性纺织技术

TECHNOLOGY FOR FUNCTIONAL TEXTILES

辛斌杰 陈卓明 刘岩 著

东华大学出版社
·上海·

图书在版编目(CIP)数据

功能性纺织技术＝TECHNOLOGY FOR FUNCTIONAL TEXTILES/辛斌杰,陈卓明,刘岩著. —上海:东华大学出版社,2017.12

ISBN 978-7-5669-1284-8

Ⅰ.①功… Ⅱ.①辛… ②陈… ③刘… Ⅲ.①功能性织物-研究 Ⅳ.①TS1

中国版本图书馆 CIP 数据核字(2017)第 232704 号

本著作受上海工程技术大学学术著作出版专项资助

责任编辑:张　静
封面设计:魏依东

出　　版:东华大学出版社(上海市延安西路1882号,200051)
本 社 网 址:dhupress.dhu.edu.cn
天猫旗舰店:http://dhdx.tmall.com
营 销 中 心:021-62193056　62373056　62379558
印　　刷:句容市排印厂
开　　本:787 mm×1092 mm　1/16
印　　张:25.75
字　　数:708 千字
版　　次:2017 年 12 月第 1 版
印　　次:2017 年 12 月第 1 次印刷
书　　号:ISBN 978-7-5669-1284-8
定　　价:79.00 元

Preface

The use of functional textiles is growing at a tremendous rate in all fields of daily life and industry. An increasing number of consumers are realizing that it is necessary for textile to not only possess visual and tactile features, but also to have practical functional properties, such as anti-ultraviolet radiation, flame retardation, antistatic ability and water repellency, etc. Therefore, it is becoming considerably important to explore efficient and environmental friendly technology for the development of functional textiles.

In order to provide reliable and useful information for further study on functional textiles, this book therefore represents an attempt to collect all the relevant and up-to-date information in the field and to introduce the preparing technologies based on the knowledge gleaned from eight years of research on functional textiles, such as wet-spinning, electro-spinning, spin-coating and magnetron sputtering. Not many people would associate the manufacture of functional clothing with high technology. The wide range of coverage in this book, along with the systematically summarize and scientific research experiment involved, may consequently attract their attention. We hope, therefore, that this book may also arouse wide interest of scientists from other non-textile fields who intend to discover topics suitable for interdisciplinary collaboration and future study.

At the beginning of this book, a brief introduction on definition, classification and application of functional textile is given systematically. Different kinds of functional fibers and yarns are also introduced. Advanced finishing techniques are described, including chemical and mechanical finishing, which are the ways about imparting functions to textiles. In addition, this book focuses on the preparation of some promising fiber-forming polymer fibers by wet-spinning and electro-spinning, such as polyaniline

(PANI), Polysulfonamide (PSA) and cellulose, especially their composite strengthened by functional materials, including carbon nanotube, titanium dioxide and graphene, etc. Moreover, Al-film coated PSA and aramid, and Al/SiO_2 multi-layer film and Ag/TiO_2 film deposited on fabrics are prepared by magnetron sputtering for the fabrication of functional textiles.

We are very grateful to all the contributors of this book for their time and effort in collection and writing in each subject area. We also hope that this book can serve not only the scientific research field but also the industries on functional textile investigation and development.

Binjie Xin
June 2017, Shanghai, China

前　言

随着我国的科技进步和经济发展，功能性纺织品在人们的日常生活中和各个工业领域的应用正在以惊人的速度增长。越来越多的消费者意识到，纺织品不仅需要拥有良好的视觉和触觉效果，而且更应该具备如抗紫外线辐射、阻燃、抗静电和防水等实用功能。因此，如何通过高效环保的制备技术开发功能性纺织品，变得尤为重要。

为了能够对功能性纺织品的进一步研究提供可靠有用的信息，本书致力于整合最新的相关信息，并根据笔者所在课题组近十余年的研究成果，对功能性纺织品的制备技术进行较为系统的介绍，如湿法纺丝、静电纺丝、旋涂成膜和磁控溅射表面功能化等技术。本书涉及范围较广，包含对功能性纺织品技术的系统总结和相关科学研究试验的介绍。希望通过这些内容引起本领域内技术类研究人员的关注，也希望本书能够激发其他非纺织领域的科学家的广泛兴趣，期望他们能从中发现适合跨学科合作和未来学习的主题。

本书开篇部分，系统地介绍功能性纺织品的定义、分类和应用，并介绍不同种类的功能性纤维和纱线，描述现阶段赋予纺织品功能性的先进后整理技术，包括化学和机械整理。此外，本书着重介绍通过湿法纺丝和静电纺丝制备的具有良好应用前景的成纤聚合物纤维，如聚苯胺（PANI）、聚砜酰胺（PSA）等，以及由这些成纤聚合物纤维与功能性材料混合而成的复合材料，常用的功能性材料包括碳纳米管、二氧化钛和石墨烯等。另外，通过磁控溅射技术在聚砜酰胺和芳纶纤维表面沉积金属铝（Al）膜，以及在其他纺织材料表面沉积 Al/SiO_2 多层薄膜和 Ag/TiO_2 膜的方式，均能制备功能性纺织材料。

非常感谢这本书的所有贡献者在撰写每个章节时所耗费的时间和精力。希望这本书不仅能有益于科学研究，也能为功能性纺织品行业的发展提供一定的技术支持。

辛斌杰
2017 年 6 月，中国 上海

Contents

Chapter 1　Overview ·· 1
  1.1　Introduction ··· 1
  1.2　Definition of Functional Textiles ·· 3
  1.3　Classification of Functional Textiles ··· 3
  1.4　Classification of Finishing Technology for Functional Textiles ········· 4
  1.5　Scope of Application of Functional Textiles ································· 5
  1.6　Future of Functional Textiles ·· 8
  References ·· 9

Chapter 2　Functional Fibers ·· 10
  2.1　Introduction ··· 10
  2.2　Flame Retardant Fibers ··· 10
  2.3　UV-resistant Fibers ·· 12
      2.3.1　Classification ··· 12
      2.3.2　Preparations and Applications ·· 12
  2.4　Hygiene and Medical Fibers ·· 13
      2.4.1　Typical Hygiene and Medical Functional Fibers ············· 13
      2.4.2　Applications of Hygiene and Medical Fibers ··················· 14
  2.5　Breathable Fibers ··· 14
  2.6　Conductive Fibers ·· 15
  2.7　Shape Memory Fibers ·· 16
  2.8　Phase Change Fibers ··· 17
  References ·· 17

Chapter 3　Functional Yarns ·· 19
  3.1　Introduction ··· 19
  3.2　Characteristics of Functional Yarns ··· 19
      3.2.1　Monofilament Yarn ·· 19

  3.2.2 Multifilament Yarn ·· 19
  3.2.3 Intermingled/Commingled Yarn ···························· 20
  3.2.4 Tape Yarn ·· 20
  3.2.5 Staple Yarn ··· 20
 3.3 **Conductive Yarns** ··· 21
  3.3.1 Intrinsically Conductive Polymer Fiber/Yarn ················ 21
  3.3.2 Polymer Yarns Twisted/Embedded with Metallic Wires ············· 22
  3.3.3 Polymer Yarns with Metallic Coatings ·································· 22
 3.4 **Shape Memory Yarns** ··· 22
  3.4.1 Shape Memory Alloy Yarn ··· 23
  3.4.2 Shape Memory Polymer Yarn ··· 23
 3.5 **Antimicrobial Yarns** ·· 23
 3.6 **Auxetic Yarns** ·· 24
 3.7 **Anti-static Yarns** ·· 25
 3.8 **Soluble Yarns** ··· 25
 3.9 **Biodegradable Yarns** ··· 26
 References ··· 26

Chapter 4 Advanced Finishing Techniques ·· 28
 4.1 **Introduction** ··· 28
 4.2 **Foam Finishing** ··· 29
  4.2.1 Types of Foam ··· 29
  4.2.2 Preparation of Foam ·· 29
  4.2.3 Foam Generation ··· 31
  4.2.4 Foam Characterize ·· 32
  4.2.5 Application Systems of Foam ·· 32
  4.2.6 Practice of Foam Finishing ·· 35
 4.3 **Plasma Treatment** ·· 36
  4.3.1 Introduction ·· 36
  4.3.2 Classification of Plasma ··· 36
  4.3.3 Textile Treated by Plasma ·· 37
  4.3.4 Summary ·· 40
 4.4 **Supercritical Fluid Technology** ··· 41
  4.4.1 Introduction ·· 41
  4.4.2 Application ··· 42
 4.5 **Ultrasonic Technology** ··· 43
  4.5.1 Introduction ·· 43
  4.5.2 Application ··· 43
  4.5.3 Summary ·· 45

4.6 Sol-gel Technology for Functional Finishing — 45
4.6.1 Introduction — 45
4.6.2 Application — 46
4.7 UV Treatment — 46
4.7.1 Introduction — 46
4.7.2 UV Treatment on Wool Fabric for Anti-felting Finishing — 47
4.7.3 UV Treatment on Silk and Wool Fabric for Hydrophobic/Hydrophilic Finishing — 47
4.8 Finishing Treatment for Improving Comfort — 48
4.8.1 Thermal Regulation — 48
4.8.2 Moisture Management — 50
4.9 Enzyme — 54
4.9.1 Introduction — 54
4.9.2 Enzyme Classification — 54
4.9.3 Textile Application — 56
References — 58

Chapter 5 Wearable Smart Textiles — 62
5.1 Introduction — 62
5.2 Heat Responsive Textiles — 62
5.2.1 Insulation Textiles — 63
5.2.2 Cooling Textiles — 65
5.3 Shape Memory Textiles — 67
5.3.1 Shape Memory Alloy Textiles — 67
5.3.2 Shape Memory Polymer Textiles — 67
5.3.3 Shape Memory Gel Textiles — 68
5.4 Color-changing Textiles — 70
5.5 Waterproof Breathable Textiles — 71
5.5.1 Densely Woven Fabrics — 72
5.5.2 Membranes — 72
5.5.3 Microporous Coatings — 73
5.6 Electronic Textiles — 74
5.6.1 Introduction — 74
5.6.2 Application — 74
References — 77

Chapter 6 Nanospinning and Nanowrapping Technologies — 80
6.1 Introduction — 80
6.1.1 Fluid Movement Parameter Simulation of Wet-spinning — 81

  6.1.2 Electric Field Simulation of Electro-spinning ················ 83
 6.2 Preparation and Characterization of PSA Fiber by Wet-spinning and Electro-spinning ················ 86
  6.2.1 Experiments ················ 86
  6.2.2 Results and Discussion ················ 87
  6.2.3 Summary ················ 92
 6.3 Preparation and Characterization of Coaxial Electrostatic Spinning of PSA/PU Composite Fiber ················ 93
  6.3.1 Experiments and Simulation ················ 95
  6.3.2 Results and Discussion ················ 97
  6.3.3 Summary ················ 106
 6.4 Preparation and Characterization of Polyester Staple Yarn Wrapped with Electro-spun PSA Nanofiber ················ 106
  6.4.1 Experiments and Simulation ················ 108
  6.4.2 Results and Discussion ················ 111
  6.4.3 Summary ················ 122
 References ················ 123

Chapter 7 Development of Conductive Polymer Materials ················ 127
 7.1 Introduction ················ 127
  7.1.1 Synthesis Method of PANI ················ 128
  7.1.2 Development of Conductive PANI Fiber ················ 132
  7.1.3 Application of PANI ················ 136
 7.2 Structure of Electrochromic Device ················ 138
  7.2.1 Transparent Conductor (TC) ················ 139
  7.2.2 Electrochromic Layer (EC) ················ 139
  7.2.3 Ion Storage Layer (ISL) ················ 139
  7.2.4 Ion Conductor (IC) ················ 139
  7.2.5 Models of Electrochromic Reaction ················ 140
  7.2.6 Classification and Selection of Electrochromic Material ················ 143
  7.2.7 Application of Electrochromic Technology ················ 147
 7.3 Preparation and Characterization of PANI Emulsion ················ 150
  7.3.1 Experimental ················ 151
  7.3.2 Results and Discussion ················ 153
 7.4 Preparation of PANI/PVA Composite Fiber by Electro-spinning ················ 156
  7.4.1 Experimental ················ 157
  7.4.2 Results and Discussion ················ 160
 7.5 Study on Electrochemical Behavior of PANI Composite ················ 175
  7.5.1 Experimental ················ 175

    7.5.2 Results and Discussion 177
 7.6 Conclusions 185
 References 187

Chapter 8 Development of Graphene Based Textiles 193
 8.1 Introduction 193
    8.1.1 Introduction of Graphene 193
    8.1.2 Graphene Dip Coating Method 202
 8.2 Preparation and Characterization of Graphene/PSA Composite Membrane
    203
    8.2.1 Experimental 204
    8.2.2 Results and Discussion 207
    8.2.3 Summary 217
 8.3 Effects of Plasma Treatment on Properties of Graphene/PSA Composite Membrane 217
    8.3.1 Experimental 217
    8.3.2 Results and Discussion 219
    8.3.3 Summary 229
 8.4 Preparation and Characterization of Graphene/PSA Yarn 230
    8.4.1 Experimental 230
    8.4.2 Results and Discussion 230
    8.4.3 Summary 235
 8.5 Preparation and Characterization of Graphene/PSA Fabric 236
    8.5.1 Experimental 236
    8.5.2 Results and Discussion 236
    8.5.3 Summary 241
 References 242

Chapter 9 Sputtering Technology for Textiles 246
 9.1 Introduction 246
    9.1.1 Protective Fabric for Robot 248
    9.1.2 Heat Insulating Material 249
    9.1.3 High-temperature Protective Fabric 250
    9.1.4 Finishing of High-temperature Protective Fabric 251
 9.2 Preparation and Characterization of High-temperature Protective Fabric by Coating with Silicone 253
    9.2.1 Experimental 254
    9.2.2 Characterization 254
    9.2.3 Results and Discussion 255

9.2.4 Summary ……… 259
9.3 Study on Al-film Coated PSA via Magnetron Sputtering ……… 259
　9.3.1 Experimental ……… 259
　9.3.2 Results and Discussion ……… 262
　9.3.3 Summary ……… 267
9.4 Study on Al-film Coated Aramid via Magnetron Sputtering ……… 267
　9.4.1 Experimental ……… 268
　9.4.2 Results and Discussion ……… 270
　9.4.3 Summary ……… 276
9.5 Preparation and Characterization of Al/SiO_2 Multi-layer Film Coated via Magnetron Sputtering ……… 276
　9.5.1 Experimental ……… 276
　9.5.2 Property Test and Characterization of Film ……… 278
　9.5.3 Results and Discussion ……… 279
　9.5.4 Summary ……… 284
9.6 Preparation and Characterization of Ag/TiO_2 Film Coated PSA Fabric via Magnetron Sputtering ……… 284
　9.6.1 Experimental ……… 285
　9.6.2 Characterization ……… 287
　9.6.3 Results and Discussion ……… 288
　9.6.4 Summary ……… 292
References ……… 293

Chapter 10　Development of Functional PSA Nano-composites ……… 295
10.1 Preparation of PSA Composite Fibers by Wet-spinning ……… 295
　10.1.1 Experimental ……… 296
　10.1.2 Results and Discussions ……… 298
　10.1.3 Summary ……… 309
10.2 Preparation of PSA Fiber by Electro-spinning ……… 309
　10.2.1 Experimental ……… 310
　10.2.2 Results and Discussion ……… 311
　10.2.3 Summary ……… 323
References ……… 324

Chapter 11　Functionalization of Cellulose Based Materials ……… 327
11.1 Introduction ……… 327
　11.1.1 Study on Antibacterial Treatment for Fiber ……… 330
　11.1.2 Research Status of Electro-spinning of Natural Cellulose ……… 333

11.2 Activation and Dissolution of Natural Cellulose and Preparation of Regenerated Cellulose Membrane 341
    11.2.1 Application of Regenerated Cellulose Membrane 341
    11.2.2 Experimental 342
    11.2.3 Results and Discussion 347
11.3 Preparation of Natural Cellulose/PAN Blended Nano-fiber 357
    11.3.1 Electro-spinning Device 357
    11.3.2 Experimental 358
    11.3.3 Analysis of Difficulty in Electro-spinning 360
    11.3.4 Analysis of Effects of Single-factor 361
11.4 Characterization of Cellulose/PAN Blended Nano-fiber 371
    11.4.1 Test of Mechanical Properties 372
    11.4.2 Thermogravimetric Analysis (TG and DTG) 373
    11.4.3 Test and Analysis of Differential Scanning Calorimetry (DSC) 374
    11.4.4 Analysis of Fourier Transform Infrared Spectroscopy (FTIR) 375
    11.4.5 Test and Analysis of Dynamic Contact Angle 375
11.5 Antibacterial Treatment of Cellulose/PAN Nano-fiber 381
    11.5.1 Purpose of Antibacterial Treatment 381
    11.5.2 Process of Antibacterial Treatment 384
    11.5.3 Experimental Equipment and Reagent 386
    11.5.4 Preparation and Antibacterial Treatment of Cuprammonium 386
    11.5.5 Analysis of Experimental Results 388
11.6 Conclusions 392
References 395

Chapter 1　Overview

Binjie Xin* and Shixin Jin

1.1　Introduction

As a traditional manufacturing industry, conventional textile products of textile industry have satisfied people's basic requirements after years of development. Nevertheless, higher requirements on textile products have been put forward by customers with the improvement of living standards. In this case, conventional textile products with single variety and less function become less able to fulfill the market requirements. *National Long-term Scientific and Technological Development Plan* proposed that: "The manufacturing industry could be improved and optimized by high-tech; the product grade, technical content and added value could be significantly improved; the whole technical level of manufacturing industry could be comprehensively enhanced."[1]. Functional development of textile products promotes the product level and competitiveness to a large degree, therefore, investigation and industrialization of functional textiles have been considered as one of the main tasks, which was proposed by *Textile Industry Restructuring and Revitalization Plan*[2].

People's requirements on functions of textiles have changed radically with the improvement of science and technology and continuously-improved quality of life. In this case, diverse new fiber materials and textile products have come forth continuously[3-4]. In addition, the structure of textile products undergoes profound changes as well, transferring from "economical and practical" to "functional, decorative, and hygienical". Resultantly, the functional, decorative and hygienical features instead of durability of textile products have aroused the potential consumers' interest and become the main con-

---

Corresponding Author:
Binjie Xin
School of Fashion Technology, Shanghai University of Engineering Science, Shanghai 201620, China
E-mail: xinbj@sues.edu.cn

sideration when they choose textile products[5].

The self-contained high-tech content and high added value of functional textiles have become a new economic growth point of domestic textile industry, which drives the recent advance of conventional textile industry and is considered as the leverage for increasing export and strengthening competent advantages in the international markets. Design and development of functional textiles not only satisfy the continuous increasing life demands, but also provide people with a healthier, more comfortable and more environment-friendly life[6].

Compared with the "Eleventh Five-Year Plan" period, more attention has been paid on the sustainable development of environment and economy, where more opportunities to textile industry can be presented, rapid processes on management and science and technology can be achieved by the enterprises, and unlimited substitution of products can be also enhanced continuously in the "Twelfth Five-Year Plan" period. Therefore, the "Thirteen Five-Year Plan" period is considered as a period with most opportunities for functional textiles.

The gross industrial output value of functional textile industry by the end of 2015 (nearly a trillion, expressed in RMB Yuan) increased 63% when compared to Year 2010[7-8]. The total export amount of functional textiles reached 20.74 billion (expressed in US dollars) with an average annual growth rate of 10.3%. The profit margin of above-scale enterprise was 5.8% with an increasing of 1%. 13 experimental industrial clusters had been built up, and 6 of them were newly-established. In addition, the labor productivities of the top 20 enterprises presented an average annual growth rate of 15%, and a batch of small and medium-sized enterprises characterized by "professional, proficient, unique, excellent, novel" were sprouted up[7-8]. As shown in Fig. 1-1, the total processing amount of fiber (13.4 million tons) in Year 2015 increased 46% when compared to Year 2009, which occupied 25.5% of the total processing amount of fibers in textile industry and presented an increasing of 5.5%.

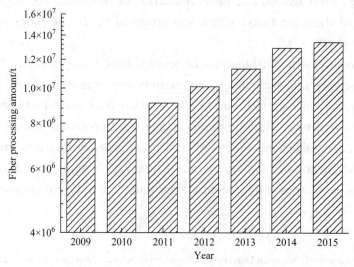

Fig. 1-1 Processing amounts of fiber in functional textiles in 2009—2015

1.2 Definition of Functional Textiles

Functionality refers that more functions than that of ordinary products can be provided except for the self-contained properties. Functional textiles refer to the textiles which are designed with one or several additional properties such as anti-bacterial, anti-acne, mildew-proofing, anti-viral, anti-mosquito, moth-proofing, flame retardant, crease-resist and no-ironing, water-repellent and oil-repellent, anti-ultraviolet, anti-electromagnetic radiation, aroma, magnet therapy, infrared physiotherapy, and negative-ion healthcare, etc. All above-mentioned additional functions contribute to adapting to other usage conditions, which not only widens the scope of application of textile product but also satisfies the customer's requirements of different levels and fields. So far, diverse functional textiles characterized by such functions from far infrared, anti-ultraviolet, no-ironing, flame retardant, anti-fouling, anti-static to micro-magnetic field, negative ion, moisture-absorption and quick-drying, moisture-absorption and heat-generating have been recognized by the market sufficiently[9].

1.3 Classification of Functional Textiles

The functionality of textile products can be classified into three categories by analyzing existing functional textiles and their standard features. Figs. 1-2~1-4 presented several representative fabrics respectively.

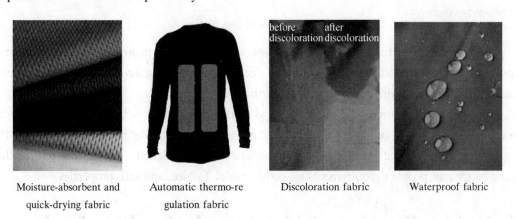

Moisture-absorbent and quick-drying fabric Automatic thermo-regulation fabric Discoloration fabric Waterproof fabric

Fig. 1-2 Comfort and aesthetics

Category 1: comfort and aesthetics, e.g., moisture penetrability, air permeability, heat resistance and moisture resistance, liquid wicking, moisture-absorbent and quick-drying, automatic thermo-regulation, coolness, aroma, no-ironing, wash-and-wear, shape memory, anti-pilling, discoloration, waterproof, oil-proof, anti-fouling, etc[10].

| Flame retardant fabric | Bullet-proof vest | Stab-resistant textile | NBC protection clothing |

Fig. 1-3 Safe and protective textiles

Category 2: safety and protection performance, e. g., flame retardant, anti-ultraviolet, thermal protection, low temperature resistant, anti-static, electromagnetic shielding, anti-alcohol-penetration, anti-blood permeability, bullet-proof, stab-resistant, NBC protection, light shielding, insect-resistant, compression resistance, etc[11].

  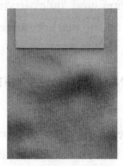 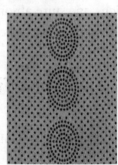

| Anti-bacterial and mildew-proofing fabric | Anti-mosquito fabric | Negative-ion fabric | Far infrared fabric |

Fig. 1-4 Health care textiles

Category 3: health care, e. g., anti-bacterial, mildew-proofing, anti-acne, anti-mosquito, deodorization, medical circular decompression, negative ion, micro-magnetic field, far infrared, formaldehyde absorption, etc[10-11].

Because of application of above-mentioned functional textiles[12], the demands of different consumer groups could be satisfied, the added value of products has been increased, and the scope of application and applicability of textiles have been widened. The textiles and clothes have better met people's requirements on comfortable, fashion and safe properties.

1.4 Classification of Finishing Technology for Functional Textiles

The functional textiles can be fabricated via weaving with functional fiber or conducting functional finishing.

Functional fiber includes: modified fiber, e. g. , such functions as anti-pilling, anti-static, hydrophilic, flame retardant can be achieved via modification in the raw material stage; such fibers as hollow fiber, special-shaped composite fiber, superfine fiber can be obtained via modification in the fiber formation stage; furthermore, far-infrared fiber (containing new ceramic ultrafine powders), insulation fiber (able to absorb the visible light and infrared ray in the solar radiation and change into heat energy, able to reflect heat radiation of human body as well), anti-ultraviolet fiber (polyester containing ceramic ultrafine powders), anti-bacterial and deodorization fiber, deodorant fiber, heat-generating fiber, magnetic fiber, etc. can be fabricated.

Functional finishing includes: comfort finishing, e. g. , air permeability, moisture penetrability, lightness, smoothness, anti-static properties, hydrophily, moisture absorption, quick drying, automatic thermoregulation, etc; wholesomeness finishing, e. g. , anti-bacterial & deodorization, etc; protection finishing, e. g. , flame retardant, anti-ultraviolet, radiation protection, etc; hygienical finishing, e. g. , cosmetology, far infrared; safekeeping finishing, e. g. , crease resistant, moth-proofing; environment-friendly finishing, i. e. , no pollution to the environment[13-14].

1.5 Scope of Application of Functional Textiles

Functional textiles have been widely applied in various fields, from basic necessities of life (food, clothing, shelter, transportation) to successful launching of manned spacecraft. The main applications can be classified into following categories: (1) textile appearance, including discoloration, bending and shaping, decorating and packing, anti-pilling, etc; (2) wearability, including warmth retention and cold protection, moisture-absorption and ventilation, anti-bacterial and mildew-proofing, crease-resistant and no-ironing, water-repellent and pollution-repellent, shrink-proof and aroma, soft and comfortable; (3) protection function, including flame retardant, anti-static, anti-ultraviolet, anti-electromagnetic radiation, thermal protection, camouflage protection, noise protection, anti-viral, bullet-proof, stab-resistant and anti-cutting, low-speed impact resistant, athletic protection, etc; (4) production function, including filtering, transmission, sealing, dust-proof, shielding, sewing, etc; (5) environment-friendly function, including ecological and environmental-protection fiber, clean textile production, ecological pre-treatment, ecological dyeing, ecological finishing, ecological textile inspection, etc[14].

The functional fibers and yarns are further introduced in **Chapter 2**, the textile composites are introduced in **Chapter 3**, and investigations of functional textiles in certain fields are introduced in **Chapters 4~11**.

Chapter 4 is about finishing technology for textiles. This part mainly introduces some advanced finishing techniques such as foam finishing, plasma treatment, supercritical fluid technology, ultrasonic technology and UV treatment. The above arrangement gives the fabric a unique performance and meets the requirements of all kinds of working conditions.

Wearable smart textiles are introduced in **Chapter 5**. This part includes heat responsive smart textiles (insulation textiles and cool textiles), shape memory textiles (shape memory alloy textiles, shape memory polymer textiles and shape memory gels functional textiles), color-changing textiles, waterproof breathable textiles (classification of waterproof breathable textiles and processing methods of waterproof breathable textiles), intelligent textiles with electronic information(information and communications, health care and medical applications, fashion, leisure and home applications, military and industrial applications).

Chapter 6 is about technologies for nanospinning and nanowrapping. In this part, comparison and analysis of polysulfonamide(PSA) molding technology are conducted and the wet-spinning and electro-spinning devices are used to prepare two series of PSA fibers under different force systems. The two kinds of fibers are characterized systematically by using SEM, FTIR, XDT and TG. The fluid motion model of wet-spinning and electrostatic field model of electro-spinning is established by using COMSOL software. The experimental results show that the PSA fiber prepared by wet-spinning has higher crystallinity and better mechanical properties and thermal stability. The electro-spun PSA fiber has smaller fiber diameter and relatively larger specific surface area. In addition, four kinds of PSA composites are prepared by using electro-spinning and the properties of the composites are systematically investigated: (1) Blended PSA/PU nano-fiber fabricated via single-axis electro-spinning; (2) PSA/PU shell-core nano-fiber fabricated via coaxial electro-spinning; (3) Blended PSA/PU yarn fabricated via electro-spinning; (4) PSA/PET wrapped-yarn fabricated via electro-spinning.

Chapter 7 is about the development of conductive polymer materials. The high-performance PSA/poly(3,4-ethylene dioxythiophene)(PEDOT) composite conductive yarn is fabricated via in-situ polymerization of PEDOT on the surface of PSA yarn, with PSA yarn as the substrate and vapor deposition polymerization as the method. The morphological analysis, chemical structure, thermal stability and electrothermal properties are investigated comprehensively by using SEM, FTIR, TG, etc.

Development of graphene based textiles is introduced in **Chapter 8**. The conductivity and anti-ultraviolet properties of PSA textiles can be improved and optimized by using the unique structure and properties of graphene, including: (1) Series of PSA/graphene composite films with different mass fraction of graphene can be fabricated via physical blending and spin coating, in addition, the chemical composition, macromolecular structure, thermal properties, mechanical properties, and anti-ultraviolet properties of the PSA/graphene composite film are characterized by using SEM, optical microscope, FTIR, XRD, surface specific resistance test, TGA, UV spectrum, etc; (2) The plasma etching to PSA film and PSA/graphene composite film is performed via plasma vapor deposition, and the samples before/after plasma treatment are investigated by using SEM, contact angle tester, moisture absorption test, etc; (3) The PSA/graphene size-impregnated yarn and PSA/graphene composite fabric are fabricated by applying the graphene-impregnated coating on the surface of PSA yarn and PSA fabric, in addition, their properties are characterized by using optical microscope, FTIR, surface specific resistance test, and TGA.

Chapter 9 is about sputtering technology for textiles. The magnetron-sputtered coating on PSA/aramid fabric is investigated, with aluminum paste as material and conventional technology as coating method. In addition, its thermal properties and physical properties are discussed. Subsequently, Al film is coated on the surface of PSA/aramid fabric for the purpose of changing the original thermal properties of the fabric. The films with different membrane structures (single-layer Al film, double-layer SiO_2/Al film, three-layer SiO_2/Al/SiO_2 film) are selected to investigate the effects of different membrane structures on the heat-insulating properties of the fabric. Finally, the double-layer Ag/TiO_2 film is fabricated on the PSA fabric via orthogonal experiment to achieve the optimum coating parameters and characterize the film.

Properties of functional PSA composites are introduced in **Chapter 10**. In this part, the preparation and characterization of PSA composites and fibers prepared by wet spinning, spin-coating and electro-spinning were introduced: (1) PSA/CNT or PSA/TiO_2 composite solutions with different mass fraction of CNT or TiO_2 were prepared by physical blending method. Corresponding composite fibers were extruded and fabricated through a set of wet spinning units and electro-spinning, and composite membranes were prepared by the spin-coating method. (2) PSA/(CNT/TiO_2) composite membranes and fibers were also prepared and characterized, whereas PSA/(CNT/TiO_2) composite spinning solutions with different mass fraction of CNT/TiO_2 were prepared by using the physical blending method before the wet spinning, electro-spinning and spin-coating. In this chapter, the thermal behavior of PSA/MWCNT (multi-wall carbon nanotube) composites was studied, and a heat conducting model was established to explain its thermal conductive behavior. The experimental results showed that the thermal behavior of MWCNT reinforced PSA composites was influenced by the mass fraction of MWCNT in the polymer matrix. MWCNT in PSA does have much influence on the thermal decomposition ratio of PSA/MWCNT composites.

Functionalization of cellulose-based materials is introduced in **Chapter 11**. Cellulose and PAN are selected as two components to fabricate the blended nano-fiber via electro-spinning, which is because, the natural celluloses fabricated via both phase transformation and electro-spinning are considered as a hotspot and difficulty in current material field. The following three items are considered when PAN is added to obtain the blended yarn: (1) PAN is characterized by excellent acid resistance, heat resistance, bacteria resistance, and it is suitable for preparing the membrane with features of ultrafiltration, nano-filtration and reverse osmosis; (2) PAN exhibits superior solubility and can be dissolved in DMAC solution together with cellulose, which increases the spinnability of cellulose spinning solution sufficiently; (3) Quite a lot of strong polar groups can be found in PAN macromolecular structure, e. g., Cyano (—CN), the complexing ability between —CN and copper ion is superior than that of —OH on cellulose macromolecular when the anti-bacterial treatment is carried out, and it presents stable chemical constitution after the anti-bacterial treatment. The advantages of natural cotton fiber and PAN can be maximized and the disadvantages of them can be minimized by the two-component fiber at nano-scale fabricated via blending of natural polymer and synthetic polymer.

1.6 Future of Functional Textiles

Domestic varieties of textiles and clothes have become increasing rich and remarkable effects on investigation, and the industrialization of functional textiles has been achieved with the development of new materials and new technologies. The development of functional textiles strengthens the level and competiveness of textile to a large extent, quite a lot researches have been done by scholars in textile industry of diverse countries for the development of new functional textiles via the high-tech as they have recognized the importance of functionality of textile in improving the added value of product[15].

The development of functional textiles has become the international trend and hotspot. The functional textile was investigated in Japan in very early years and has attracted more attention of customers. In the textile enterprises in Japan, the development and production of various functional materials have predominated and quite a lot of patents have been applied. Meanwhile, the functional textiles with features of healthcare and comfort have widely accepted by customers. Researches focusing on combustibility, anti-ultraviolet properties and anti-static properties have been done in developed countries and districts such as U. S. and Europe, which complies with local laws and regulations[16]. So far, the functional textiles with features of safety and comfort have captured significant shares in the international high-end market.

Since the 1990s, developments of various functional textiles in domestic have presented a continuously rising trend. The clothing textile, household textile and protective textile have been focused, with both after finish and fiber process being developed, resultantly, product developments have coincided with the functional applications of product. Compared with the technology-leading products manufactured by developed countries, the R & D ability of domestic functional textiles is far lower; in this case, the pace of development and innovation must be quickened.

References

[1] Liang S. National Long-term Scientific and Technology Development Plan. China Legal Publishing House, 2006.

[2] Cai Q. Industrial textiles draw a grand blueprint. Textile Apparel Weekly, 2016, 20: 36-36.

[3] Luo Y F. Novel functional fibers and their textiles. China Textile Leader, 2013, 3: 53-58.

[4] Zhang Y H, Zhang L Y. Development and application of several new fibers. China Textile Leader, 2012, 9: 51-53.

[5] Yao M. Textile industry prospect and development of detection technology. Consumer Guide, 2014, 3: 43-44.

[6] Liu L Z, Zhao Y, Liao X H, et al. Functional textiles. Textile Dyeing and Finishing Journal, 2013, 35(1): 7-11.

[7] Cao X J. Future planning and development prospect of textile industry. The Marketing of Cotton and Jute of China, 2016, 4: 14-17.

[8] Zhang J R. Discuss the application and development trends of functional textile in the clothing. Jiangsu Textile, 2014(2):51-52.

[9] Zhang Y T. Industrial Textiles. China Textile & Apparel Press, 2009:5-8.

[10] Yan X. Industrial Textiles. Donghua University Press, 2003:11-15.

[11] Wei X. Industrial Textiles Design and Production. Donghua University Press, 2009:7-11.

[12] Wei Q F. Functional Nanofibers and their Applications. Woodhead Publishing, 2012:53-58.

[13] Tian J Y. Textiles Functional Finishing. China Textile & Apparel Press, 2015:22-28.

[14] Jiang H. Development and Application of Functional Textiles. Chemical Industry Press, 2013: 16-23.

[15] Luo Y F. Recent developments of functional fibers abroad. Hi-Tech Fiber & Application, 2011: 36 (6).

[16] Cristina L, Luís D D S, Osvaldo C, et al. Functional textiles for atopic dermatitis: a systematic review and meta-analysis. Pediatric Allergy & Immunology 2013, 24(6): 603-613.

Chapter 2 Functional Fibers

Fujun Xu[*], Wei Liu and Zhuoming Chen

2.1 Introduction

Since the nineteenth century, revolutionary of science and technology changes have been occurring at an unprecedented rate in many fields, including the textile and clothing industry. Solid foundations of scientific understanding have been laid to guide the improved usage and processing technology of natural fibers and the manufacturing of synthetic fibers. The functional fibers are defined as the fibers with certain functions or special properties such as thermal conductivity, electrical conductivity, antibacterial, flame retardant and so forth. These functional fibers are widely applied to military personnel, or civilian police, firemen, and emergency responders. In this chapter, the properties, preparation methods and applications of some typical functional fibers are introduced.

2.2 Flame Retardant Fibers

Adding flame retardants during polymer processing is the most widespread and efficient method of FR protection of polymeric materials, since this method does not require new equipment and is economically efficient. Halogen-containing organic compounds and organo-phosphorous compounds are the most commonly used flame retardant additives for poly (ethylene terephthalate) (PET) fibers. Halogen compounds are inert to polyesters under thermal conditions. All phosphorous organic compounds can produce phosphor-

---

Corresponding Author:
Fujun Xu
College of Textiles, Donghua University, Shanghai 201620, China
E-mail: fjxu@dhu.edu.cn

ous-containing acids like H_3PO_4, H_3PO_3, H_3PO_2, which can cause degradation of polyester by hydrolysis, leading to reduction of mechanical stability. The melt viscosity of PET is lowered by addition of these compounds. Flame retardant polyester fibers are manufactured mostly from reactive additives or monomers containing a phosphorus component[1].

Typical flame retardant fibers include:

Aromatic polyamide fibers or aramids, which were first developed in 1960s, are one of the most commonly used polymers for developing flame retardant textiles now. These fibers find usage in manufacturing firefighter's uniforms, industrial worker's uniforms, military clothing and accessories. DuPont has developed some products, such as Kevlar® and Nomex®. Nomex® is registered trademark for flame resistant meta-aramid material. Meta-aramid fibers are characterized by long-term heat and flame resistance, and are favored for applications such as heat-resistant filters and other industrial materials and fire proof clothing for firefighters.

Melamine fibers are manufactured from fiber-forming synthetic polymer composed of at least 50% by weight of a cross-linked melamine polymer. These fibers can be operated at high temperatures; they have high LOI (limited oxgen index) values, and their typical target markets are hot gas filtration and safety and protective apparel. The major application of melamine fibers is in areas of bedding industry, upholstery manufacturing of seats for automobiles and aircrafts, protective clothing for firefighters and industrial workers, and in the filtration industry involving high temperature applications.

Polybenzimidazole (PBI) is an organic fiber with excellent thermal resistant and a good hand. The federal Trade Commission definition for PBI fiber is a manufactured fiber in which the fiber-forming substance is a long chain aromatic polymer having recurring imidazole groups as an integral part of the polymer chain. PBI does not burn in the air and does not melt or drip. The high LOI value coupled with its good chemical resistance and good moisture regain makes PBI an excellent fiber for fire blocking end uses such as safety and protective clothing and flame retardant fabric.

A silica-containing example like VISIL® is a specialty cellulosic/viscose fiber for flame retardant applications. It is permanently fire resistant because of the high polysilicic acid complex content (30% ~ 33%) built into the fiber in the manufacturing process. A fire-blocking barrier made by using VISIL® fibers can be the fire blocking layer in complete seating units. The use of this inherently fire resistant fiber, either alone or in blends, reduces the need for flame retardants in upholstery fabric of foam. It does not melt or flow when it is in contact with heat or flame, and emits essentially no smoke or toxic fumes[1-3].

2.3 UV-resistant Fibers

In the recent years, consumers have become increasingly aware of the need for sun protection, which is related to the incidence of sun induced skin damage and its relationship with increased exposures to UV light. UV radiation can lead to acute and chronic reactions and damages, such as acceleration of skin ageing and sunburn. Textile and clothing are the most suitable interface between environment and human body. They can reflect, absorb and scatter solar wave, but in most cases, they do not provide full sun screening properties. UV protection ability highly depends on large number of factors such as type of fiber, fabric surface and construction, type and concentration of dyestuff, fluorescent whitening agent (FWA), UV-B protective agent, as well as nanoparticles if applied[4].

2.3.1 Classification

(1) Fibers naturally with UV-resistant property. Acrylic fiber is an excellent UV-resistant fiber; —CN group can absorb UV energy and convert the UV energy into heat loss.

(2) Fibers contained with anti-ultraviolet additives. Most of synthetic fibers have poor anti-ultraviolet ability; anti-ultraviolet additives can be blended into synthetic polymer to fabricate UV-resistant fibers. Anti-ultraviolet additives can be mainly categorized as two kinds:

Inorganic anti-ultraviolet additives: Inorganic substances, contained titanium dioxide, zinc oxide, talc, clay, calcium carbonate, can scatter or eliminate ultraviolet, which have a higher refractive index and can prevent UV radiation from getting into skin.

Organic anti-ultraviolet additives: Those organic compounds are able to absorb UV-R range of 290~360 nm, known as an ultraviolet absorber.

2.3.2 Preparations and Applications

There are many ways to prepare UV-resistant fibers by adding UV-resistant additives into composite solution; it can be summarized as the following ways: (1) UV-resistant copolymer is prepared to produce UV-resistant fibers via copolymerization with a suitable UV absorber and fiber-forming polymer monomers. (2) Organic UV absorbers or inorganic UV scattering agents are used in single or mixed way to prepare UV-resistant fibers by dipping, printing, spinning or exhaustion methods through combining the agents with natural fiber or synthetic fiber.

The clothing made by UV-resistant fibers are particularly suitable for people who work for a long time in summer, such as military, police, geologists, construction workers, etc, and they put on the developed costume to prevent UV from penetrating. Therefore, UV-resistant material is a great development prospect with a protective function of clothing materials[1-2].

2.4　Hygiene and Medical Fibers

Hygiene and medical fibers are two important kinds of high-tech fibers. They are closely related with human body: the former is mainly about health care, improving the quality of life, reducing disease, and reducing patients' pain; the latter is concerned with people's life, repairing illness and injury. The industry generally puts the two parts together and refers to them collectively as medical functional fibers. Fibers used in relation to health care and surgery must be non-toxic, non-allergenic, and non-carcinogenic, and must be able to be sterilized without imparting any change in their physical or chemical characteristics. The fiber with medical application characteristics itself is called the medical functional fiber[2].

2.4.1　Typical Hygiene and Medical Functional Fibers

Antibacterial synthetic fibers can be produced by bonding silver or a combination of silver and either copper or zinc ions with ion-exchange group of the fiber. The bonding does not produce a chemical reaction between the metal and the fiber and facilitates control of the ion to be released on the surface of fiber to provide the antibacterial effect. Polyester, acrylic, and polyamide fibers were made antibacterial by treating them with a copper compound containing a borate, a carbonate, or a mixture of the two.

Superabsorbent polymers (SAPs) are employed to produce superabsorbent (SA) fibers. Superabsorbent polymers are normally not used alone but are combined with other materials, such as a liquid-absorbent component. Compared with particulate products, superabsorbent fibers show many advantages, such as a high surface area, fast absorption, flexible handle, and ease of formation of soft products with different shapes to fit the surface of wound or body.

Alginate fiber is such an example where naturally occurring, high molecular weight carbohydrates or polysaccharides obtained from seaweeds have found uses in medical textiles. Chemically, alginate is a copolymer made from α-L-guluronic acid and β-D-mannuronic acid. It is made into fibers by extruding sodium alginate into a calcium chloride bath where calcium alginate filaments precipitate. Upon contact with wound fluid, these fibers are partially converted to a water-soluble sodium alginate that swells to form a gel around wound, thus keeping the wound moist during healing period.

Chitin is another polysaccharide which, after cellulose, is the most abundantly available natural polymer. It is found in outer shells of shrimps and crabs in combination with protein and minerals. Medicinal and medical use of this polymer has been realized seriously since the early 1970s. High purity chitin, from which protein, heavy metals and pyrogens have been removed, can be used for a range of applications from food additives for controlling cholesterol levels in blood to artificial skins into which tissues can safely grow[1, 5-7].

2.4.2　Applications of Hygiene and Medical Fibers

Textile materials and products that have been engineered to meet particular needs are suitable for any medical and surgical application where a combination of strength, flexibility, and sometimes moisture and air permeability are required. The number of applications is huge and diverse, ranging from a single thread suture to the complex composite structures for bone replacement, and from the simple cleaning wipe to advanced barrier fabrics used in operating rooms.

Textile materials used in the operating theatre include surgeon's gowns, caps and masks, patient drapes, and cover cloths of various sizes. Surgical gowns should act as a barrier to prevent the release of pollutant particles into the air. Surgical masks consist of a very fine middle layer of extra fine glass fibers or synthetic microfibers covered on both sides by either an acrylic bonded parallel-laid or wet-laid nonwoven. Disposable surgical caps are usually parallel-laid or spun-laid nonwoven materials based on cellulosic fibers. Surgical drapes and cover cloths are used in the operating theatre either to cover the patient (drapes) or to cover working areas around the patient (cover cloths). The second category of textile materials used for healthcare and hygiene products are those commonly used on hospital wards for care and hygiene of patients and include bedding, clothing, mattress cover, incontinence product, cloth and wipe[6-7].

2.5　Breathable Fibers

Water vapor transfer is one of the key physical properties of fabrics in today's active wear, as the loss of water vapor in clothing is fundamental for the heat balance of the body and comfort. A theoretical study of heat and moisture exchange of liquid water with water vapor, moisture diffusion into fiber and liquid water and vapor transfer within the inter-fiber void space is presented, together with a study of water vapor transfer from different aspects or areas.

Each fiber in a fibrous media continually exchanges heat and moisture with the air in the microclimate immediately surrounding it. In addition, there will be radiation heat exchanges with other fibers and other surfaces. When there is a temperature difference between a fiber and the air in the surrounding microclimate, a net heat flow results; this exchange is generally well understood, at least in principle. Similarly, if there is a difference between the water vapor pressure at the fiber surface and the water vapor pressure in the air in the surrounding microclimate, there will be a net exchange of moisture. For a given fibrous material, the vapor pressure at the surface depends upon the amount of moisture adsorbed onto that surface and the temperature of the fiber. The amount of moisture on the fiber is not limited by adsorption, however. When the fiber becomes saturated with respect to the adsorption state, i. e. it has adsorbed as much moisture as it can, additional moisture may condense as a liquid onto the surface of the fiber. Depending on the nature of the fibrous media, large amounts of water condensate may be held on the surface of the fiber[2].

In recent years, design engineers of synthetic fibers have taken a leap in realizing and utilizing fiber cross sectional shape as a powerful functional parameter. For example, the C-slit cross sectional shape was developed to entrap the air for thermal insulation, while simultaneously improving elastic behavior. The C-shaped sheath originally has an alkali-soluble polymer core reaching the external surface through a narrow longitudinal slit. After drawing, the filaments can be textured by false twisting or commingling. The core is then removed using alkali finishing, giving a hollow C-shaped cross section with a longitudinal slit. The combination of texturing and cross sectional shape provides void fraction exceeding 30% and a springy feeling. Different fibers including polyester or polyamide based can be used. In another derivative, the sheath can be made of a blend of polyester and hydrophilic polymer. In this case, alkali finishing introduces micro-crazes and pores throughout the sheath, which allows liquid sweat absorption and transportation from the skin to the hollow core[8].

2.6 Conductive Fibers

Conductive fiber consists of a non-conductive or less conductive substrate, which is then either coated or embedded with electrically conductive elements, often carbon, nickel, copper, gold, silver, or titanium. Substrates typically include cotton, polyester, nylon, and stainless steel to high performance fibers such as aramid and PBO. Electrically conducting fibers can be used for anti-static, anti-microbial, anti-odour, shielding and other applications.

The development of conductive or semi-conductive materials, which are flexible and soft, is essential for fiber-based wearable electronics because of their unique electronic, chemical and mechanical properties. Materials like conductive polymers, metals and metal oxide nanoparticles/nanowires, carbon based micro/nano materials, such as carbon particles (CPs), carbon nanotubes (CNTs), carbon fiber and graphene, have been used and investigated. These materials are promising for a variety of applications including flexible optical and electronic devices, and chemical and biological sensors.

Conductive polymers or, more precisely, intrinsically conducting polymers (ICPs) are organic polymers that conduct electricity. Such compounds may have metallic conductivity or can be semiconductors. The biggest advantage of conductive polymers is their process ability, mainly by dispersion. Conductive polymers are generally not thermoplastics, i.e., they are not thermoformable. However, they are organic materials like insulating polymers. They can offer high electrical conductivity but do not show similar mechanical properties to other commercially available polymers. The electrical properties can be fine-tuned using the method of organic synthesis and by advanced dispersion techniques. Additionally, manufacturing techniques have been developed that embed carbon onto the surface of the fiber. Some companies use a suffusion to incorporate carbon particles into nylon or polyester. This process chemically saturates the outer skin of the fiber with carbon particles and makes them an integral part of the structure of the fiber. This characteristic makes them both highly conductive and durable. The durability makes them useful in mechanical, chemical and thermal applications[2, 9-11].

2.7 Shape Memory Fibers

Shape memory polymers can be processed in fiber form. Several methods, i.e., wet-spun, melt-spun, dry-spun, reaction, and electro-spun, are employed for their production to fiber form as other synthetic fibers do. In addition, shape memory polymer profile fibers such as hollow fibers were also achieved by melt spinning process. In comparison with other forms of smart polymers, fibers exhibit higher molecular orientation with lower shape fixity, higher shape recovery, and higher recovery stress due to stress-induced orientation during their production. Shape memory polymer fibers (SMPFs) in fibrous form allow one to design and control its mechanical and shape memory properties like elongation, switch temperature, shape fixity, shape recovery, etc. SMPFs can be knitted directly or with other common fibers and yarns such as cotton, polyester, and nylon. It is now known that shape fixity and shape recovery can be designed from above 95% to 20 %. It is also possible to obtain required transition temperature for the specific applications based on various recipes[2].

Their applications in everyday used textiles include long sleeve shirts where the sleeves shorten with increasing temperature (i.e., making the wearer cooler) or reduce the need to iron clothes after washing when a breeze of dry air from dryers or the like could trigger the shape memory effect and remove creases. More precisely engineered clothings include body suits intended for sailors and deep sea divers when body temperature could be maintained with self-adjustable shape memory weaves/functions that would modify the thermally insulating or conducting properties. Attempts have also been made to produce bicomponent fibers with shape memory effects whose individual fibers in the given structure curl up based on their different thermal expansions, thereby regulating the passage of the air and thus controlling the temperature. In medical applications, upon application of the trigger mechanism, shape memory sutures would tighten the stitches where inserting sutures in hard-to-reach places cannot provide adequate tension to close cuts and wounds[1, 12-13].

2.8 Phase Change Fibers

Phase change materials (PCMs) are a type of important functional materials. They react immediately to environmental temperature and body temperature change by releasing/storing heat without changing their own temperature. The first and most immediate application of PCMs is in textile. Phase change textiles can regulate and keep human body at a constant and comfortable temperature upon temperature fluctuation by absorbing or releasing heat. PCMs change from one physical state (solid or liquid or gas) to another physical state without significant change in temperature. During this phase changing process, PCMs release or store a large amount of heat. In nature, four fundamental phases of matter are commonly observed: solid, liquid, gas, and plasma. Several other states of matter also exist such as Bose-Einstein condensates, neutron degenerate matter, etc. The states only occur in extreme situation such as ultracold or ultra-dense matters. Other states of matter are also possible such as quark-gluon plasma which has not been demonstrated experimentally[12, 14-15].

The temperature regulating effect of phase change textiles can keep the human body's temperature at a constant and comfortable stage. They can bring good benefits to the individual wearer on both physical and physiological aspects. Phase change textiles can be used widely in casual wear, sportswear, protective clothing, underwear, jacket, firefighter uniform, skiwear, bulletproof vest, diver's coverall, space suit, airman suit, field uniform, sailor suit, curtain, quilt, quilt cover, sleeping bag, special glove, shoe lining, building material, automotive interior, and medical and hygiene application[14].

References

[1] Horrocksand A R, Anand S C. Handbook of Technical Textiles, 2000.

[2] Horrocks R, Anand S. Handbook of Technical Textiles (2nd ed.): Technical Textile Processes, 2015.

[3] Zhang S. Horrocks A R. A review of flame retardant polypropylene fibres. Progress in Polymer

Science, 2003, 28:1517-1538.
[4] Rosace G, Migani V, Guido E, et al. Flame Retardant Finishing for Textiles, 2015.
[5] Qin Y, An Overview of Medical Textile Products, 2016.
[6] Yuan S S, Luo J, Dong Y J. Simply analysis on the nonwoven materials for hygiene and medical products. Chemical Fiber & Textile Technology, 2015.
[7] Qin Y. Medical Textile Materials, 2016.
[8] Mattila H. Intelligent Textiles and Clothing, 2006.
[9] Cork C R. 1 - Conductive fibres for electronic textiles: An overview. Electronic Textiles, 2015: 3-20.
[10] Dias T. Electronic Textiles: Smart Fabrics and Wearable Technology, 2015.
[11] Kumar L A, Vigneswaran C. Electronics in textiles and clothing: Design, products and applications. Ifip Tc8 Open Conference on Business Process Re-Engineering: Information Systems Opportunities and Challenges, 2015:1-30.
[12] Tao X. Handbook of Smart Textiles. Springer Singapore, 2015.
[13] Hu J, Lu J. Shape Memory Fibers. Springer Singapore, 2015.
[14] Meng Q, Li G, Hu J. Phase Change Fibers and Assemblies, 2015.
[15] Raoux S. Phase Change Materials. Springer US, 2009.

Chapter 3 Functional Yarns

Fujun Xu[*], Wei Liu and Zhuoming Chen

3.1 Introduction

In **Chapter 2**, the functional fibers are introduced. The functional yarns are the assembly of the functional fibers, which can be designated as filament, tape, spun, core spun, plied, braided, etc. Therefore, in this chapter the structures and characteristics of functional yarns are presented first, and then some typical functional yarns are introduced as well.

3.2 Characteristics of Functional Yarns

3.2.1 Monofilament Yarn

Technical monofilament yarn consists of a single, solid filament having a diameter in the range of 100~2000 mm. The cross-sectional shapes of filaments can be varied depending upon the end uses (Fig. 3-1). It could be circular (Fig. 3-1(a)) or profiled, solid or hollow. The surface could be smooth or structured. A non-circular cross-section encourages wicking. The surface area of the fiber increases as the cross-sectional shape becomes more and more non-circular. Monofilaments have high bending rigidity and more resistance to abrasive damage. The diameter of the monofilament depends upon its application.

3.2.2 Multifilament Yarn

A multifilament yarn (Fig. 3-1(b)) is a bunch of thin continuous monofilaments of infi-

---

Corresponding Author:
Fujun Xu
College of Textiles, Donghua University, Shanghai 201620, China
E-mail: fjxu@dhu.edu.cn

nite length. The filaments are assembled together to form a coherent strand through incorporation of a nominal amount of twist known as producer's twist. The cross-sectional shape of the filament decides how closely the fibers can be brought together in the yarn. A non-circular cross-section inhibits close proximity between the fibers in the yarn, and hence bulky or voluminous yarns are produced from them. Porous polyester fibers with pore radii in the range of 5~1500 mm enable the fibers to absorb water and dry rapidly. Circular fibers promote closeness and therefore give the yarn a lean look. The yarn is smooth, compact, dense and uniform, with maximum fiber strength exploitation. Multifilament yarns are much more flexible than the equivalent monofilament yarns.

3.2.3 Intermingled/Commingled Yarn

This is essentially a filament yarn. However, instead of twist holding the fibers together, the filaments are intermingled or entangled in order to avoid their separation during processing. When filaments of the same type are entangled, the yarn is known as an intermingled yarn; when filaments of two or more types, e.g., carbon and polyester, are mingled together, the yarn is known as a commingled yarn. The yarn looks tight at the mingle points which are distributed at regular intervals along the yarn length (Fig. 3-1(c)). The mingle points hold the filaments together[1].

3.2.4 Tape Yarn

A tape yarn is basically a thin, narrow, and ribbon-like film produced from a synthetic polymeric material such as polyethylene, polyamide or polyester (Fig. 3-1(d)). A flat polymeric sheet or film is sliced into a large number of narrow tapes 20~40 mm in width with a thickness of 60~100 microns for technical applications. A tape may be further split or fibrillated (Fig. 3-1(e)) mechanically to produce a regular network of interconnected fibers which gives it a multifilament yarn-like texture.

3.2.5 Staple Yarn

Spun or staple yarns are linear assemblies of short discontinuous fibers. Synthetic or natural fibers (cotton, wool, jute, coir, etc) are used as raw material for these yarns (Fig. 3-1(f)). When synthetic fibers are used, they are cut into shorter length (staples) so as to make them compatible in physical dimension to their natural fiber counterparts. This makes synthetic fibers processable on machines that were designed primarily for natural fibers. Furthermore, it makes them suitable for blending if necessary. The fibers are held together by twist or fiber/filament wrapping. Hence, the magnitude of the twist or wraps is the most important parameter for these yarns. The yarn surface shows a helical arrangement of fibers with a lot of project indents.

Core spun yarn: Core spun yarns have a distinct core and sheath fiber assembly (Fig.3-1(g)). This can be either elastic or non-elastic. Non-elastic core: A filament (mono or multi) is placed at the core of the yarn and wrapped by staple fibers. Such a combination leaves the opportunity to select appropriate fibers for the core and sheath to suit a specific application[2].

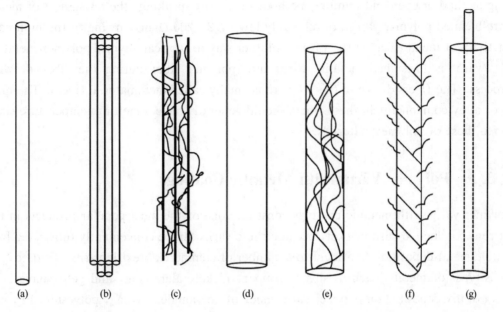

Fig.3-1 Schematic diagram of various yarns: (a) Monofilament yarn; (b) Multifilament yarn; (c) Commingled yarn; (d) Tape yarn; (e) Fibrillated tape yarn; (f) Spun yarn; (g) Core spun yarn[2]

3.3 Conductive Yarns

3.3.1 Intrinsically Conductive Polymer Fiber/Yarn

Intrinsically conductive polymers (ICPs), also known as conjugated polymers, are organic polymers that can conduct electricity. ICPs have been under intense investigation during the latest 50 years; however, fibers or yarns based on the pure ICPs were not reported until the 1980s, because it was believed that ICPs were intractable from a conventional polymer processing viewpoint. In fact, the majority of ICPs are non-thermoplastic materials that decompose at a temperature lower than their melting point, such that these ICPs cannot be melt-processed. Therefore, it is necessary to resort to a solution-spinning method for fabrication of ICP fibers/yarns.

3.3.2 Polymer Yarns Twisted/Embedded with Metallic Wires

One of the simplest methods of fabricating polymer conductive yarns is to blend metal filaments or wires directly into traditional textile polymer yarns. Such conductive yarns can be readily fabricated by conventional yarn spinning techniques such as the ring spinning method or open-end spinning method. Generally speaking, the resistances of metal-wire-blended polymer yarns are relatively low (0.2~200 Ω/m), owing to the high conductivity of metal fibers in the yarns. Another way to fabricate hybrid polymer-metal fibers/yarns is to embed metallic wires into polymer fiber during the fiber drawing process. The fibers drawn in this way are normally called metamaterial fibers. The metal or alloy components in these fibers should generally have a melting temperature similar to those of polymer materials.

3.3.3 Polymer Yarns with Metallic Coatings

Polymer yarns with metallic coatings constitute one of the most popular products in the current conductive yarn market, because these yarns can be conveniently fabricated by a simple metal-deposition treatment to a number of commercial textile yarns. To date, various types of metals, such as silver, copper, nickel, aluminum, and gold, have been successfully deposited on polymer yarns made of polyamide (nylon), polyester, PP, and so on. A variety of coating techniques have been suggested and commercially used, including polymer-metal lamination, physical vapor deposition, metallic-paint brushing, and electroless plating[1, 3-4].

3.4 Shape Memory Yarns

Shape memory materials are those materials that have the ability to "memorize" a macroscopic (permanent) shape, be manipulated and "fixed" to a temporary and dormant shape under specific conditions of temperature and stress, and then later relax to the original, stress-free condition under thermal, electrical or environmental command. Shape memory textiles are a wonderful innovation that offers great opportunities for smart products. They are having a significant impact in textile, clothing and other industries like defense and aerospace. These "intelligent" textiles have the capability of remembering their original shape. No matter what happened to them during their process of change, such as washing or steam treatment, they can still recover their original shape or state under suitable conditions. Shape memory textile is often divided into shape memory alloy (SMA) fiber/yarn and shape memory polymer (SMP) fiber/yarn.

The shape memory effect of SMA fiber stems from the existence in such materials of two stable crystal structures: a high temperature-favored "austeniti" phase and a low temperature-favored (and "yieldable") "martensitic" phase. Deformations of the low temperature phase, occurring above a critical stress, are recovered completely during the solid-solid transformation to the high temperature phase[5-6].

3.4.1 Shape Memory Alloy Yarn

Shape memory alloy fibers such as nickel-titanium (NiTi) can be incorporated into textiles such as polyester, acrylic, cotton, etc, during textile finishing, fiber and film making. Italy Luotaliyani designed "lazy shirt" fabric combining nickel, titanium and nylon fiber, these shape memory alloy fibers having shape memory function. In hot weather the wearer can roll up the sleeve from wrist to elbow; when the temperature drops and the sleeve is unrolled, it automatically returns to its original shape. The clothing also has super wrinkle-free capacity, regardless of massive pressures, and can return to its original status in 30 s[7]. Despite the demonstrated merits, SMA fibers also show some downsides that limit their applications, such as limited recoverable strains of less than 8%, inherently high stiffness, high cost, a comparatively inflexible transition temperature, and demanding processing and training conditions.

3.4.2 Shape Memory Polymer Yarn

Compared with shape memory alloys, polymeric shape memory materials possess the advantages of high elastic deformation (strain up to more than 200% for most of the materials), low cost, low density, and potential biocompatibility and biodegradability. They also have a broad range of application temperatures that can be tailored, have tunable stiffness, andante easily processed. These two materials (polymers and metal alloys) also possess distinct applications due to their intrinsic differences in mechanical, viscoelastic and optical properties. Shape memory yarns are likely to be increasingly significant in future for sutures and stents. For instance, shape memory implants can be brought into the body in a compressed temporary shape, through a small incision. A suitably constructed implant, on reaching body temperature, would then change to its "remembered" permanent shape. These materials can also be biodegradable, so that repeat surgery for removal of the implant would not be required.

3.5 Antimicrobial Yarns

Micro-organisms or microbes are microscopic organisms that are usually too small to be seen by the naked eye. Microbes are very diverse, and they include a variety of micro-organisms like bacteria, fungi, algae and viruses. Textile fibers and the structure and chemical processes of textile substrates provide room for the growth of micro-organisms, especially in suitable conditions of humidity and the warm environment in contact with the human body. The growth of microbes on textiles during use and storage negatively affects the textiles and causes potential health risks to the wearer. Though the use of antimicrobials has been known for decades, it is only in the last couple of years that several attempts have been made to manufacture antimicrobial textiles, especially by finishing. An antimicrobial finish is a recent innovation in finishes. It prevents the growth of bacteria, and the products finished with it have been proved to be environmentally friendly and health protecting, preventing diseases.

The antimicrobial agents can be introduced during fiber formation, or deposited during the finishing processes of yarns or textile fabrics by exhaust, pad-dry-cure, coating, spray and foam techniques. The bioactive substance deposited on the surface of textiles may be easily washed out, and may also disturb the fabric's usability and comfort. The antimicrobial yarn can be produced by incorporating biologically active substances into the yarns, such as: internal antimicrobial release, a viable option for synthetic fibers, where antimicrobials can be incorporated into the fibers when they are spun; surface application, applicable to all fibers; chemical bonding, the best way to achieve durability.

Textile fibers with built-in antimicrobial properties will serve the purpose alone or in blends with other fibers. Bioactive fiber is a modified form of the finish which includes chemotherapeutics in its structure, i.e., synthetic drugs of bactericidal and fungicidal qualities. These fibers are used not only in medicine and health prophylaxis applications but also for manufacturing textile products in daily use and technical textiles. The field of application of bioactive fibers includes sanitary materials, dressing materials, surgical threads, materials for filtration of gases and liquids, air conditioning and ventilation, constructional materials, and special materials for the food, pharmaceutical, footwear, clothing and automotive industries, etc[6, 8-9].

3.6 Auxetic Yarns

A recent exciting development in textile technology is auxetic fibers that exhibit the unusual property of getting fatter when stretched and narrower when compressed. In fact, these fibers, in contrast to conventional fibers, swell on stretching with consequent increase in their internal void volume. The processing of making auxetic multifilament yarn from polymers such as polytetrafluoroethylene, polypropylene and nylon into knitted and woven textile constructions has been demonstrated by Alderson in Bolton University, UK[10].

One application for auxetic fabrics is in wound bandages that contain a wound-healing agent. As the infected wound swells, so does the auxetic bandage. The internal voids in the bandage expand and release the wound healing agent. Once the wound starts to heal, the swelling goes down, the bandage contracts, and release of the wound-healing agent ceases. Thus, the auxetic fibers provide a means of controlled drug delivery. Auxetic fabrics are also envisaged in compression bandages and arterial prostheses. Auxetic fabrics are particularly desirable in applications that require highly curved hard surfaces, such as those found in the body parts of aircraft and cars. Auxetic textiles are extensively used also for making personal protection clothing, filtration, mechanical lungs, ropes, cords and fishnets, fibrous seals, etc[11].

3.7 Anti-static Yarns

Anti-static yarn is designed to make anti-static textiles to allow the discharge of accumulated static electricity from a person's skin as well as acting as a barrier against electromagnetic radiation. Anti-static textile is required to prevent damage to electrical components or to prevent fires and explosions when working with flammable liquids and gases. If not controlled, static electricity can cause product damage and lead to machinery downtime, lost man hours, returned products and warranty costs, particularly in the semiconductor and electronics industry.

Anti-static properties can be imparted into textiles by either using anti-static yarn by combining the conductive fibers with natural and/or synthetic fibers like cotton, polyester, etc., or enhancing the electro-conductivity of yarns by coating with metal or by conductive polymers such as carbon, polypyrrole, polyaniline, etc. The characteristics of the fabric include an associated earth connection based on textile fibers and conductors and being sufficiently long for the fabric to be electrically connected to the ground, thereby generating a discharge. Such anti-static fabrics are used in cleanroom garments, workwear, carpets, sewing threads, wool garments, brushes, blankets, wrist-straps, anti-static shoes, etc[6].

3.8 Soluble Yarns

Soluble fiber is a newly developed environmental protection fiber. They possess the property of dissolving in water at a particular temperature depending on their composition. Water-soluble PVA fiber has a unique specific property in that it is soluble in hot water below 93 ℃, and it has high strength, 1.5 to 3 times that of cotton. Water-soluble fiber degrades naturally in the soil. Blended yarns can be produced with cotton or other fibers and made into cloth. Since the water-soluble fiber in fabric can be dissolved, the style and grade of the fabric will certainly be greatly improved. After it is processed to water-soluble non-woven fabric, it can be used as a disposable cloth such as a computer embroidery cloth, medical clothing, etc[6].

Water-soluble garments are sterile hygienic materials that are used to help protect patients and medical staff from dangers of infections to which they are exposed in hospitals. Specific uses include surgical garment and drape, facemask and shoe cover. Other than these, different kinds of water-soluble non-woven fabrics are used as backing or bearing on embroidery fabrics, and dissolve totally in hot water after the embroidery is finished. Water-soluble high-strength fiber can be spun into high-strength yarn for production of high quality and recoverable sacks for cement or fodder, etc.

3.9 Biodegradable Yarns

Biodegradable polymers are classified according to their synthesis method, chemical composition, processing method, application, etc. Biodegradable polymers can be divided into two types: those from natural origins and those from mineral origins. The subgroups of the former include polysaccharides (e.g., starch, cellulose), proteins (e.g., casein, silk and wool), polyesters produced by microorganisms or plants (e.g., polyhydroxylalcanoates, poly-3-hydroxybutyrate), and polyesters synthesized from bio-derived monomers (e.g., polylactic acid). There is another sub-class of biopolymers, derived from mineral origins, consisting of aliphatic polyesters (e.g., polyglycolic acid, polycaprolactone), aromatic polyesters (e.g., polybutylenesuccinate terephthalate) and polyvinyl alcohols[6].

Biodegradable fibers and yarns have applications in non-wovens, fabrics, bedclothes, wipes, wet tissues, medical items, interlinings, etc. Hydrolytically degradable polymers are generally preferred for implants due to their minimal site-to-site and patient-to-patient variations compared to enzymatically degradable polymers. Major applications for PLA yarn and non-wovens include clothing and furnishings such as drapes, upholstery and covers. Some exciting potential applications include household and industrial wipes, nappies, feminine hygiene products, disposable garments, and UV-resistant fabrics for exterioruse (awnings, ground cover, etc) among others. PLA polymers, which have shown much promise, could be used in a number of unexplored applications by replacing the conventional polymers, where they can contribute a significant role in the form of composites, copolymers and blends for different applications[12-13].

References

[1] Mattila H. Intelligent Textiles and Clothing, Cambridge: Woodhead Publishing, 2006: 1-4.

[2] Chattopadhyay R. 1-Introduction: Types of technical textile yarn. Technical Textile Yarns, 2010: 3-55.

[3] Watson D L. Electrically Conductive Yarn, 1999.

[4] Cork C R. 1-Conductive fibres for electronic textiles: An overview. Electronic Textiles, 2015: 3-20.

[5] Liu C, Qin H, Mather P T. Review of progress in shape memory polymers. Journal of Materials Chemistry, 2007, 17(16): 1543-1558.

[6] Horrocks R, Anand S. Handbook of Technical Textiles (2nd ed.): Technical Textile Processes, 2015.

[7] Yan L, et al. Shape memory behavior of SMPU knitted fabric. Journal of Zhejiang University-SCIENCE A, 2007, 8(5): 830-834.

[8] Horrocks A R, Anand S C. Handbook of Technical Textiles, 2000.

[9] Uddin A J. 9-Novel technical textile yarns. Technical Textile Yarns, 2010: 259-297.

[10] Alderson K L. et al. Auxetic polypropylene fibres. Part 1-Manufacture and characterisation. Plastics Rubber & Composites, 2013, 31(8): 344-349.

[11] Sloan M R, Wright J R, Evans K E. The helical auxetic yarn-A novel structure for composites and textiles: Geometry, manufacture and mechanical properties. Mechanics of Materials, 2011, 43(9): 476-486.

[12] Zupin Z, Dimitrovski K. Mechanical properties of fabrics made from cotton and biodegradable yarns bamboo, SPF, PLA in weft. Woven Fabric Engineering, 2010: 25-46.

[13] Netravali A N. Biodegradable natural fiber composites. Biodegrable & Sustaintable Fibres. Cambrige: Woodhead Publishing, 2005: 271-309.

Chapter 4 Advanced Finishing Techniques

Yan Liu* and Yaqian Xiao

4.1 Introduction

Technical textiles are usually manufactured not only for aesthetic purposes, but also because its functions can benefit for most of fields. Different physical and chemical treatments could be used to impart the fibers or fabrics the required functional properties, and some of the treatments are in the wet processing which involves three stages, including pretreatment, coloration and finishing. Finishing is the final step of the fabric treatment process which can add value to the fabrics. Textile finishing can be divided into two distinct types, including chemical and mechanical finishing.

Chemical finishing(namely, wet finishing) is a method which uses chemicals to achieve desired properties of fabrics. The purpose of the finishing is to change the chemical composition of the fabrics. In other words, an elemental analysis of a fabric treated with a chemical finish will be different from the same analysis done prior to the finishing[1]. The chemicals usually include anti-biometric, anti-UV, waterproof and breathable chemicals and so on.

Mechanical finishing(namely, dry finishing) mainly uses physical (especially mechanical) methods, such as wet, heat, pressure and other mechanical processes, to improve the appearance (e.g., luster, smoothness), performance (e.g., residual shrinkage), or hand feeling of the fabric. The mechanical finishing is an innovation technology to add new values to high performance textiles products.

Corresponding Author:
Yan Liu
College of Chemistry and Chemical Engineering, Shanghai University of Engineering Science, Shanghai 201620, China
E-mail: liuaynxin@hotmail.com

Although new finishing techniques are still needed to be improved, physical and chemical treatments have been widely used in textile industries, including foam finishing, plasma treatment, supercritical fluid technology, ultrasonic technology, sol-gel technology, UV treatment, comfort related finishing and biological control finishing[2].

4.2 Foam Finishing

4.2.1 Types of Foam

Foam is an agglomeration of gaseous bubbles, usually of air, dispersed in a liquid and separated from each other by thin films of liquid or lamellae[3]. Generally, foams can be solid or gaseous. Solid foams are produced from gas and solid, and are mainly used in the preparation of insulation and in upholstery and arc found in pumice stone. Gaseous foams are produced from gas and liquid, and this type is applicable to textile and classified into two types, as follows:

(1) Condensation foams are similar to the gas of the liberated beer and soft drinks which cannot be used in textile because the amount of gas liberated is not large enough to give the appropriate blow ratio needed, and the foams are unstable.

(2) Dispersion foams are produced by the introduction and mixing of gas from an external source into a liquid phase; whipped cream is a familiar example. Foamsare applied to textiles which are predominantly of the dispersion type, in which the gas phase is air and the liquid phase is water, which contains a surface-active agent (surfactant) as the foaming agent.

4.2.2 Preparation of Foam

Preparation of the liquor to be foamed requires careful attention with regard to the selection of the individual components and the way in which they are mixed. The composition of the liquor is similar to that for conventional processing, but higher concentrations of reagents are used, and the addition of other components, such as a foaming agent, a foam stabilizer, and a viscosity modifier, may be necessary for the foam system to work. The essential ingredients of foam include functional reagent, foaming agent (surfactant), viscosity modifier (thickener), foam stabilizer, emulsion polymer or copolymer, and other additives (catalyst, wetting agent, filler, pH buffer, or dyes).

Functional reagents could be applied to the foam system, such as those used for resin-finishing, softening, filling, stiffening treatment, hydrophobic, hydrophilic, oleophilic effects, flame-resistant and non-slip finishing. Usually, those foaming agents are chemical products, and they can promote generation of foam with mechanical treatment, as shown in Tab. 4-1. Generally, a wide range of surfactants can be considered as foaming agents, but there is no single surfactant that can satisfy all requirements in textile finishing. Hence, the selection of a foam surfactant should involve some prerequisites, as follows:

(1) The foam surfactant must be foamed readily.

(2) Exerting a fast and uniform wetting action.

(3) Having little or no effect on color-fastness.

(4) Be compatible with other additives in the finishing liquor.

(5) Be less sensitive to temperature changes.

(6) Be tolerant of water hardness.

(7) Causing no yellowing in white textiles.

(8) Imparting stability to the foam.

Tab. 4-1 Common Surface-active Agents

Class	Surface-active Agent
Anionic	Sodium/ammonium stearate Sodium oleate Sodium dodecyl sulphonate Sodium dodecyl benzene sulphonate
Cationic	Dodecylamine hydrochloride Hexadecyltrimethyl ammonium bromide
Non-ionic	Polyethyleneoxide Spans(sorbitan esters) Tweens
Amphoteric	Dodecylbetaine

Thickeners are used to modify the viscosity and hold up the compound particles in the feed stock, so that they remain be separate and homogeneous. Thickeners increase the viscosity of the compound and therefore slow down the drainage of the inter lamellar liquid, prolong the activities of foam. Examples of suitable thickeners are locust-bean gum, guar gum, methylcellulose, hydroxyethyl cellulose, polysaccharide, alginate, and xanthan gum. Stabilizers are products of fatty-acid alkylolamides, which can improve foam stability further with no thickeners. Typical examples are sodium polyphosphate and dodecyl[4]. Emulsion Polymers or Copolymers are colloidal aqueous dispersions used to modify the physical properties of the finished fabric and partly or completely compensate for the adverse effects of a cross-linking agent[5]. The polymer emulsion in the treating liquor acts as a carrier and binder of the ingredients to the fabric. Common types of polymer emulsion are polyacrylate and polyvinyl compounds. The selection of the polymer emulsion is mainly governed by the type of finish and the fabric properties needed.

Poly(polyvinyl chloride) is important to flame-retardant finishes because it is a chlorine donor. Some additives are also important and they can be used for promoting liquor penetration into the fabric, dispersing the fillers, and controlling pH.

4.2.3 Foam Generation

To generate foam, air has to be dispersed in water. When pure water is agitated, the entrapped air bubbles rise to the surface and burst, so that no foam is generated. But, if the water contains a surfactant, this is adsorbed on the surface of the bubbles lo form a film. As shown in Fig. 4-1, the bubbles break through the liquid-air interface, a double film is formed. The film lamellaes tend to be rigid since they are made up of three layers; two layers that form the walls of adjoining bubbles and a layer of the liquid between them. Thus the presence of the surfactant confers rigidity to the bubble walls and hence stability to the foam.

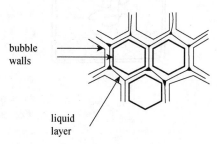

Fig. 4-1 The diagram of structure of polyhedron-foam

Foam generation needs two dynamic systems: air-blowing system and stirring system. In principle, foam generation is achieved by vigorous mechanical action, the air being injected under pressure (air-blowing)[6] or captured from the atmosphere by the turbulence of the liquid (stirring); a combination of these two methods may also be used. Foam generation by using the principle of air-blowing can he achieved under static conditions. In the static foam generator (Fig. 4-2), a pressurized air stream is introduced into the liquor to produce the foam. Alternatively, the liquor and air are fed through a mixing head containing a number of closely packed glass balls, stainless-steel shavings, or an internal series of baffles. Both air and liquor are metered into the head, where they are mixed to produce the foam. The foam produced is then pushed out by the air-liquor pressure.

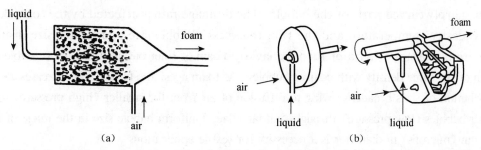

Fig. 4-2 Air-blowing foam generators: (a) Static foam generator; (b) Dynamic foam generator

A stirring generator consists of a generating tank and a stainless-steel steering shaft. The steering shaft is mounted in the middle of the tank and fitted at its lower end with at least three blades (Fig. 4-3). Fig. 4-3 (b) shows that the air is injected so that it can strike directly at the bottom of the tank and raise the bubbles, which in turn are stirred by the blades.

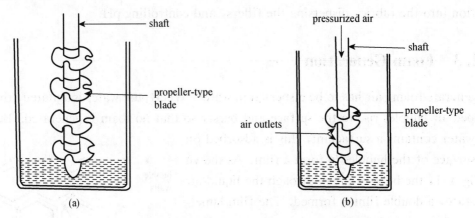

Fig. 4-3   Stirring foam generators: (a) A stringe foam generator;
(b) Combined with air-blowing foam generator

4.2.4   Foam Characterize

The performances of foam for the textile finisher are foam degree, foam stability, viscosity, wetting power, bubble size and distribution.

The foam degree is normally expressed in terms of the foam density (g/mL), which is calculated from weighing a known volume of foam. The foam stability is defined as a measure of the time that foam will maintain its initial properties. Foam viscosity is a measure of the foam stiffness or resistance to flow when the bubble walls are still in the liquid state. Textile substrates are not readily wettable by foam. The foam has to spread out on the substrate at first and then break, release the liquid to penetrate the fibers. The wetting power can be assessed by the spreading or penetration rate, which is governed by the foam-drainage rate. Liquid can drain out of the lamellae films between foam bubbles owing to gravity and surface-tension differences between the least and the most sharply curved parts of the bubble. The drainage rate is affected by the foam density, viscosity, temperature and the film thickness. Bubble size and bubble-size distribution are important criteria of foam behavior. Foam of finer bubble size is more stable than foam of the same density with coarser bubbles. As foam ages, the bubble size increases and the distribution of sizes broadens owing to diffusion of air from the smaller (high pressure) to the larger bubbles (low pressure) through thin lamellae. Uniform bubble size in the range of 50~100 μm (microns) in diameter is a necessity for textile applications.

4.2.5   Application Systems of Foam

There are several systems for applying foam to textile substrates (Tab. 4-2). These can be classified as direct (pressurized and non-pressurized) in direct systems and indirect systems[7].

Tab. 4-2 Foam Application

System	Pressure Mode	Type of Applicator	Manufacturer
Direct	Pressurized	Slot	Gaston Country Dyeing Machine Company
		Screen	Zimmer, Milter Foam System
	Non-Pressurized	Knife-on-air	Sonic Air System
		Knife-over-roller	Webcon
		Knife-over-blanket	DMfit
		Horizontal pad	Holy Moly
		Kiss roller	Morrison Textile Machinery
		Pad box	Hoechst AG
Indirect		Doctor roller	Stork
		Doctor knife on a blanket	Monforts

In the direct pressurized system, the foam held under pressure in the distribution box or manifold. Foam is applied to the fabric is done through a variable-dimensional slot applicator or through a rotary screen with the fabric pressed against a backing roller. Fig. 4-4 illustrated direct pressurized foam applicators, the slot applicator as shown in Fig. 4-4 (a) and screen applicator as shown in Fig. 4-4 (b).

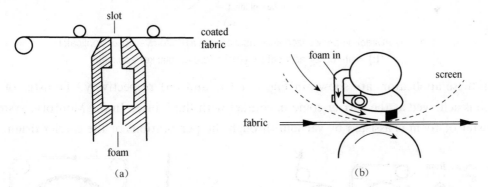

Fig. 4-4 Direct pressurized foam applicators: (a) FFT slot applicator; (b) RSF stork-brabant screen applicator

In the direct non-pressurized system, the reservoir of foam is not maintained under pressure, but this does not imply that there is no pressure involved during application. The foam is applied to the fabric by a horizontal pad or a knife-over-roller, knife-on-air, knife-over-blanket application system (Fig. 4-5, Fig. 4-6).

Indirect systems of the foam are metered by some means on a carrier before being transferred to the fabric. The carrier can be a drum (Janus) or blanket (Monforte), and

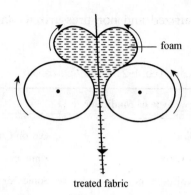

Fig. 4-5 Direction non-pressurized horizontal pad applicator

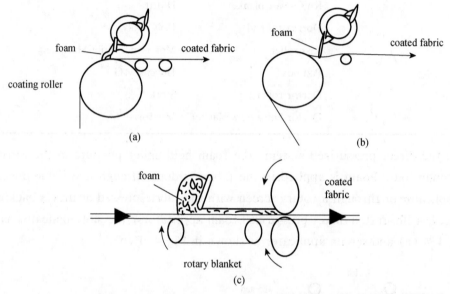

Fig. 4-6 Direction-pressurized knife applicator: (a) Knife-over-roller applicator; (b) Knife-on-air applicator; (c) Knife-over-blanket applicator

both these applicators are shown in Fig. 4-7 (a) and (b) respectively. Transfer of the foam is achieved as the carrier come in contact with the fabric. In the Monforte system, transfer of foam is assisted by vacuum through the perforations of the carrier drum.

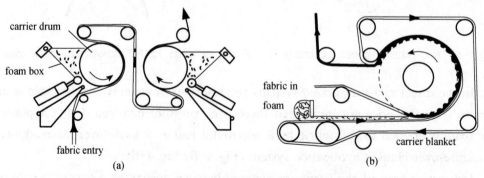

Fig. 4-7 Indirection foam applicators: (a) The Janus Mini-foam applicator; (b) The Monforts Vacu-foam applicator

Foam should be destroyed on the fabric shortly after it is applied, releasing the finishing liquor transport medium. This allows the reagents to penetrate and spread on the fabric before any fixation treatment is applied. Foam destruction is carried out mainly by the squeeze rollers of a conventional padder installed in front of a stenter, or by vacuum application, or by a combination of these methods.

Finally drying and cure fixation of foam-impregnated fabrics are usually carried out on the same traditional machinery (curing chambers and setters). As the amounts of water to be evaporated are much lower than those with a conventional pad-mangle finish-application system, then higher speeds and lower temperatures are used.

4.2.6 Practice of Foam Finishing

Foam finishing is by far the most popular among the wet processes in which foam is used. Successful foam application of most finishes has been well documented in the literature, and this will be explored next.

Elbadawi[8] studied the effects of process variables on the performance of antistatic and crease-resist finishes by using both conventional pad-mangle and foam systems. The foamed antistatic and crease-resist finishes were applied by using a knife-over-roller applicator. In the antistatic-finish experiment, the effects of the antistatic-agent concentration and the addition of polymer emulsion were investigated. The results showed that the antistatic treatment was very effective in reducing the amount of accumulated static electricity on the finished fabric. It was also found that the application of antistatic agent by using the foam system produced better results than using the conventional pad-mangle system in terms of antistatic performance

Carpet and holstery fabrics could be endowed the flame-retardancy, soil-resistant and stain-release function by foam finishing. It allows the foam to penetrate deeply into the carpet pile. The system used less energy and chemicals, so it is more environmentally friendly.

Monforts[9] described a new high-speed denim-shrinkage range based on the Vacu-foam system. The new range combines foam-finishing with twin compressive-shrinkage devices. It is claimed that the costs of processing by using the new high-speed range are no higher than those of the traditional denim-shrinkage range, and the achievable residual shrinkage factor is the same for both systems.

The wool fabrics finished by mothproofing agents (e.g., Permethrin) have been reviewed. The report highlighted the effectiveness of low-wet-pick-up applications such as the auto-foam and also dry-powder techniques in reducing the concentration residual liquor by improving the fixation of Permethrin to wool fibers. Such methods provide a reduced amount of water in the application process, enable easier disposal of effluent to be achieved, and thus improve the Permethrin eco-toxicological profile.

4.3 Plasma Treatment

4.3.1 Introduction

Plasma was first identified by William Crookes in 1879, and it was named as "plasma" by Irving Langmuir in 1928. Plasma is mainly composed of free electrons and charged ions, and it is often considered as the fourth state of matter (plasma, solid, liquid and gas). It can be used for surface modification of polymeric/textile substrates by generating polar groups with surface activation. In addition, it can improve the surface roughness and functional properties of materials by plasma etching and plasma polymerization, respectively.

4.3.2 Classification of Plasma

4.3.2.1 Hot and Cold Plasma

Plasma can be classified into hot plasma and cold plasma based on the temperature of the plasma zone. Hot plasma occurs when the temperatures of the electrons, atomic and molecular species are extremely high and remain near to the thermal equilibrium state. In that condition, molecules remain almost fully ionized (100%). So the temperature of the plasma zone is around $10^8 \sim 10^{10}$ K with an electron density more than 10^{20} m^{-3}. For the cold plasma, the electrons remain at a significantly higher temperature, however, the ions and neutral molecules remain near to ambient temperature ($T_e \geqslant T_{ion} \approx T_{gas}$ = 25–250 ℃, $T_e \approx 727$ ℃). In cold plasma, the electron density is significantly lower ($n_e \approx 10^{10}$ m^{-3}) and only a small fraction of gas molecules (0%~1%) is ionized[10]. Therefore, cold plasma is only suitable for surface modification of heat-sensitive polymeric and textile substrate.

4.3.2.2 Low and Atmospheric Pressure Plasma

Low pressure plasma is easy to ionize a gaseous molecule by electrical breakdown under a low pressure condition, and it has been extensively studied for material processing. The low pressure plasma processing has some advantages, such as providing a vast array of chemically reactive species, uniform glow plasma, low temperature (below 250 ℃) of plasma and lower breakdown voltages. However, there are some limitations of low pressure plasma processing, such as longer processing time, limit of sample size and reactor size, and mostly batch processing.

It is difficult to obtain atmospheric pressure plasma due to the presence of high voltage in a narrow electrode gap, and it is also difficult to ionize gaseous molecules and generate uniform cold plasma. However, if the stable cold plasma can be generated at atmospheric pressure, it can overcome the limitations of low pressure technology. It can be also easily integrated into existing textile processes for continuous treatment of textiles. The three major types of atmospheric pressure cold plasma commonly used in textile processing can be described as below:

(1) Dielectric barrier discharge is produced by an arrangement consisting of two parallel flat electrodes. The textile sample is kept between the two electrodes having a gap of 1mm to more, and gas or a mixture of gases is then injected for uniform treatment of the textile. The DBD (dielectric-barrier discharge) plasma is used in plasma-assisted chemical vapor deposition, surface etching, surface cleaning, surface activation, and plasma polymerization.

(2) Corona discharge is the characteristic of an asymmetric electrode pair either powered by a continuous or pulsed by an AC/DC electrical supply. It is used in electrostatic precipitators in dust collection and activation of polymer/textile substrates to improve the hydrophilic property.

(3) The atmospheric pressure plasma jet (AAPJ) consists of two concentric electrodes, through which a flow of a mixture of gases is supplied. The discharge is ignited by a RF (radio frequency) power between the outer and the inner electrodes, thus producing a high velocity of highly reactive species for the downstream processing of textiles.

Low pressure plasma has been extensively studied, but the processing technology has not been commercialized in textiles due to its inherent techno-economical limitation. On the contrary, atmospheric-pressure cold plasma can overcome this limitation and it is explored for similar applications as low-pressure plasma. Hence, it attracts an increasing attention in research and has a wide application in commercial field of textiles. Atmospheric pressure plasma is an emerging technology and several challenges associated with plasma generation for in situ plasma reactions with textile substrates are still not fully addressed. Surface modification for a desired function could be conducted by selecting appropriate plasma processing parameters. Fragmentation of a precursor followed by plasma reaction on textile substrates is the best way of surface engineering to develop high-valued apparel and home textiles.

4.3.3 Textile Treated by Plasma

Plasma treatment on textiles can modify material surface without altering the bulk properties and impart the textiles a unique function. Different value-added functionalities, such as water resistance, oil repellent, antibacterial properties, flame-retardance, ultraviolet protection and antistatic properties, can be obtained by modifying the fiber surface at a nanometer level (as shown in Fig. 4-8). The plasma processing of textiles has some advantages when compared with traditional textile processing, such as no wet chemistry involved, complex and multifunctional reaction type, highly surface specific, low energy and water consumption, environment-friendly and rapid industrial developments. The application of plasma treatment on different textiles can be described as below.

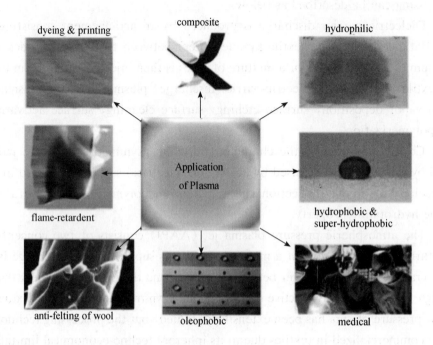

Fig. 4-8 Application of plasma for textile processing and finishing

4.3.3.1 Wool

Felting shrinkage is the most undesirable feature of wool fibers. When wool is agitated or washed in aqueous liquor, the fibers become closer to form a more compact and denser assembly due to its instinctive characteristics and directional frictional effect. Therefore, anti-felting treatment should be conducted on wool and it is usually processing in a chlorine-containing solution. However, chemical methods will produce harmful substances, such as high-temperature wastewater with acidic pH and chlorine discharges during the operation. Compared to the chemical treatment, plasma pretreatment is environmental friendly and energy-efficient process for modifying the surface chemistry of materials.

Shahidi et al[11] used low pressure plasma with non-polymerizing reactive gases to treat wool, such as O_2, N_2 and Ar. The results proved that the morphology of the wool was modified, and the wettability and dye ability of the wool could be also increased. In addition, plasma treatment can endow the wool fabrics with effective shrink-resistance and anti-felting effect. Moreover, studies on the anti-felting plasma treatment on wool fabrics in non-continuous vacuum and batch condition have been reported by researchers.

Kan et al[12] improved the anti-felting property of wool fibers and fabrics by treating with nitrogen plasma jet under atmospheric pressure. The results revealed that the anti-felting effect of wool textiles can be improved by atmospheric plasma treatment, but the hand feeling of the wool was decreased. Therefore, a polysiloxan polymer deposition process on the fiber surface should be conducted to improve the hand feeling of wool textiles.

Kim et al[13] applied three different functions of silicone polymers, such as amino functional, epoxy-functional and hydrophilic epoxy-functional silicone polymers, to plasma pre-treatment on wool fabric to improve its dimensional stability. The results showed that the anti-felting shrinkage of wool fabric treated by plasma and silicone was improved. However, tear strength and hand feeling of the treated wool were decreased; these properties were favorably restored after polymer application. The results also revealed that plasma pre-treatment can modify the cuticle surface of the wool fiber and increase its reactivity with silicone polymers.

Xu et al[14] investigated the effect of environment relative humidity (RH) on the etching of wool fibers by atmospheric plasma treatment, and the subsequent anti-felting properties of wool fabrics treated under the same conditions. In the shrinkage testing, plasma-treated wool fabrics preconditioned in RH of 100% showed the lowest shrinkage ratio of 5% which is below the ISO standard (8%) for machine-washable wool fabrics.

4.3.3.2 Cotton

Plasma induced by chemical vapor deposition (CVD) polymerization process of a flame retardant monomer is easily integrated in a cotton fabric as finishing process, totally compatible with reactive dyeing process and water repellent treatment. Moreover, the flame retardant properties of the treated fabric are not affected by the plasma treatment and the fabric can exhibit excellent wash-fastness properties after 50 cycles of laundry.

In addition, atmospheric plasma can be also applied to cotton fabrics. Carneiro et al[15] reported that the wrinkle recovery angle of cotton fabrics can be improved by dielectric barrier discharge (DBD) plasma at atmospheric pressure with a crosslinking resin of low-formaldehyde content. Horrocks et al[16] discovered that the flame retardancy of the cotton fabrics can be enhanced by atmospheric plasma with a nanoclay and hexamethylenedisiloxane (HMDSO). Lam et al[17] treated the cotton fabrics with oxygen plasma. The wicking rate, contact angle and wettability of the treated cotton were studied and the results indicated that the hydrophilicity of cotton fiber could be significantly improved after atmospheric plasma treatment. Such improvement can enhance the effectiveness of post-finishing processes during the production of cotton. Alongi et al[18] studied the impregnation of two nanoparticles suspensions (Hydrotalcite and SiO_2) on cotton fabrics to improve their thermal stability and flame retardancy. In order to significantly increase these properties by increasing the amount of nanoparticles on the cotton fiber, cold oxygen plasma discharge was introduced. The results showed that the thermal stability and flame retardancy of cotton can be modified by the addition of nanoparticles.

4.3.3.3 Silk

Zheng et al[19] modified degummed silk fabrics by cold oxygen plasma (COP) and titania sols (TSs). The results demonstrated that processing sequence of COP and TSs, and treatment conditions had significant influence on the crystalline, thermal stability and UV resistance of silk fabrics. Applying TSs before COP treatment or treating at a low temperature was beneficial to obtain high crystallinity.

4.3.3.4 Polyester

Plasma modification has a great ability to retain moisture and then increases the dissipation of the static charge. For instance, it can be used to improve the antistatic properties of polyester. It is also found that both chemical and plasma antistatic finishing treatment can endow fabrics with antistatic property by different values of recovering moisture. It is reported than there is difference in the mechanism between plasma treatment and chemical antistatic agent on the improvement of antistatic property of polyester. Some researchers also investigated the influence of atmospheric air plasma treatment on the performance of silicone nano emulsion softener on polyester fibers. Results indicated that the plasma pre-treatment modified the surface of fibers and increased the reactivity of substrate toward nano emulsion silicone. Moisture regain and microscopic tests showed that the combination of plasma and silicone treatments on polyester could decrease moisture absorption due to uniform coating of silicone emulsion on the surface of fibers.

4.3.3.5 Blended Fabrics

Hydrophobic property of blended fabrics, such as cotton/polyester (50/50) fabric, can be improved through treating by plasma CVD deposition of a fluorocarbon on fabric surface. It is also reported that durable antimicrobial properties of nylon/cotton (50/50) standard military fabric can be increased by using glow discharge plasma in atmospheric pressure through grafting and polymerizing a quaternary ammonium salt monomer onto the fabric surface. Results showed that almost 100% reductions in the bacterial activities of the fabric can be obtained after treatment.

4.3.4 Summary

The studies on plasma treatment in post-finishing process prove to be effective for the improvement of a great variety of effects and durability of many types of textiles. The application of atmospheric-pressure plasma, especially DBD plasma, has been proposed as a suitable alternative method in textile finishing. Nowadays, DBD plasma has been widely applied in the commercial and industrial fields[20].

4.4 Supercritical Fluid Technology

4.4.1 Introduction

In 1822, Baron Charles Cagniard de la Tour discovered the critical point of a substance in his famous cannon barrel experiments. A supercritical fluid is defined as a substance at a temperature and pressure above its critical point, where distinct liquid and gas phases do not exist[21].

It can effuse through solids like a gas, and dissolve materials like a liquid. Moreover, close to the critical point, small changes in pressure or temperature can lead to large changes in density, allowing many properties of a supercritical fluid to be "fine-tuned". Supercritical fluids are suitable as a substitute for organicsolvents in a range of industrial and laboratory processes. Carbon dioxide and water are the most commonly used supercritical fluids, being used in textile dyeing and finishing.

Many pressurized gas are actually supercritical fluids. Diagrams of pressure-temperature and density-pressure phases of carbon dioxide are shown in Fig. 4-9 and Fig. 4-10, respectively. As illustrated in Fig. 4-9, the critical temperature of carbon dioxide is approached at 300 K. The boiling separates the gas, liquid region and ends in the critical point, where the liquid and gas phases disappear to become a single supercritical phase. This can be observed in the density-pressure phase diagram for carbon dioxide, as shown in Fig. 4-10. At well below the critical temperature, e.g., 280 K, as the pressure increases, the gas compresses and eventually at just over 4 MPa (40 bar) condenses into a much denser liquid, resulting in the discontinuity in the line (vertical dotted line). The system consists of two phases in equilibrium, a dense liquid and a low density gas. As the critical temperature is approached (300 K), the density of the gas at equilibrium becomes higher, and that of the liquid lower. At the critical point, that is 304.1 K and 7.38 MPa (73.8 bar), there is no difference in density, and the two phases become one fluid phase. Thus, above the critical temperature, a gas cannot be liquefied by pressure. At slightly above the critical temperature (310 K), in the vicinity of the critical pressure, the line is almost vertical. A small increase in pressure causes a large increase in the density of the supercritical phase. Many other physical properties also show large gradients with pressure near the critical point, e.g., viscosity, the relative permittivity and the solvent strength, which are all closely related to the density. At higher temperatures, the fluid starts to behave like a gas, as can be seen in Fig. 4-10. For carbon dioxide at 400 K, the density almost linearly increases with pressure.

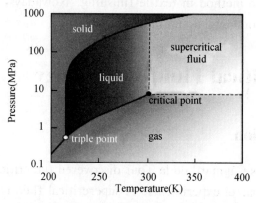

Fig. 4-9　Diagram of pressure-temperature phase of carbon dioxide

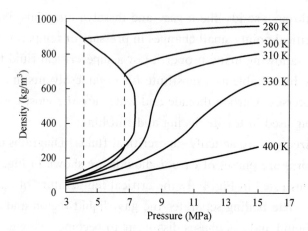

Fig. 4-10　Diagram of density-pressure phase of carbon dioxide

4.4.2　Application

Supercritical fluid technology is widely used for textiles dyeing because it is an inexpensive, nontoxic, nonflammable, environmentally friendly and chemically inert gas under many conditions. Successful pilot plant machines have been publicly exhibited at ITMA 1995 (Milan, Italy), Otemas 1997 (Osaka, Japan) and elsewhere for the past few years. At present, there is not machine especially for the finishing processing, so it is realized by using the supercritical fluid dyeing machine in laboratory.

Ma et al[22] claimed that biomacromolecules (sericin orchitosan) were immobilized onto the surface of synthetic fibrous fabrics (PET and PP) via a cross-linking agent of cyanuric chloride. Prior to the immobilization of biomacromolecules, cyanuric chloride was introduced to the fabric surface by an impregnation treatment in SCCO (2). The results showed that the treatment can improve the surface hydrophilicity and water absorption ability in the modified fabrics of PET and PP, especially for the CC-PET fabrics.

Wen et al[23] studied the UV resistance of polyester in supercritical carbon dioxide with benzotriazole absorber UV-234. In this study, the influences of treatment temperature and time on absorption, UV resistance and physical properties of polyester fabric were discussed. Results showed that ultraviolet protection factor (UPF) of polyester fabric was above 100 under conditions of temperature at 393 K, pressure of 20 MPa and treating time for 90 min.

4.5 Ultrasonic Technology

4.5.1 Introduction

Ultrasonic or ultrasound is a member of sound wave, and refers to such sound wave whose vibration frequency is larger than 20 kHz and people can't feel or hear in natural environment. Ultrasonic is a kind of wave-moving form, which can be used as a carrier or medium to detect or load information (for example, B-SCAN for medical judgment), ultrasound wave is also a kind form of energy. When its intensity reaches a certain value, it can influence the touched medium, e. g., to change or damage medium's status, characteristic, or structure (used for medical treatment). Since the 19th century, after discovering piezoelectric effect and converse piezoelectric effect in physics, people can utilize electronic technology way to produce ultrasound wave which witnesses a new chapter of developing and spreading application based on ultrasonic technology.

4.5.2 Application

With unique characteristics and sound cavitation, ultrasonic wave canaffect the macromolecule of fiber and further influence the properties of the treated fiber. In addition, the conditions of ultrasonic bath used in the treatment can also influence the properties of the fiber. In the recent decades, ultrasonic technology has been applied in the field of textile, such as dyeing finishing for the improvement of the fabric properties.

Wide-track ultrasonic spray systems (as shown in Fig. 4-11) apply performance textile finishing, including flame retardant coatings, water and oil repellent coatings, antimicrobial, durable press, moisture management, and anti-stain finishes. Uniform thin films of nano-suspensions or solutions are sprayed onto wide webs of fabric for medical textiles, flooring, carpeting, automotive or other industrial textile manufacturing applications. Wide-track textile finishing systems feature can obtain a dramatic reduction in water, chemical and energy usage. Ultrasonic spray replaces inefficient dip or padding methods with uniform thin film ultrasonic textile coatings.

Fig. 4-11　Ultrasonic spray equipment

Ge et al[24] studied the influence of ultrasonic conditions on the anti-ultraviolet performance of pure cotton fabric. In the study, ZnO nano particles were made by mechanical and cavitations effect of ultrasonic wave in liquid. The particles were treated on fabric surface, and then the anti-ultraviolet textiles were made. Some parameters were tested, including Zn^{2+} concentration in liquid, ultrasonic wave treatment time, solvent ratio, processing temperature, effect of liquid pH value on the anti-ultraviolet performance of pure cotton fabric. The results showed that the suitable concentration of Zn^{2+} is 10 mol/L, volume ratio of anhydrous ethanol and distilled water is 5 : 5, the treatment time of ultrasonic wave is 30 min, the pH value of reaction solution is about 8~9, and the processing temperature is 130 ℃. The result showed that the anti-ultraviolet performance of pure cotton fiber can be improved by ultrasonic treatment and the UPF value could reach 103.

Perincek et al[25] studied on the design requirements for industrial size ultrasound bath for textile treatments. For this purpose, effects of sound pressure, bath temperature, bath volume, textile material type and hydrophility degree of fabric were examined extensively. Finite element analysis (FEA) was used to investigate spacing and alignment of the ultrasound source transducers to reach effective and homogenous acoustic pressure distribution in the bath. It was found that textile material type, bath temperature and volume led to significant changes at sound pressure level. These parameters should be taken into consideration in designing of industrial size ultrasound bath for textile treatments. Besides, wettability of textiles is highly depended.

Sadr et al[26] prepared TiO_2 loaded cotton fabric through synthesizing TiO_2 in low temperature by ultrasonic irradiation and then loading the nanoparticles onto the cotton fabric. The results showed that the formation of anatase TiO_2 nanoparticles with 3~

6 nm crystalline size loaded onto the cotton fabric at low temperature (75 ℃) can lead to good self-cleaning and UV-protection properties. The excellent UV-protection rating of the treated fabric maintained even after 25 home launderings indicated that an excellent washing durability of the treated fabric can be obtained after loading with TiO_2. In addition, it was discovered that the sonochemical method had no negative influence on the structure of cotton fabric. The statistical analysis indicated a significant effect of both the concentration of titanium tetra is opropoxide (TTIP) used as precursor as well as the ultrasonic time on the content of the loaded TiO_2 on the fiber and self-cleaning properties of the fabric.

4.5.3 Summary

Ultrasound technology is an effective method to improve the dispersion of materials by cavitation when compared with the traditional treatments. It can reduce the consumption of chemicals and save water and energy (temperature reduction). Ultrasound technology is also an efficient finishing technology for the properties enhancement of textiles.

4.6 Sol-gel Technology for Functional Finishing

4.6.1 Introduction

Sol-gel technology refers to metal organic or inorganic compounds through the solution, sol-gel and curing, and then heat treatment to form oxides or other solid compounds. In 1840, Ebelmen reported the formation of a transparent solid by slow hydrolysis of silicic esters. In the 1930's, Geffcken verified that could obtain oxide thin film by hydrolysis of metal alkoxide. In 1971, Dislich reported that prepared SiO_2-B_2O-Al_2O_3-Na_2O-K_2O multicomponent glass by hydrolysis of metal alkoxide. The sol-gel reaction mainly includes two processes, hydrolysis (alcoholysis) and polymerization. The raw materials are dispersed in a solvent, and then generate the active monomer through the hydrolysis, active monomer polymerize, the sol becomes form, and then generates the gel with spatial structure. After drying and heat treatment, the nanoparticles and required materials are prepared.

Two processes are given as below:

(1) Hydrolysis reaction: $M(OR)n + H_2O \longrightarrow M(OH)x(OR)n - x + xROH$

(2) Polymerization: $-M-OH + HO-M- \longrightarrow -M-O-M- + H_2O$

$-M-OR + HO-M- \longrightarrow -M-O-M- + ROH$

4.6.2 Application

Sol-gel technology was applied to textiles refers to the use inorganic or organic modification of metals form a sol by the hydrolysis of metal alkoxide. Under certain conditions such as (heating, Ultraviolet radiation), the sol formed the gel by condensation and condensation, and finally formed a solid, transparent, porous oxide film in textiles. The fabric could obtain self-cleaning, antibacterial, antistatic, anti UV function etc.

Zhen et al[27] prepared the solid-phase microextraction (SPME) fibers by using sol-gel technology. An electron microscopy experiment suggested a porous structure for Superox-4 (polyethylene glycol, PEG) coating. SPME-GC analyses provided evidence that the sol-gel fibers have some advantages, such as high velocities of mass transfer, efficient extraction rates, high thermal stability, long life span, and spacious range of application for both polar and non-polar analytes.

Yin et al[28] prepared a new dyeing solution of containing silica and direct dyes by sol-gel treatment to improve color strength (K/S value) and dyeing fastness. In this case, EtOH, TEOS, H_2O and GPTMS were added in turn. The molar ratio of TEOS : H_2O : EtOH was 1 : 5 : 8 and the concentration of GPTMS was 0.05 mol/L. Fabric was dyed at 90 ℃ for 40 min. The concentration of NaCl added into the dyeing solution was 10 g/L. The dyed fabrics were baked at 150 ℃ for 5 min. With this process, the results indicated that the K/S value was enhanced by more than 10%, the rubbing fastness and the washing change fastness were improved by one grade, and the washing staining fastness was improved by half a grade. The fastness and the color strength were even better than the fabric fixed with fixing reagent MMF-1. The fiber surface became smoother by using a video microscope. The calculated sol-gel weight gain on the fabric was 4.6%. As a nonpolluting process, the sol-gel technology shortens the dyeing process and brings a better fixing property, meeting the needs of energy-saving and pollution-free processes.

4.7 UV Treatment

4.7.1 Introduction

Similar to plasma processing of textiles, UV treatment can also be used for increasing surface area, hydrophilic property, anti-felting, antistatic, dyeing, printing, and adhesion strength. Actually UV eximer and UV lamps can launch wavelengths in the 172~400 nm which are widely used for the treatment of polymer/textile substrates to improve their functionality in UV treatment. The UV excimer lamp with 172 nm photons is most commonly used for surface modification of heat-sensitive textile substrates to alter physicochemical properties. Mostly, the textile samples are UV treated in the presence of air, oxygen (O_2), and nitrogen (N_2) gases for surface etching, activation, oxidation, and radical generation.

4.7.2 UV Treatment on Wool Fabric for Anti-felting Finishing

Wool fabric is felted in an aqueous medium by mechanical agitation due to the presence of scale on the fiber surface. Dood et al[29] reported that UV treatment is effective for shrink proofing of wool fabric treated with silicone polymer. In this process, UV treatment helps to achieve adequate shrink proofing (only 3% shrinkage) of wool fabric treated with only 3% silicone-containing polymer compared to the 20% shrinkage of untreated wool fabric. UV treatment helps to adhere silicone polymer on the wool surface so that it can maintain its shrink-resistance properties even after washing cycles. There is only a marginal change in tensile strength and weight loss properties.

Sayed and Khatib[30] reported that the UV treatment can modify wool fabric to enhance its felting shrinkage and pilling resistance properties. In this study, freshly scoured wool fabric was first irradiated with 254 nm UVC radiations for 10 to 100 min followed by treatment with an ecofriendly oxidizing agent such as H_2O_2 or proteolytic enzyme. It was found that UV/H_2O_2-treated wool fabric showed lower area shrinkage of 13.7% compared to 32.10% in the control wool fabric. Shrinkage and pilling resistance results of the treated fabric were compared with harsh and non-eco-friendly chlorination treatment. The results were attributed to the fact that UVC radiation could partially oxidize wool fiber surface (—S—S into —SO_3H), oxidize the thin fatty lipid layer of the upper epicuticular, and break the waxy smooth scales responsible for felting. These physicochemical surface changes facilitate the adhesion and penetration of H_2O_2 inside the modified wool fabric. Treatment of wool fabric with these systems was found to be effective in reducing pilling and shrinkage without severe loss in weight and strength of the fabric.

4.7.3 UV Treatment on Silk and Wool Fabric for Hydrophobic/ Hydrophilic Finishing

Periyasami et al[31] studied that one side of mulberry silk fabric was irradiated by using 172 nm UV excimer lamp. It had been elucidated that after one side irradiation for 5 min in air atmosphere, wetting and wicking properties could be improved significantly on the irradiated side. The UV irradiation showed an average wetting time of 7.2 s, which was 100% lower than the control silk fabric. However, the other side, that is the nonirradiated side of the fabric showed behavior similar to the control silk. In the treated surface, the wetting/wicking of water was better due to the improvement in the hydrophilic property in addition to formation of nano pores of 100 nm×10 nm and the surface morphology of the other side (not treated) remaining unchanged. Then this process endowed silk fabrics got the multifunction which had the hydrophobic property on one side and the hydrophilic property on the other side.

Later on based on this concept, Basak et al[32] developed hydrophilic-hydrophobic wool fabric by 172 nm UV excimer lamp irradiation treatment. Scoured wool fabric was initially treated with a fluorocarbon chemical to make the sample hydrophobic (both sides). Then one side of the hydrophobic sample was exposed to UV irradiation for 5~30 min. It was found that with increasing UV irradiation time, hydrophilic properties in terms of wetting, wicking, and contact angle improved significantly. The contact angle of the scoured wool (control sample) was 90° and it increased to 140° in the fluorocarbon-treated sample. It was seen that the contact angle on the irradiated side of the fabric decreased exponentially with increasing irradiation time. The contact angle decreased to 100° within 5 min of UV exposure and it was as low as 10° after 30 min of irradiation. The other side of the fabric (the unexposed side) remained completely hydrophobic (water-repellent) and there was no visible change in contact angle. It might because the high-energy UV irradiation (7.2 eV) on the fluorocarbon finished fabric caused photo-induced oxidation involving defluorination of the surface and incorporation of oxygen by forming $CF-O-CF_2$, CF_2-O-CF_2, and $CF-O-CF_3$ moieties[33].

4.8 Finishing Treatment for Improving Comfort

In its broadest terms, clothing comfort is often described as consisting of thermal and sensorial comfort. Thermal comfort is determined by the ability of the body to have balanced heat exchange with the environment. The thickness of fabrics and the ability of entrapping air are the most important factors in determining heat transmission or insulation value.

4.8.1 Thermal Regulation

Clothing creates a buffer layer which alters the heat transfer rate from the human body to the environment. The heat transfer rate is determined by the nature of the clothing fabrics and air layers (microclimate) created. Phase change materials (PCMs) are a relatively new technology that can be widely used to aid the process of thermal energy transfer. PCMs are applied in textiles and clothing that can create an additional buffering effect and minimizes changes in skin temperature to maintain the comfort of the wearer[34]. The cooling and insulating effects of PCMs provide a thermos regulating function for wearers. These materials have been described as "thermal shock absorbers" in that they are able to slow the rate of temperature change within a wearer's personal microclimate. Thus, PCMs in textiles and clothing are able to reduce the rate at which a person will overheat or get cold.

4.8.1.1 Introduction

PCMs were initially developed by the United States' National Aeronautics and Space Administration's (NASA) research program in the 1910s to protect instruments from the extreme temperature fluctuations they encountered in space, and more than 100 different PCMs were discovered. Now PCMs have been widely applied in many fields, such as protective clothing, sportswear, medical textiles, footwear, and interior textiles among others.

Phase change textiles and apparel are unique in that they are able to actively combat changes in a person's microclimate. PCMs work by absorbing heat energy and changing from a solid state to a liquid state when the temperature reaches the PCMs melting point. This phase change will produce a temporary cooling effect in the clothing layers as the PCMs absorb heat from the surface of the body, giving more comfort to the functional finishes to improve the comfort and protection of apparel wearer in a warm or hot climate. On the other hand, when a wearer's body or the external environment is cold, the PCMs will release stored heat to warm the wearer. Ideally, PCMs regulate or adjust the wearer body temperature through modifying heat transfer to create and ensure a constant comfort zone next to the skin.

PCMs can be incorporated into textiles by using coatings with embedded PCMs microspheres. Such coatings may be based on acrylic or polyurethane, and can be applied using processes such as knife-over-roll, knife-over-air, pad-dry-cure, gravure, dip coating, and transfer coating[35]. Microencapsulation is another approach. Before inclusion into textiles, PCMs are made into microcapsules which are a tiny package with a polymer wall less than 2 μm thick and less than 40 μm in diameter. The capsules contain solid or liquid PCMs, usually paraffin, which is released under certain conditions of heat, abrasion, diffusion, degradation, or dissolution of the outer polymer. Capsules of paraffin are produced when the PCMs are in solid form, so the polymer can be easily applied over the surface of the paraffin. The microencapsulated paraffin can be embedded in acrylic fibers or polyurethane foams, and also be coated onto the surface of a textile structure by using one of these polymers as the carrier in the finish.

4.8.1.2 Application

Textiles finished with PCMs are considered smart textiles. Static thermal resistance is greater in coated PCMs fabrics than in untreated fabrics, but this can be attributed to the binder material used to coat the fabric, which closes pores in the textile. The key role of PCMs coated fabrics is to serve as a buffer when a person goes from a neutral environment to a cold one.

Ghali et al[36] investigated the effects of PCMs on body heat loss during exercise by means of a numerical, three-mode fabric ventilation model. The results imply that the heating effect lasts approximately 12.5 minutes depending on PCMs percentage and outdoor conditions. It is also found that overall heat released by PCMs decreases the clothed-body heat loss by an average of $40 \sim 55 \text{ W/m}^2$ for a one-layer suit depending on the frequency of oscillation and crystallization temperature of the PCMs. However, after the initial transition, when the PCMs are in the solid phase, it shows no effect on the thermal performance of the fabric.

The thermal performance of microencapsulated PCMs incorporated into a textile through coating was evaluated by Kwok et al. In their study, three indices and related test methods were used to characterize the thermal functional performance of the textiles. The fabric thermal regulating capability was found to be strongly related to the amount of PCMs. A similar study of the effects of PCMs applied in cold weather protective clothing was advances in the dyeing and finishing of technical textiles conducted by Wang et al[37]. It was found that conductive fabric significantly changed the temperature distributions in clothing assemblies to increase the thermal-protective performance of the clothing, and that clothing assemblies with PCMs saved around 30% of energy in the temperature control process. PCMs were tested on thermal manikins moved from a warm environment to a cold one, and the effects were compared to those of a test where the manikin was dressed in a comparable non-PCMs outfit. The effect of the PCMs was that it buffered the wearer from the cold for approximately 15 min. These results indicate that it would only be possible to keep the wearer warm if they move back and forth between warm and cold environments so the heating effect would not have time to wear off.

The US Army uses a type of PCMs called Shape Memory Polymers (SMPs), which include a membrane with a flexible moisture barrier. It is used in a dry suit that aims to keep the wearer dry in cold water, while still preventing discomfort from sweating in warmer environments[38]. The molecular structure of the SMPs is rigid at temperatures below activation point and prevents permeation of water molecules, yet relaxes at warmer temperatures, ensuring the user to be comfortable in a wide variety of environments. "Air Force One Operators" also rely on PCMs in thermos regulated gloves, which ensure comfort and safety in many different conditions[39].

4.8.2 Moisture Management

4.8.2.1 Introduction

Moisture transfer is an important aspect of clothing comfort, contributing to both thermal and sensorial comfort. The human body supplies sensible and insensible moisture to the surface of the skin in the form of perspiration. Evaporation of the moisture from the skin serves to cool the body; it must be moderated to prevent over drying or overcooling of skin. On the other hand, significant discomfort may occur if the body moisture condenses in the fabric or next to the skin. Achieving balance in moisture transfer between the body and the environment, and keeping skin dry, are particularly important for people in high-activity situations and/or exposed to severe thermal conditions where they may be sweating heavily, such as during sports activities or firefighting. Often clothing impedes that balance by trapping moisture next to the skin or by becoming saturated with the body's moisture. Recent approaches to this problem involve the development of functional fabrics designed to transport moisture through the fabrics to the environment where it can be rapidly released through evaporation. Moisture management finishes are one method for achieving that function.

Moisture management can be defined as the engineered or inherent transport of water vapor or aqueous liquid (perspiration) through a textile. It is an integrated function of wetting, wicking, liquid moisture transport and surface evaporation. Moisture vapor or liquid absorbed from the skin side needs to move from the fabric inner surface to its exterior surface to vaporize. Moisture management systems often make use of finishes to improve wetting, wicking or absorbency and also that impart hydrophobicity.

4.8.2.2 Application

Fabrics made of superfine ormicrodenier fibers promote a wicking effect through capillary interstices in the fabric, and this effect can be further enhanced by hydrophilic finishes. Sampath et al[40] introduced a moisture management finish (MMF) on microdenier polyester fabric based on aluminosilicate polyether copolymer. Fabric testing indicated the MMF treatment improved absorbency and wicking characteristics, and this change led to quicker evaporation of moisture from the fabric. Gibson[41] reported the use of a nano-based water-repellent finish on the outer surface of military uniforms to obtain a water-repellent surface, while maintaining wicking characteristics in the inner fabric. These finished fabrics developed a moisture management property that absorbed liquid and delivered it to the outer surface where it could easily evaporate. The uniforms made of these fabrics demonstrated improved comfort properties.

Yoo et al[42] explored moisture management properties of thermal protective fabrics. The protective fabrics were treated with hydrophilic finishes on the surfaces of fabrics which were produced using yarns blended with hygroscopic fibers. The moisture management performance was assessed by liquid moisture absorption and evaporation properties, including absorption capacity, absorption rate and drying rate (evaporation). The results revealed that wicking finishes on aramid fabrics noticeably increased the rate of liquid absorption, but did not change the total amount of water absorbed by the fabric or its water vapor absorption. FR rayon blends showed improved vapor absorption compared to aramid fabrics.

Ma et al[43] introduced a novel MMF by using supercritical carbon dioxide. The treatment included a series of immobilization processes, a pad-dry cure process and a solution-additive process to finish PET fabric with natural functional agents (sericin, collagen, or chitosan). Test results showed improved fabric surface wettability and moisture regain of the PET fabrics. The new approach was promoted as simple, low-cost and exhibiting a stronger immobilization of functional agents on the fiber surface.

Nanocomposites have also proved effective in functional finishing of textiles. Dong et al[44] applied a TiO_2 finish to polypropylene fibers and found hydrophilicity was improved significantly after sputter coating, as would be expected from better surface coverage by TiO_2 nanoparticles. Mikotajczyk et al[45] explored the effects of the structure of polymers by applying nanosilica additives on various alginate fibers. Regardless of the differences in the chemical structure or the presence of nanosilica, the types of alginate fibers examined showed similar values of moisture absorption. The amount of moisture absorbed by the fiber-forming polymer of alginate fiber exerted a strong influence on its electrical properties. Similar research was reported by Viswanath et al[46] using fluoro-alkyl nano lotus finish on cotton fabrics. The finish was applied in conjunction with a silicone softener and the results exhibited optimum performance with respect to thermal insulation value, moisture transport and air permeability as compared to controlled fabrics.

A hybrid pattern with hydrophobic and hydrophilic areas on the inner fabric surface was formed as illustrated in Fig. 4-12. The dual functionality of the surface enhanced capillary effects. The researchers reported that wicking windows were formed between the two surfaces to allow the passage of liquid moisture. The chemically finished fabric exhibited a one-way moisture transport property. Fig. 4-13 displays the moisture transport process of the finished fabric.

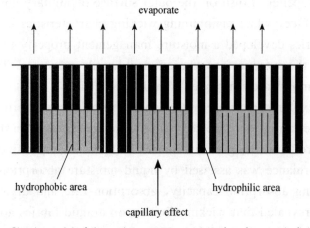

Fig. 4-12 Simple model of the moisture transport transfer of sweat in finished fabric

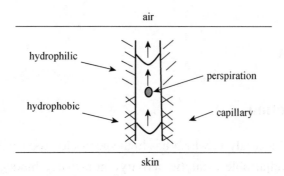

Fig. 4-13 The one-way moisture capillary process

4.8.2.3 Odor Control

Odor retention in textiles is a common occurrence and a determining factor in the overall comfort and sanitation of a fabric. To combat the incidence of odor-producing and odor-retaining biological hazards such as fungi and bacteria, the use of antimicrobial finishes on textiles has become a common occurrence in the textile industry, especially in sports and active wear, in which higher levels of sweat and body odor are present. Sweating is a normal body mechanism, most useful for cooling off the body through evaporation. Sweat glands produce a mixture of water, salt and other trace components which are then secreted onto the skin[47]. Once secreted, Gram-positive bacteria metabolize the nutrients present in the sweat, creating body odor[48]. The moist, warm, dark environment often found in shoes provides the ideal environment for the growth of bacteria; therefore, the presence of odor on feet and socks are common. Preventing odor build-up on clothing has become a prevalent trend in textiles and is a complex. Odor varies immensely from one individual to the next, and can depend on a combination of factors including (but not limited to) diet, smoking habits, leisure pursuits and hygiene. There are several types of finishes and methods of application for imparting odor reduction or other antimicrobial outcomes in textiles, and these are discussed by the protective finishes section. Some odor-resistant finishes have limitations with their use and application. Antibacterial finishes are generally not durable, and after a certain number of washings they are rendered ineffective[49]. Another limitation of anti-odor finishes is their interactions with the skin. Some anti-bacterial finishes or metallic additives can be skin irritants, and some ingredients such as triclosan have been investigated for the possibility of carcinogenicity. Therefore, further work with anti-odor finish needs to be completed to ensure such finishes are durable to repeated laundering, as well as safe for continual human exposure. Recently, an interesting development is the incorporation of fragrance within microencapsulated finishes, allowing gradual release of an appropriate fragrance. In addition to social comfort and aesthetic considerations, these finishes can be applied to fabrics to promote health and wellbeing[50].

4.9 Enzyme

4.9.1 Introduction

Enzyme is a kind of biocatalyst, which has high specificity and catalytic efficiency, mild reaction condition, adjustable catalytic activity, non-toxic, biodegradable, less sewage and environmental pollution. Enzymes are a sustainable alternative to the harsh toxic chemicals in the textile industry. As enzymes operate in moderate conditions of temperature and pH, energy consumption is reduced, thereby reducing greenhouse gas emissions. Both water consumption and waste generation during textile manufacturing are also minimized by enzyme use. As the enzymes are very specific in their actions, minimum by-products are generated. The risks to humans, wildlife, and the environment are minimized. Enzymes can be utilized in several steps of textile processing namely desizing, scouring, bleaching, and biopolishing of cotton, degumming of silk, bleaching and shrink proofing of wool, biostoning (of denim), and for treatment of textile effluents.

4.9.2 Enzyme Classification

4.9.2.1 Amylases

The amylases are starch degrading enzymes. There are basically four groups of amylases: endoamylases, exoamylases, debranching enzymes and transferases. α-amylase is a well-known endoamylase that widely distributes in animals, plants and microorganisms. α-amylase catalyzes the hydrolysis of internal α-1,4-glycosidic to low molecular weight products, such as glucose, maltose and maltotriose. Industrial use of α-amylase for the production of dextrose powder and dextrose crystals began in 1959, since then, amylases have been used for various purposes[51]. Amylases are used in the textile industry for the desizing process, which involves the removal of starch. Under the specific catalysis of the amylase, the starch chain is quickly hydrolyzed by catalytic cracking, than the starch slurry is dissolved and removed. The use of enzyme desizing does not damage the fiber; desizing efficiency; environmental protection; fabric hair effect is better, high whiteness, good dyeing performance.

4.9.2.2 Pectinases

Pectinolytic enzymes or pectinases are a group of enzymes that hydrolyzepectin substances, which mostly present in plantsfungi, yeasts, insects, nematodes, and protozoa[52]. The pectinolytic enzymes may be divided into three groups: pectin esterase, depolymerases and protopectinases. Over the years, pectinases have been used in several conventional industrial processes such as bioscouring of cotton in textile processes. Alkaline pectinase is often considered as the most suitable enzyme for cotton scouring. Despite many researches have done, there is still demand for a pectinase with higher activity and stability at high temperatures and alkaline conditions. Studies of the optimal fermentation medium have been performed to reduce the cost of an alkaline thermostable pectinase. Thermostability of pectinase can be improved by using direct evolution technique.

4.9.2.3 Cellulases

Cellulases hydrolyze cellulose (β-1, 4-D-glucan linkages) produce primary products glucose, cellobiose and cello-oligosaccharides[53]. Natural cellulases, described as total crude cellulases, are complex mixtures of three major kinds: cellobiohydrolase (EC 3.2.1.91), endo-β-1,4glucanase (EC 3.2.14) and β-glucosidase (EC 3.2.1.21). Natural cellulose macromolecules are composed of a large number of crystalline regions and a small amount of amorphous regions. Hydrolysis occurs first in the amorphous region and then in the crystalline region. The fibers are degraded by three cellulases synergistically treatment.

4.9.2.4 Proteases

Peptidases, commonly designated as proteases (EC 3.4), are the enzymes that catalyze the cleavage of peptide bonds in other proteins. They are degradative enzymes which catalyze the total hydrolysis of proteins. Proteases can be classified intoexopeptidases that attack the ends of protein molecules, and endopeptidases which cleave peptide bonds within polypeptide chains. Depending on the nature of the functional group at the active site, they are divided into serine proteases, aspartic proteases, metallo proteases, cysteine proteases and endopeptidases of unknown catalytic mechanism. In the textile industry proteases are used in wool finishing and in a smaller scale, in silk degumming.

4.9.2.5 Carboxylic ester hydrolases

Carboxylic ester hydrolyzing enzymes are a group of enzymes that are able to catalyze the hydrolysis, synthesis or transesterification of an ester bond. There are two well-known groups within carboxylic ester hydrolases: lipases and esterase[53]. Esterases differ from lipases by showing a preference for short-chain acyl esters (shorter than 10 carbon atoms) and that they are not active on substrates that form micelles. Lipases are the most important group of biocatalysts for biotechnological applications. In the textile industry their applications are diverse, including the hydrolysis of synthetic fibers and bioscouring in combination with pectinase. Recently, lipases and esterase has been applied in the textile industry because of they can hydrolyze synthetic fibers such as polyester, polyamide and polyacrylonitrile and its potential application in wool processing.

4.9.2.6　Nitrilases

In nature, three different groups of enzymes are involved in the microbial hydrolysis of nitriles. Nitrilases (EC 3.5.5.1 and 3.5.5.7) hydrolyze nitriles to the corresponding carboxylic acids, forming ammonia; and nitrile hydratases (EC 4.2.1.84) form amides from nitriles which can be subsequently hydrolyzed by amidases (EC 3.5.1.4). In the textile industry, nitrile hydrolyzing enzymes are used in the hydrolysis of PAN surfaces toincrease hydrophilicity of fabrics.

4.9.2.7　Oxidases, Peroxidases and Laccases

The oxidase is that enzyme catalyzes oxidation-reduction where oxygen is reduced to water (H_2O) or hydrogen peroxide (H_2O_2). Glucose-1-oxidase (EC 1.1.3.4) is an enzyme which catalyzes the oxidation of β-D-glucose to D-gluconolactone and hydrogen peroxide. Its main application in the textile industry is a bleaching agent, as it only needs glucose and oxygen to generate H_2O_2. Peroxidases catalyze the oxidation of a wide variety of substrates, using H_2O_2 or other peroxides (EC 1.10.1.7). In the textile industry, they are used in the decoloration of dye baths and in the bleaching of textiles. Laccase (EC 1.10.3.2) catalyzes the oxidation of ortho and para diphenols, aminophenols, polyphenols, polyamines, lignin and aryl diamines as well as some inorganic ions coupled to the reduction of molecular dioxygen to water. The use of laccase in the textile industry is growing very quickly. It is mainly used to bleach textiles, synthesize dyes and decolorize textile wastewater.

4.9.3　Textile Application

4.9.3.1　Denim Biowash

Cellulases have been widely used in textile because of their ability to modify cellulosic fiber in a controlled and desired manner, so as to improve the quality of fabrics. Bio-stoning and bio-polishing are the best-known current textile applications ofcellulases.

　　Blue jeans and other denim garments have gained remarkable popularity in recent years. It is estimated that over 800 million pairs of blue jeans are produced worldwide every year. In denim fabrics, the indigo dye is mostly attached to the surface of the yarn and the most exterior short cotton fibers. In the late 1970s and the early 1980s, industrial laundries developed methods for producing faded jeans by washing the garments with pumice stones, which partially removed the dye revealing the white interior of the yarn, which leads to the faded, worn and aged appearance. This was designated as "stonewashing". The conventional denim wash process consists of treatment with pumice stones. About 1 kg stone is required for 1 kg denim fabric. A large amount of pumice sludge is produced in this process. For example, a denim finisher processing 100 000 garments a week with stones typically generates 18 ton of sludge[8]. This may cause several problems including rapid wear and tear of washing machines, unsafe working conditions, large numbers of second class garments, environmental pollution, and the need for manual removal of pumice from pockets and folds of garments. In the mid-1980s, biotechnology provided a perfect alternative for stone-washing using microbial cellulases, later known as "bio-stoning".

During the bio-stoning process, cellulases act on the cotton fabric and break off the small fiber ends on the yarn surface, thereby loosening the indigo, which is easily removed by mechanical abrasion in the wash cycle. The advantages in the replacement of pumice stones by a cellulose based treatment include: (1) reduced wear and tear of washing machines and short treatment times; (2) increased productivity of the machines because of high loading; (3) substantial decrease of second quality garments; (4) less work-intensive and safer working conditions; (5) safe environment, since pumice powder is not produced; (6) flexibility to create and consistently reproduce new finished products; and (7) the possibility to automate the process with computer-controlled dosing devices when using liquid cellulases preparations.

4.9.3.2 Modification of Wool

The wool modification mainly uses proteases. Proteases are the catalyst for hydrolysis of peptide chains, which can promote the hydrolysis of peptide bonds (CO—NH—) on wool. It can strip the scales or lipid layers at different extents[54].

The catalytic reaction of protease to wool belongs to heterogeneous catalysis, that is, the catalyst (enzyme, liquid) and the matrix (wool, solid) are not in the same phase. First, the wool surface attracts the surface of the enzyme medium, and then diffuses into the fiber, generates biocatalytic reaction. The dyeing ability of wool is greatly improved through enzyme treatment that is much higher than untreated at low temperature. Because the scaly layer was destructed this was favorable for the diffusion of the dye after enzyme treatment. Therefore, the enzyme treatment can improve the initial dyeing rate of wool, shorten the dyeing balance time, achieve low-temperature dyeing (80 ℃), and avoid wool fiber damage in the high temperature. The scaly layer of the wool was partially softened or peeled after enzymatic treatment, and the directional friction effect of the wool was reduced, so that the anti-felting ability was improved and the anti-felting effect was obtained. Meanwhile, the hand feeling of wool became soft and slippery which has the characteristics of cashmere. After the enzymatic treatment, the wool is refined and the degree of bending is increased, and the spinnability is improved. Because the protease has catalyzed hydrolysis on the wool surface, more carboxyl and amino groups were exposed after hydrolysis, significantly improved the hydrophilic surface of the wool.

4.9.3.3 Surface Modification of Synthetic Fibers

Synthetic fibers have some disadvantages such as high hydrophobicity and crystallinity, which affect not only decrease wearing comfort, but also imped the application of compounds and coloring agents in textile finishing. The traditional methods such as alkaline or acid hydrolysis are used to improve fiber hydrophilicity, but lead to the deterioration of fiber properties such as irreversible yellowing and loss of resistance. Recent studies show that surface modification and hydrolysis of polyester and polyamide with enzyme are environmentally benign methods. The major advantages of polymer modification with enzyme as compared with chemical methods are milder reaction conditions and highly specific nondestructive transformations, targeted to less fiber damage.

Yoon et al[55] reported that polyesterase modified the surface of PET and PTT. The authors thought that the formation of terephthalic acid (a hydrolysis product) could be monitored at 240 nm. The enzymatic treatment resulted in significant, efficient desizing, increased hydrophilicity and reactivity with cationic dyes, and improved oily stain release.

For acrylic and cellulose acetate fibers enzymes can be used to accomplish the formation of reactive or hydrophilic groups at the surface by hydrolysis. Their pendant groups without affecting, in theory, the integrity of the main chain of the polymer. The pendant group in polyacrylonitrile (PAN) is the nitrile group. The ester group of the polysaccharide substituent is a potential supplier tohydrolyze the cellulose acetate fibers in a controlled manner, creating hydroxyl groups at the surface. The modification of PAN and cellulose acetate with enzymes results in two types of products: soluble compounds and new chemical groups attached to insoluble fiber substrate. For PAN fiber, only nitrilase and amidase generate a soluble product, ammonia; nitrile hydratases generate amide groups as new sidechains of the PAN main chain. For cellulose acetate, the hydrolysis of its sidechains releases acetic acid to the reaction media and the hydroxyl group is located on the polymer backbone.

The surface of PAN was modified by nitrile hydratase and amidase enzymes obtained from different sources. After enzymatic treatment the fabric became more hydrophilic and the adsorption of dye was enhanced[56].

References

[1] Schindler W D, Hauser P J. Chemical finishing of textiles. Cambridge: Woodhead Publishing, 2004.
[2] Gulrajani M L. Advances in the dyeing and finishing of technical textiles. Cambridge: Woodhead Publishing, 2013.
[3] Bickerman J J. Foams. Heidelberg: Springer, 1973.
[4] Evans J O, Chase R L. Dyers woad control [Isatis tinctoria, weeds, cultural and chemical control].

Leaflet-Utah State University, 1981.

[5] Nakamichi T. Cross-linking Agent. Shikizai Kyokaishi, 1992, 65(8): 511-525.

[6] Anooshehpoor A, Brune J N. Frictional heat generation and seismic radiation in a foam rubber model of earthquakes. Pure & Applied Geophysics, 1994, 142(3-4):735-747.

[7] Baker K L, Bryant G M, Camp J G, et al. Foam finishing technology. Textile Research Journal, 1982, 52(6):395-403.

[8] Pearson J, Elbadawi A. Tandem wet-on-wet foam application of both crease-resist and antistatic finishes. Ecotextiles, 2007: 195-199.

[9] Lennox-Kerr P. Germans develop high-speed denim shrinkage. Textile World, 1996,146(6):90.

[10] Nehra V, Kumar A, Dwivedi H K. Atmospheric non-thermal plasma sources. International Journal of Engineering, 2008, 2(1): 53-68.

[11] Shahidi S, Rashidi A, Ghoranneviss M, et al. Plasma effects on anti-felting properties of wool fabrics. Surface & Coatings Technology, 2010, 205(7):S349-S354.

[12] Kan C W, Yue M, et al. Plasma pretreatment for polymer deposition-improving antifelting properties of wool. IEEE Transactions on Plasma Science, 2010, 38(6):1505-1511.

[13] Kim M S, Kang T J. Dimensional properties of low temperature plasma and silicone treated wool fabric. Fibers and Polymers, 2001, 2(1): 30-34.

[14] Xu H, Peng S, Wang C, et al. Influence of absorbed moisture on antifelting property of wool treated with atmospheric pressure plasma. Journal of Applied Polymer Science, 2009, 113(6): 3687-3692.

[15] Carneiro N, Souto A P, Foster F, et al. A DBD plasma machine in textile wet processing. 21st IF-ATCC International Congress, 2008.

[16] Horrocks A R. Textiles-fire retardant materials. Fire Retardant Materials, 2001, 81(2):128-181.

[17] Lam Y L, Kan C W, Yuen C W M. Physical and chemical analysis of plasma-treated cotton fabric subjected to wrinkle-resistant finishing. Cellulose, 2011, 18(2):493-503.

[18] Alongi J, Tata J, Frache A. Hydrotalcite and nanometric silica as finishing additives to enhance the thermal stability and flame retardancy of cotton. Cellulose, 2011, 18(1):179-190.

[19] Zheng C, Chen G, Qi Z. Compound finishing of bombyxmori silk: a study of cold oxygen plasma/titania sols treatments and their influences on fiber structure and performance. Plasma Chemistry and Plasma Processing, 2012, 32(3): 629-642.

[20] Mollah M Y A, Schennach R, Patscheider J, et al. Plasma chemistry as a tool for green chemistry, environmental analysis and waste management. Journal of Hazardous Materials, 2000, 79(3): 301-320.

[21] Keskin S, Kayrak T D, Akman U, et al. A review of ionic liquids towards supercritical fluid applications. Journal of Supercritical Fluids, 2007, 43(1):150-180.

[22] Ma W X, Okubayashi S, Hirogaki K, et al. Impregnation of cyanuric chloride into synthetic fabrics (PET and PP) by supercritical carbon dioxide and its application in immobilizing natural biomacromolecules. Fiber, 2010, 66(10):243-252.

[23] Wen H B, Dai J J. UV protection of polyester in supercritical carbon dioxide fluid. Dyeing & Finishing, 2006.

[24] Ge J, Xu B, Cai Z,et al. Research of ultrasonic wave anti-ultraviolet textiles.Cotton Textile Technology, 2013.

[25] Perincek S, Uzgur A E, Duran K, et al. Design parameter investigation of industrial size ultrasound

textile treatment bath. Ultrasonics Sonochemistry, 2009, 16(1):184.

[26] Sadr F A, Montazer M. In situ sonosynthesis of nano TiO₂ on cotton fabric. Ultrasonics Sonochemistry, 2014, 21(2): 681-691.

[27] Wang Z, Xiao C, Wu C, et al. High-performance polyethylene glycol-coated solid-phase microextraction fibers using sol-gel technology. Journal of Chromatography A, 2000, 893(1): 157-168.

[28] Yin Y, Wang C, Wang C. An evaluation of the dyeing behavior of sol-gel silica doped with direct dyes. Journal of Sol-gel Science and Technology, 2008, 48(3): 308-314.

[29] Dodd K J, Carr C M, Byrne K. An investigation into the application of a UV-curable silicone for the shrinkproofing of wool fabric. Journal of the Textile Institute, 1993, 84(4):619-630.

[30] El S H, El K E. Modification of wool fabric using ecologically acceptable UV-assisted treatments. Journal of Chemical Technology and Biotechnology, 2005, 80(10): 1111-1117.

[31] Periyasamy S, Gupta D, Gulrajani M L. Modification of one side of mulberry silk fabric by monochromatic VUV excimer lamp. European Polymer Journal, 2007, 43(10):4573-4581.

[32] Basak S, Samanta K K, Chattopadhyay S K, et al. Development of dual hydrophilic wool fabric by Q172 nm VUV irradiation. Journal of Scientific & Industrial Research, 2016, 75(7):439-443.

[33] Fu Y Z, Viraraghavan T. Removal of Cl acid blue 29 from an aqueous solution by Aspergillus niger. Am. Assoc. Text. Chem. Color Review, 2001, 1(1): 36-40.

[34] Mansfield E, Yohe G. Microeconomics: Theory and Applications. 7th ed. Norton Company, 2004.

[35] Mondal S. Phase change materials for smart textiles-An overview. Applied Thermal Engineering, 2008, 28(11-12):1536-1550.

[36] Ghali K, Ghaddar N, Harathani J, et al. Experimental and numerical investigation of the effect of phase change materials on clothing during periodic ventilation. Textile Research Journal, 2004, 74(3):205-214.

[37] Wang S X, Li Y, Hu J Y, et al. Effect of phase-change material on energy consumption of intelligent thermal-protective clothing. Polymer Testing, 2006, 25(5):580-587.

[38] Meinander H. Smart and intelligent textiles and fibre. Textiles in Sport, 2005:120-133.

[39] Zhang X. Heat-storage and thermo-regulated textiles and clothing. Smart fibers and clothing. Cambridge: Woodhead Publishing, 2001: 34-57.

[40] Sampath M B, Senthilkumar M. Effect of moisture management finish on comfort characteristics of microdenier polyester knitted fabrics. Journal of Industrial Textiles, 2009, 39(2):163-173.

[41] Gibson P. Water-repellent treatment on military uniform fabrics: Physiological and comfort implications. Journal of Industrial Textiles, 2008, 38(1):43-54.

[42] Yoo S, Barker R L. Moisture management properties of heat-resistant workwear fabrics-effects of hydrophilic finishes and hygroscopic fiber blends. Textile Research Journal, 2004, 74 (11): 995-1000.

[43] Ma W X, Zhao C, Okubayashi S, et al. A novel method of modifying poly(ethylene terephthalate) fabric using supercritical carbon dioxide. Journal of Applied Polymer Science, 2010, 117(4): 1897-1907.

[44] Dong W G, Huang G. Research on properties of nano polypropylene/TiO₂ composite fiber. Journal of Textile Research, 2002, 23(1):22-23.

[45] Wojtczak M, Kuzminiski H, Dobosz S, et al. Milt characteristics in European whitefish (Coregonus lavaretus) in relation to season and hormonal stimulation with a gonadotropin-releasing hormone analogue, azagly-nafarelin. Fundamental & Applied Limnology, 2007: 171-185.

[46] Viswanath C S, Ramachandran T. Comfort characteristics of cotton fabrics finished with fluoro-alkyl nano lotus finish. Indian Journal of Fibre Textile Research, 2010(35):342-348.

[47] Robinson R A, Stokes R H. The variation of equivalent conductance with concentration and temperature. Revista Da Associao MédicaBrasileira, 1954, 76(7):1-2.

[48] Mcqueen F M, Gao A, Ostergaard M, et al. High-grade MRI bone oedema is common within the surgical field in rheumatoid arthritis patients undergoing joint replacement and is associated with osteitis in subchondral bone. Annals of the Rheumatic Diseases, 2007, 66(12):1581-1587.

[49] Gao Y, Cranston R. Recent advances in antimicrobial treatments of textiles. Textile Research Journal, 2008, 78(1):60-72.

[50] Specos M M, García J J, Tornesello J, et al. Microencapsulated citronella oil for mosquito repellent finishing of cotton textiles. Transactions of the Royal Society of Tropical Medicine & Hygiene, 2010, 104(10):653-658.

[51] Antranikian G, Vorgias C E, Bertoldo C. Extreme environments as a resource for microorganisms and novel biocatalysts. Adv. Biochem. Eng. Biotechnol. , 2005, 96:219-262.

[52] Jayani R S, Shukla S K, Gupta R. Screening of bacterial strains for polygalacturonase activity: Its production by bacillus sphaericus. Enzyme Research, 2010(3).

[53] Levisson M, Van J D O, Kengen S W. Carboxylic ester hydrolases from hyperthermophiles. Extremophiles Life under Extreme Conditions, 2009, 13(4):567.

[54] Middlebrook W R, Phillips H. The Application of enzymes to the production of shrinkage-resistant wool and mixture fabrics. Coloration Technology, 2010, 57(5):137-143.

[55] Yoon M Y, Kellis J, Poulose A J. Enzymatic modification of polyester. Aatcc Review, 2002,2(6): 33-36.

[56] Colbrie G, Herrmann M, Heumann S, et al. Surface modification of polyacrylonitrile with nitrile hydratase and amidase from agrobacterium tumefaciens. Biocatalysis and Biotransformation, 2006, 24(6): 419-425.

Chapter 5 Wearable Smart Textiles

Wei Liu* and Na Meng

5.1 Introduction

Smart textiles is a term used to define the materials which have advanced responsive properties covering an extensive range of technologies and products. It could be also defined as the flexible fibrous structures which are expected to sense, actuate and control the external environment. The micro-electronic textile is one of the popular smart textiles which can be integrated or assembled by micro-electronic devices into a composite system. In addition, some of the new materials such as phase change, shape memory, chromic and conductive materials play important roles in the innovation, design and production of intelligent textiles. Smart textiles could be also applied for the development of multifunctional wearable products used for medical, health care, assistive, protective and other functional purposes.

Smart textiles have a promising future based on the development of smart materials. The products of smart textiles will be introduced systematically in this chapter, including heat responsive textiles, shape memory textiles, color-changing textiles, waterproof breathable textiles and electronic textiles.

5.2 Heat Responsive Textiles

Usually, heat responsive textiles are defined as fiber assembles (woven fabric based, knitted fabric based, nonwoven fabric based, and composites of the related textiles), which have stimulus-active function of sensoring the environmental temperature. Tem-

---

Corresponding Author:
Wei Liu
School of Fashion Technology, Shanghai University of Engineering Science, Shanghai 201620, China
E-mail: wliu@sues.edu.cn

perature is one of the most commonly monitored factor regarding functional textiles, such as sportswear, medical and rescue products. The measured temperature could be used as a stimulus-active element of alert, heating, lifesaving or analysis system, which aims to cool, heat and/or maintain body temperature.

5.2.1 Insulation Textiles

Currently, insulation clothes utilizing heating systems can be summarized into the following four types according to the heat source, including electric-heating clothing, solar thermal clothing, chemical energy thermal clothing, and phase change material thermal clothing.

5.2.1.1 Electric-heating Clothing

Electric-heating clothing is a class of clothing in which the heating element is driven by electrical power. It generally consists of power, heating element, automatic temperature control device, security device, etc. These components are connected to each other through wires. Considering the safety of the human body, an electric power source should be DC limited to 0~5 V. The heat carriers of electric-heating fabric are distributed on heat sensitive parts of the human body, such as chest, preabdomen, lower back, back, joints and so on[1]. Automatic temperature control system regulates the opening and closing of heating element according to the change of body temperature. Security device automatically turns off the power when the circuit breaks down. Currently, there are lots of brands of electric-heating clothing on the market, such as Gerbing, Tour Master, ProSmart and so on, as shown in Fig.5-1. The products could be electric jacket, vest, pant, shoe, sock, glove, etc. They are very useful and helpful for soldiers, workers, outdoor athletes who work in a cold environment for a long time, and these wearable devices could provide a better thermal comfort condition, as well as avoid frostbite. In addition, electric-heating clothing is useful in the medical field, for example, it provides hyperthermia for people with rheumatism, arthritis and poor blood circulation[2]; electric-heating clothing combined with other technologies, such as communication, monitoring, health care and physical therapy, can be developed into versatility smart clothing[3].

  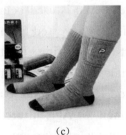 

(a) (b) (c) (d)

Fig.5-1 Various electric-heating clothing: (a) Electric-heating jacket produced by Tour Master company; (b) Carbon fiber electric-heating fabric produced by Gerbing company; (c) Electric-heating sock produced byReLang electrical company; (d) Electric-heating glove produced by ReLang electrical company

5.2.1.2 Solar Thermal Clothing

Solar thermal clothing is a kind of heating clothing, and its heat source is solar energy. According to the use of the solar energy, it can be divided into two types, one is solar battery based clothing and the other is solar thermal insulation material based clothing.

Solar battery can be made of solar materials, such as silicon-based solar materials and organic solar materials. Flexible solar battery made of organic solar materials is light and flexible, and it can be used in jacket. Green dye sensitive battery made from synthetic dyes not only can be used to heat garment, but also can be used to provide electricity for mobile phone, MP3 and other digital products. Solar thermal insulation materials with good heat preservation property can absorb the visible light and near-infrared, and reflect thermal radiation from human body. Solar selective absorption film is generally made of cermet materials, such as zirconium carbide. Solar fiber is usually formed by the combination of fine particles of zirconium carbide and fiber-forming polymer, and it has the capability to absorb the solar energy. In addition, these fibers can be assembled to prevent the diffusion of body heat. As a result, the clothes made of this kind of fiber possess pretty weatherization effect.

5.2.1.3 Chemical Energy Thermal Clothing

Chemical energy thermal clothing provides heat for human body mainly relying on some chemical reactions. In the market, common products are warm paste[4] produced in Japan, consists of iron powder, diatomaceous earth, activated carbon, wood flour, salt and superabsorbent composition. It mainly uses iron oxide exothermic principle to achieve heating purpose.

5.2.1.4 Phase Change Material Thermal Clothing[5]

Phase change material thermal clothing is a new type of smart textile, which can achieve liquid-solid and solid-liquid reversible changes according to the external environment[6]. When the environment temperature or body skin temperature reaches the melting point of phase change material within textiles, the phase change material can convert its state from solid into liquid to absorb heat from outside and save it as latent heat stored in textiles. If the phase change textile is used below its crystallization temperature in a cold environment, its state will be converted from liquid to solid to release the heat stored in textiles to achieve a short-term heating effect.

Phase change materials can be applied to textiles through composite spinning and hollow fiber filling method to achieve the purpose of thermo-regulation. For example, the United States Department of Agriculture Southern Lab[7-8] carried out a series of researches with the crystal water of inorganic salts or polyethylene glycol sealing in the hollow fiber inside in 1980s. In addition, phase change materials can be applied in textiles by post-treatment of fabric, such as the coating method.

5.2.2 Cooling Textiles

Cooling clothing can be dated back to the NASA's Apollo program in the late 1960s, which is a part of extravehicular spacesuit. Because the outer space environment is very harsh, high vacuum, strong radiation and severe changes of temperature, the astronauts must be protected in extravehicular. With the development of cooling garments, they can be also applied to pilots, tank crews, steel workers, miners and medical staff. Common cooling garments include circulating air cooling clothing, liquid cooling clothing, phase change cooling suits and fan cooling clothing.

5.2.2.1 Circulating Air Cooling Clothing

Circulating air cooling clothing use the cold air as the cooling medium. This clothing can absorb heat of vaporization and make latent heat exchange by accelerating the evaporation of sweat. In addition, the formation of air convection can make sensible heat transfer between the cool air and the human body to reduce the body surface temperature. The principle is similar to that of air conditioning system. Firstly, the air is cooled by cooling device and then purified. After the purification, the cold air is exhausted into the clothing through the turbine pipe, and then flows to cool the heat of the human body, and finally discharges outside to the clothing to complete a cycle. Although the circulating air cooling clothing has good cooling effect, light weight, low energy and consumption, the mobility is limited by its dilatability. With the rapid development of microelectromechanical systems (MEMS) technology, the refrigeration system in the circulating air cooling clothing trends to be miniaturized and lightweight.

5.2.2.2 Liquid Cooling Clothing

The main cooling mediums of liquid cooling clothing are water, ice-water mixture, sub-zero coolant composed of water and ethylene glycol, and microencapsulated phase change material (MEPCM) suspension liquid[9]. The liquid cooling clothing is mainly composed of the basic garment, PVC tube road network, liquid inlet and outlet tube, refrigeration equipment, pump, electron flux control equipment and a thin ninon lining, etc. It can be divided into liquid cooling helmet, vest and somatotype liquid-cooled clothing based on its requirement. Full body cooling clothing includes slices, single-tube rotary cooling clothing and multi-pipe straight cooling clothing. For the multi-pipe straight cooling clothing, because of its low flow resistance, good permeability, high cooling efficiency and reliability, it is widely used in spacesuit cooling system.

5.2.2.3 Phase Change Cooling Clothing

Phase change cooling clothing can be divided into different kinds depending on the cooling medium, including ice-cold clothing, dry ice cooling clothing, cold gel cooling clothing and phase change material cooling clothing. The principle of phase change cooling clothing is uncomplicated, using latent heat for heat exchange to achieve the purpose of cooling. Phase change cooling clothing is typically designed in the form of armor with several pockets in the front or back of the cooling clothing. Each pocket has a seal, and then the cooling medium can be placed in pockets for cooling purpose, as shown in Fig. 5-2.

Fig. 5-2　Phase change cooling clothing[10]

5.2.2.4　Fan Cooling Clothing

The principle of fan cooling clothing is that it strengthens airflow through the rotation of the fan promoting sweat evaporation and heat output. Early fan cooling clothing was developed by a Japanese company, which made the ordinary fan sewn into the fabric to achieve lower body temperature[11], as shown in Fig. 5-3. For the realization of cooling function of clothing, Liu Jing from the Chinese Academy of Materialization Sciences developed a new kind of air conditioning suit which used micro-fan-array to cool human body[12] (Fig. 5-4), and this study was achieved based on the current level of developing micro-processing technology. These mini-fans are small size, low power consumption, light weight, which are not only used for high-temperature workers, but also for ordinary consumers going out in summer[13].

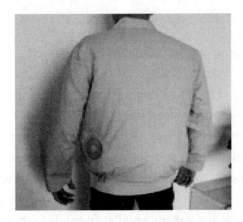

Fig. 5-3　Air conditioning clothing produced by a Japanese company[11]

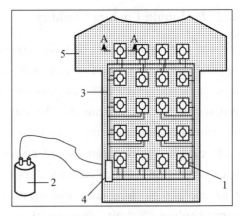

Fig. 5-4　Schematic of micro-fan-array based air conditioning suit for cooling human body[12]. (1—Mini-fan-array; 2—Micro power; 3—Circuit connection; 4—Switch array; 5—T-shirt)

5.3 Shape Memory Textiles

Materials possessing shape memory effect can be applied to textiles, such as metallic alloy, ceramic, glass, polymer and gel. The important and promising applications of shape memory materials mainly focus on biomedicine, textile and apparel, toy, packaging, national defense and industrial fields.

5.3.1 Shape Memory Alloy Textiles

Shape memory alloys (e.g. Cu-Zn-Al, Fe-Ni-Al, Ti-Ni alloys) are already used in biomedicine as cardiovascular stent, guide wire and orthodontic wire. The shape memory effect of these materials is based on a martensitic phase transformation[14].

For general metal, it is difficult for its shape to recover permanently after uploading when the strain exceeds 0.2%~0.5% in elastic zone. This is called plastic deformation. For shape memory alloy, the metal would fully restore to its original shape despite the strain outstrips the elastic limit, which is just like the metal can remember the past state when heat the metal or remove the external stress[15].

Shape memory materials can be used in apparel, bandage and so on, and most of them are mainly shape memory alloys. A smart shirt with automatic crimple sleeves, reported in Italy, was fabricated from a core yarn of nickel titanium alloy with hard-hand, and it is expensive to manufacture due to the high cost of nickel titanium alloy. The application of the shape memory alloys weaved in fabric has been reported in many patents and literature. However, the shape memory alloy fabric has the following disadvantages[16]:

(1) They have hard-hand feeling;
(2) Processing is different;
(3) Switch temperature limits;
(4) The fibers and fabrics are limited in the structure and in the use of apparel.

5.3.2 Shape Memory Polymer Textiles

Shape memory polymers (SMPs), a type of shape memory materials, are defined as polymeric materials with the ability to sense and respond to external stimuli in a predetermined shape. Compared with shape memory metals and ceramics, SMPs possess characteristics such as light weight, corrosion resistance (compared with metals), formability, workability and low cost. Some of SMPs, such as polynorbornylene, transpolyisoprene, styrene butadiene copolymer, block polyurethanes, polyethylene and polyester copolymerized with other materials, have been found to possess shape memory effect[16].

For preparation of textiles, clothing and related products, the applications of SMPs are wider than that of other shape memory materials. There are many forms of SMPs used in textiles, mainly including shape memory fibers, yarns, fabrics and chemicals. However, patents and literature are mainly focused on woven or nonwoven fabrics, in which the shape memory fibers are bonded with adhesive. SMPs can easily return to their original shapes when they are heated above the shape memory switch temperatures, even if they wrinkle or deform.

Several kinds of SMPs can be employed to prepare SMP fibers. They include shape memory polyurethanes(SMPUs, including modified SMPUs by incorporating ionic or mesogenic components into the hard segment phase), polyethylene terephthalate-polyethylene oxide copolymer, polystyrene-poly(1,4-butadiene) copolymer, polyethylene/nylone-6-graft copolymer, triblock copolymer made from poly(tetrahydrofuran) and poly(2-methyl-oxazoline), thermoplastic polynorbornene and other polymers with shape memory effect by cross-linking after spinning, such as polyethylene, poly(vinyl chloride) and polyethylene-poly(vinyl chloride) copolymer. There are many spinning methods which can be used to produce SMP fibers, including wet, dry, melt, reaction and electric.

Shape memory yarns include individual shape memory fibers, blended yarns of SMP fibers and natural, regenerated or synthetic fibers. The blended yarns can be core spun yarn, friction yarn and fancy yarn. Textiles also include sewing threads of SMP fibers and ordinary natural or synthetic fibers.

Shape memory fabrics include woven, knitted, braided and nonwoven fabric. The fabric can be applied to textiles depending on the properties and desired applications, such as collar, cuff of shirt, and any other apparels which need shape fixity; elbow, knee of apparel and any other clothes which need bagging recovery; and shape fixity of denim, velvet, cord, knitting fabric and any other fabrics. Shape memory fabrics can be obtained by finishing method through coating the shape memory emulsion or combining shape memory film onto fabrics. An U.S. patent reported a shape memory fabric which was obtained by bonding the SMPs powder onto the ordinary fabric[17]. Hong Kong Polytechnic University has studied shape memory finishing chemicals and technologies for cotton fabric, wool fabric and garment finishing[18]. This method is more environmental friendly, simple and better temperature sensitivity than other traditional finishing methods.

5.3.3 Shape Memory Gel Textiles

Shape memory gel(SMG) is one of the most promising new materials which can fulfill the future demand as a smart material[19-20]. Because of its biocompatible and biodegradable features, it is belonging to environmental friendly materials. SMG can memorize its original shape during gelation process, and this characteristic is called shape recovery property[21]. The SMG becomes soft and elastic if it is heated above its critical temperature, in this case, it can change its phase[22]. After deformation, this gel can recover its original shape by heating above the same critical temperature.

Hydrogel, a smart material, can be used to form shape memory textile. Its important feature is that hydrogel will occur volume phase transition under certain stimulus conditions, namely when the external environmental condition changes continuously, the gel volume can produce a discontinuous change. Studies have shown that the smaller the size of the gel, the faster its response to stimulation. Fiber with a high length/diameter ratio is help to improve the response rate of the gel[23]. Therefore, copolymerization, cross-linking, blending, coating, composite spinning and other methods can be used to prepare hydrogel fibers, and develop a special smart textile with specific intelligence. Hydrogel in the smart textile has two functions, one is to respond moisture and the other is to respond temperature.

One example of hydrogel fiber which can respond moisture is "Aikesu" fiber and it is developed by a Japanese company Toyobo. "Aikesu" fiber can be fabricated by introducing polyacrylic acid molecular chain which is treated by amino, carboxyl and other hydrophilic groups into common fiber. The moisture absorption performance of "Aikesu" fiber in standard state is 3.5 times higher than that of kapok fiber. The structure of "Aikesu" fiber would change after moisture absorption, in this case, hygroscopicity and exothermicity of the fiber become moderate, and that can prevent the cold sense after sweating[24]. Intelligent anti-leaching clothing is also hydrogel material which can be used to respond moisture. The anti-leaching principle is that once putting the clothes in water, anti-leaching hydrogel in the fabric will swell and cause the permeability channel quickly to be closed, and then prevent the water infiltrating into the clothes and maintain the human body with normal temperature for normal physiological activities. The hydrogel of the soaked fabric can be dehydrated by drying treatment, and then it can swell and return to the state of breathable and moisture permeable.

One example of hydrogel fiber which can respond temperature is the intellectualized thermoregulation diving suit which is developed by American researchers and it is considered to be a breakthrough in protective clothing of wet environments. The interlayer of the suit is composited by heat-sensitive hydrogel polymer and foam composite (namely SmartSkin) which is the core of the suit. Therefore, the working principle of the suit is that the gel shrinks and the foam material thins when the diver has a large amount of activity and high surface temperature, in this case, the flow of cold water outside the diving suit increases; the gel and the foam material swell when the surface temperature of the diver drops to a certain value, and then the outside cold water will be prevented to continually flow into the garment, so the heat exchange between the human body and the environment greatly reduces and the insulation can be achieved[25].

5.4　Color-changing Textiles

Color-changing textile can change its color according to the external stimulus, such as light and temperature. Discoloration textile can be mainly prepared by spinning with color-changing fiber, as well as by dyeing and printing technology.

Color fiber has a special composition or structure, which can automatically change its color after stimulated by light, heat, moisture, radiation or the other external conditions. Color fiber materials have been developed rapidly in recent years, because they are high-tech functional fibers full of vitality and have high values. There are four types of color fibers, including color photochromic fiber, thermochromic fiber, humidity color fiber and electrochromic fiber. In addition, photochromic and thermochromic fiber materials have wide applications.

Photochromic fiber can be produced by spinning technology after blending the photochromic material with polymer. Photochromic fiber will change color when it is exposed to light with a certain wavelength, and then change back to its original color when it is stimulated by another wavelength of light irradiation or the light disappears. Moreover, photochromic fabric can be prepared by staining or pigment printing technology, especially the color cotton and knitted fabrics, by using photochromic dye. For the latter, the photochromic dye powder mixes with a binder resin solution to obtain a paste, and then the paste was printted onto the fabric surface to obtain photochromic fabric. However, staining or pigment printing technology is unsuitable to prepare color fiber.

Thermochromic fiber can be prepared by filling the thermochromic agent into fiber or applying a kind of vinyl chloride polymer solution which contains thermochromic microcapsules. Heat treatment on the fiber surface should be conducted in the latter method to improve the reversible thermochronic effect.

Humidity color fiber can be produced by inorganic membrane which is made of cobalt, salts-cobalt or chloride-containing complexes with six knot water. Color changing effect is achieved by heating the inorganic membrane to lose water into cobalt chloride with two crystal water, and the shape and number of ligand complexes change, causing absorption spectrum (color) changes. It is used in printing process of textiles mixed with glue[26]. For example, when the fabric is dry, it is colorless; but it changes color in the form of presenting a pattern after wetting, as shown in Fig.5-5.

Electrochromism is a phenomenon presented by some materials with properties of reversibly changing colour by using bursts of charge to cause electrochemical redox reactions in electrochromic materials. Transition metal oxides are a large family of materials, which have a variety of interesting properties in the field of electrochromism.

Fig. 5-5 Humidity sensitive color printing paste-watermarking

Among them, tungsten oxide (WO_3) has been the most extensively studied material. It is usually applied to produce electrochromic windows or smart glasses, and can be also used in preparation of electrochromic displays on paper substrate as anti-counterfeiting systems integrated on packaging. Another example of electrochromic material is polyaniline which can be fabricated by electrochemical or chemical oxidation of aniline. Other electrochromic materials which have wide technological applications include viologens and polyoxotungstates.

Color textiles have good wearability and wide applications. It can be used in civil field for the preparation of fashionable garment, such as T-shirts, pants, swimwear, casual sportswear, clothes, and children's clothing. It can be also used in military field for military camouflage purpose, for instance, the soldier can conceal himself in landscapes when put on the color clothes which can change color in specific environment[27].

5.5 Waterproof Breathable Textiles

Waterproof breathable textiles are designed for application in garments to provide protection from wind and rain, and can be also used to prevent the loss of body heat. For example, waterproof breathable textiles can prevent the penetration of liquid water from outside to inside the clothing, and can permit the penetration of water vapor from inside to outside the clothing[28]. Waterproof breathable textiles can be applied under different conditions such as in leisure and in work as shown in Tab. 5-1.

Tab. 5-1 Applications of waterproof breathable fabrics[29-30]

Leisure	Work
Heavy duty, foul weather clothing: Anoraks, cagoules, packs, over-trousers, hats, gloves, gaiters	Foul weather clothing: Survival suits, special military protective clothing, clean-room garments, surgical garments, hospital drapes, mattress and seat covers, specialized tarpaulins, packaging, wound dressings, filtration
Fashionable weather protection: Rainwear, skiwear, golf suits, walking boot linings, panels and inserts, sport footwear linings	Domestic and transport: Non-allergic bedding, car covers, fire smoke curtains in ships, cargo wraps in aircraft

Waterproof breathable textiles can be divided into three types, including densely woven fabrics, membranes and coated fabrics. According to different kinds of waterproof breathable textiles, there are various preparation methods as described below.

5.5.1 Densely Woven Fabrics

The finest long staple cotton fibers are selected into the densely woven fabrics, so that there are very small spaces between the fibers. Before the preparation, the cotton should be processed into combed yarn to improve its regularity and ensures that the fibers are as parallel as possible to the yarn axis. In this case, it is difficult for the water to penetrate into the yarn which without large pores. The yarn is woven using an Oxford weave, which is a plain weave with two threads acting together in the warp. This gives minimum crimp in the weft, and ensures that the fibers are as parallel as possible to the surface of the fabric. When the fabric surface is wetted by water, the densely cotton fibers swell in the transverse direction, and then the pore size of the fabric reduces and a quite high pressure is required to cause penetration of water. Therefore, any water-repellent finishing treatment is not required for the densely woven fabrics[31].

Moreover, densely woven fabric can be also made from synthetic microfilament yarns. The individual filaments are less than 10 mm in diameter, so that fibers with very small pores can be designed. Microfilaments are usually made from polyamide or polyester. The latter is particularly useful because it has inherent water-repellent properties. Furthermore, water penetration resistance of the fabric can be improved by post-treatment with silicone or fluorocarbon[32].

5.5.2 Membranes

Thickness of a typical membrane is only about 10 μm, and it can provide a conventional textile fabric with necessary mechanical strength after laminating process. The membrane can be divided into two types, namely hydrophobicity with micro-pores and hydrophilic.

Expanded polytetrafluoroethylene (PTFE) polymer is one example of hydrophobic membranes and it has 1.4 billion tiny holes per square centimeter. These holes are much smaller than the smallest raindrop[33], but larger than a water vapor molecule. The hydrophobic nature of the polymer with small pore size requires a very high pressure to cause water penetration. Hydrophobic membranes with micro-pores usually have a layer of hydrophilic polyurethane to reduce contamination[34].

Hydrophilic membranes are very thin films of chemically modified polyester or polyurethane containing no holes, sometimes referred to as non-poromeric. Water vapor from perspiration is able to diffuse through the membrane in relatively large quantities. The polyester or polyurethane polymer is modified by incorporating with polyethylene oxide up to 40% by weight[35]. The poly(ethylene oxide) constitutes the hydrophilic part of the membrane by forming part of the amorphous regions of the polyurethane polymer system. It has a low energy affinity for water molecules which is essential for rapid diffusion of water vapor[36]. These amorphous regions are described as acting like intermolecular 'pores' allowing water vapor molecules to pass through but preventing the penetration of liquid water owing to the solid nature of the membrane.

5.5.3 Microporous Coatings

Microporous coatings can be prepared by two methods, including wet coagulation and thermocoagulation[37].

5.5.3.1 Wet Coagulation

Polyurethane polymer which is insoluble in water is dissolved in the organic solvent dimethyl formamide to produce a solution and then coated onto a fabric. The coated fabric is passed through a conditioning chamber containing water vapor. Because the organic solvent is miscible with water, in this case, it can be diluted and the precipitates of solid polyurethane can be formed onto the fabric. The fabric is then washed to remove the solvent, which leaves pores in the coating. Finally, the coated fabric is mangled and dried. However, the application of this method is limited by the high cost of machine and expensive solvent recovery.

Another example of wet coagulation is foam coating. For this technology, a mixture of polyurethane and polyurethane/polyacrylic acid esters is dispersed in water to form foam. The foam is stabilized with additives and then coated onto the fabric. The coated fabric is dried to form a microporous coating. It is important that the foam is microporous structure to allow penetration of water vapor, but it also has small enough cells to prevent the penetration of liquid water. The fabric is finally introduced into a low pressure to compress the coating. Because no organic solvents are used, this type of coating production is environmental friendly.

5.5.3.2 Thermocoagulation

Polyurethane is dissolved in an organic solvent and the resulting solution is mixed with water to produce emulsion. The emulsion "paste" is then coated onto the fabric followed by going through a two-step drying process. The first step employs a low temperature to remove the organic solvent to precipitate the polyurethane. In this case, the coating is a mixture of solid polyurethane and water. The second step employs a high temperature to evaporate the water and leave pores in the coating.

5.6 Electronic Textiles

5.6.1 Introduction

Electronic textiles are also called wearable electronics which can be prepared by introducing electronic devices into textiles to provide intelligent functions, such as memory, intellect, creativity, communication and physical senses[38]. In addition, the electronic textiles can be worn into the human body as implantable devices, such as pacemakers and neuroprosthetics. Wearable electronics can be also worn externally in the form of ring, badge, wristwatch, eyeglasses, jewelry, shoes or clothing. The key considerations for wearable electronics are that they have to be robust, small, consume a small amount of power and be comfortable to wear. Wearable electronics can function as sensors or computers that consist of input, output and a motherboard made up of transistors and various interconnections. The wearable computer must be powered by lightweight batteries or fuel cells and can't add too much additional burden to the wearer. Accordingly, the favorable ingredients for wearable electronics are lightweight, flexible and conductive materials. Conductive materials in fibrous form, such as yarns and fabrics, are preferred candidates for wearable electronics by serving as interconnects, functional devices and sensors.

5.6.2 Application

In the past few decades, many desk electronic appliances have been made portable because of constant miniaturization in electronics. It is reasonable to assume that, in the future, some of these portable devices can be wearable because they are small and convenient to carry. Applications of the technology will be developed and widespread[39].

5.6.2.1 Information and Communication

The evolution of lifestyles in recent years has led to increased mobility, and a strong desire for instant access to information and communications. The tendency is convenient that the devices should be more unobtrusive and convenient. Apparel integrated with wearable computer or mobile phone will have a market if it is in an appropriate price.

A wearable computer vest, constructed from commercial components in 1998, is illustrated in Fig. 5-6. The input device in the configuration presented in Fig. 5-6 is a chording keyboard called Twiddler, which allows one hand operation[40]. This wearable computer was the first prototype at Tampere University of Technology, and the later methods have been developed based on the previous experiences. The platform for the computer system is the vest, which contains several pockets for additional components. The function of the wearable computer which can move with its user is to provide a variety of data processing in daily life. The CPU (Via II) is integrated into the back of the vest[41]. A head mounted display V-Cap 1000 manufactured by Virtual Vision Inc. is connected to the CPU module with a thick and inflexible cable.

Fig. 5-6　First wearable computer assembly at Tampere University of Technology

5.6.2.2 Health Care and Medical Application

In affluent societies, people's health consciousness is getting stronger. Meanwhile, the ageing of the population in many developed societies is bringing a heavy burden to medical, especially hospital, social systems and government budgets. Wearable electronics may provide personal assistance to monitor people's physiological status. If necessary, medical advice or treatment can be given anywhere, not just in hospital to provide more mobility, efficient and effective health services.

The Wearable Cardioverter Defibrillator by LifeCor Inc. has a chest harness and a hip pack, which provides immediate emergency medical aid to people who is prone to heart attacks. As soon as electrodes of the defibrillator sense the irregular heartbeat, an audio warning is given before electricity is discharged. Then, the nearest hospital is notified. Another example is a wearable artificial kidney, which serves as a hemodialysis but has the advantage of being able to be fitted around the neck. In a form of undergarment, the wearable motherboard by Georgia Tech has sensors that are detachable so that they can be positioned at the right locations for users of different sizes. Such sensors can be used to monitor vital body signs (such as the heartbeat, respiration rate or temperature) of patients recovering from specific illnesses or to monitor patients at home rather than in hospital. The Wearable Polysomnography by Advanced Medical Corporation is a wearable ECG (electrocardiograph), which allows patients to be monitored at home with data being sent to hospital via an internet connection. Wearable medical devices can be integrated into a biosensor layer to monitor body conditions, such as heartbeat, blood pressure and temperature. Another function is called Virtual Doctor, which would assess and give advice on the overall health of the individual. In the future, the clothing may detect the user's feelings, moods, aches and pains, and monitor brain activity and change its color, pattern, shape, and even its smell. However, fashion items worn by ordinary consumers require a high standard of quality and easy care. They should be washable in addition to be wearable electronics and photonics flexible and robust, and ideally foolproof. Currently, this is a real challenge to most electronic devices[42].

Many people live with a physical handicap, such as the loss of a sense like vision and hearing, or a loss of mobility in one part of their body or in all of it. Some future wearable electronic devices could help alleviate the suffering of such people. For example, wearable devices equipped with artificial muscles can be designed to help limbs and arms, and increase the viability of theirs movement range. Personal guidance systems are being developed to help the blind to use the global positioning system (GPS), and the computer's geographical information system (GIS) can keep track of locations with the aid of a highly detailed map. Cochlear implants are being developed to help the deaf replace the lost functionality of damaged or missing hair cells by sending signals to the intact underlying nerve structure.

5.6.2.3 Fashion, Leisure and Home Application

Wearable electronics have a very exciting but challenging market in this sector. As wearable and fashionable clothing, it is important to attract people's attention and make people want to wear them and feel good when they do wear them.

The Philips/Levis ICD$^+$ can be viewed as the first generation of smart clothing because they integrate mobile phones and music players, which try to enhance the "organizer" functions of clothing. Until now, the closest wearable mobile communications have been Ericsson's Bluetooth headset. Nokia's prototype of the mobile snow jacket is attempting to have devices such as the mobile phone fully incorporated within clothing. France Telecom has developed a phone coat equipped with an extra-flat (100 g) mobile telephone integrated in its lining. The keypad is placed on the sleeve of the jacket and a microphone is discretely placed on the collar. Infleon's MP3 jacket is an application in the area of infotainment, a combination of information and entertainment.

Clothing has always acted as an interface between the body and the external world. They offer ample opportunities for the creation of intelligent clothing that performs functions according to the body's needs and requirements, and adapts to the environment. Some of the functions are described as follows. Intelligent clothes can give reminders to people, identifying and memorizing different objects and taking with them, such as keys and wallets. They may perform temperature feedback and control mechanisms in a smart jacket, and adjust its interior temperature[7, 43].

5.6.2.4 Military and Industrial Application

This is a very promising market where earlier penetration of the technology is expected to take place. In combat, soldiers must use their hands at all times to control weapons and machinery. Wearable devices would be very useful to assist them. The soldiers may be connected to navigation systems via wearable computers to guide them through difficult terrain and unknown areas. The systems may also let them know the positions of enemy and allied soldiersby satellite systems. Soldiers would also constant touch with their superiors. Others in nearby areas can be notified if a soldier falls down. Soldiers can look up on a stored database of information that how to fix any damaged equipment and even how to apply first aid to injured soldiers. Other people, such miners and mountaineers, may need navigation and detection systems to guide them to avoid dangerous areas.

Initially, the wearable electronic and photonics are expensive, and only the military and industrial sectors that especially required such performance can afford. However, the functions of the wearable electronic and photonics are dominant and bring more efficiency to consumers.

References

[1] Kukkonen K, Vuorela T, Rantanen J, et al. The design and implementation of electrically heated clothing. Proceedings Fifth International Symposium on Wearable Computers, Zurich, 2001.
[2] Koscheyev V S, Leon G R, Paul S, et al. Augmentation of blood circulation to the fingers by warming distant body areas. Eur. J. Appl. Physiol., 2000, 82 (1): 103-111.
[3] Ren P. Research and application of wearable human thermal comfort system technology. Graduate School of Chinese Academy of Sciences, 2009: 1-10.
[4] http://www.zgsunny.com/.
[5] Zhou L J, Xu J, Lu L N. The application of smart materials in textiles. Proceedings of Textile Bioengineering and Informatics Symposium 2010, 2010.
[6] Zhang D X, Guo F Z. The application of phase change materials in thermoregulation clothing. Journal of Knitting Industries, 2007, 3: 28.
[7] Vigo T L, Burno J S. Temperature adaptable textile fibers and method of preparing same. WO 8707854.
[8] Vigo T L, Frost C M. Temperature-adaptable fabrics. Journal of Textile Research, 1986, 56: 737.
[9] Li L N, Qian X M, Xu J. Development status and application of cooling clothing. China's Individual Protective Equipment, 2008 (2): 24-28.
[10] http://www.labbase.net/Product/ProductItems-29-61-439-20177.html.
[11] http://www.1688.com/gongsi/-C8D5B1BEBFD5B5F7B7FE.html.
[12] Liu J. Based on nano/micron fan array driven by the human body cooling air conditioning service. CN 2005101027002.

[13] Zeng Y Z, Deng Z S, Liu J. Micro-fan array system based on human cooling air conditioning service. Journal of Textile Science, 2007, 28 (6): 100-105.
[14] Tao X M. Smart Fibers, Fabrics and Clothing. Cambridge: Woodhead Publishing,2001:280-281.
[15] Zhou M L, Gao W, Tian Y T. Hybrid control based on inverse Prandtl-Ishlinskii model for magnetic shape memory alloy actuator. Journal of Central South University, 2013, 5:1214-1220.
[16] Leng J S, Du S Y. Shape memory polymers and multifunctional composites. CRC Press, 2010: 294, 297-299.
[17] Araki E, Sugihara N, Nakao K, et al. Process for producing adhesive for fusion bonding, adhesive for fusion bonding obtained by the process, and adhesive fabric containing the adhesive for fusion bonding. US 6743741 B1, 2004.
[18] Lu J, Hu J L, Zhu Y,et al. Shape memory finishing of wool fabrics and garments. Advanced Materials Research, 2012, 441: 235-238.
[19] Varadan V K. Nanosensors, biosensors, and info-tech sensors and systems. Nanosensors, 2011, 7980: 284-318.
[20] Osada Y, Matsuda A. Shape memory in hydrogels. Nature, 1995, 376(6537):219-221.
[21] Kabir M H, Gong J, Watanabe Y, et al. Hard-to-soft transition of transparent shape memory gels and the first observation of their critical temperature studied with scanning microscopic light scattering. Mater. Lett. , 2013, 108(5): 239-242.
[22] Tanaka T, Sato E, Hirokawa Y, et al. Critical kinetics of volume phase transition of gels. Phys. Rev. Lett. , 1985, 55(22): 2455-2458.
[23] Gong J, Watanabe Y, Watanabe Y, et al. Development of a novel standard type of gel engineering materials via simple bulk polymerization. Journal of Solid Mechanics and Materials Engineering, 2013, 7(3): 455-462.
[24] Furukawa H, Horie K, Nozaki R, et al. Swelling-induced modulation of static and dynamic fluctuations in polyacrylamide gels observed by scanning microscopic light scattering. Phys. Rev. E. , 2003, 68(3):1406-1414.
[25] Yu J L, Liu J Q, Li Y Q. Shape memory function fibers and textiles. Synthetic Fiber, 2008, 37(2): 6-8.
[26] Xue L Y. Discoloration materials and their application in textiles. Shanghai Wool & Jute Journal, 2010(1): 20-22.
[27] Wang W Z. Color-changing textiles and their applications. Shandong Textile Science and Technology, 2007(2): 54-56.
[28] Holmes D A, Grundy C, Rowe H D. The characteristics of waterproof breathable fabric. Journal of Industrial Textiles, 1995, 29(29): 306-313.
[29] Krishnan S. Technology of breathable coatings. Journal of Coated Fabrics, 1992, 22(7): 71-74.
[30] Horrocks A R, Anand S C. Handbook of Technical Textiles. The Textile Institute, 2000(12):282-315.
[31] David A H. Waterproof breathable fabrics. Handbook of Technical Textiles, 2000(29):282-315.
[32] Lomax G R. Journal of Coated Fabrics. 1995,15(10): 115-126.
[33] Hosokawa T. Waterproof-breathable fabric,"Gore-Tex Laminate®". Sen-ito Kogyo, 1979, 35(3): 69-73.
[34] Lomax G R. Hydrophilic coatings. Journal of Coated Fabrics, 1990,20(10): 88-107.
[35] Lomax G R. Breathable waterproof fabrics explained. Textiles, 1991(4): 12.

[36] Krishnan S. Waterproof breathable polyurethane membranes and poroussubstrates protected therewith. US 5239037A, 1993.

[37] Perumalraj R, Lecturer S. Waterproof breathable fibre. Fiber2fashion.com.

[38] Tao X M. Wearable electronics and photonics. Cambridge: Woodhead Publishing, 2005.

[39] Bar-Cohen Y. Electroactive polymer(EAP) actuators as artificial muscles, reality, potential and challenges. SPIE Press, 2001.

[40] Handykey Corporation. http://www.handykey.com/, 2002-11-01.

[41] ViA Inc.. http://www.via-pc.com/, 2002-11-01.

[42] Tao X M. Smart fibres, fabrics and clothing. Cambridge: Woodhead Publishing, 2001.

[43] Venture Development Corporation. Smart Fabrics and Intelligent Clothing, Boston, 2003.

Chapter 6 Nanospinning and Nanowrapping Technologies

Xi Tong, Binjie Xin* and Zhuoming Chen

6.1 Introduction

PSA fiber belongs to the series of Poly Para-Phenylene Terephthalamide (PPTA) fibers. PSA fiber, as a new kind of high-performance synthetic fiber, has some excellent properties, such as heat resistance, thermal stability and flame retardancy. At present, PSA fiber is mainly prepared by wet-spinning[1-2] and electro-spinning[3].

Wet-spinning is suitable for polymer which cannot be molten and can be only dissolved in non-volatile or heat-unstable solvent. In addition, the polymer with high rigid-chain, such as acrylic, vinylon and polyvinyl chloride fibers, is generally prepared by wet-spinning. Electro-spinning process is applicable to many polymers which can be a melt or a solution. The diameter of the fiber prepared by wet-spinning is at micron scale, while the diameter of the fiber prepared by electro-spinning is at nano-scale which can meet the functional requirements of high-temperature filtration and strong absorption. Research on the influence of high-voltage electrostatic field fluid on the performance of fiber can be traced back to over 100 years ago. As early as 1882, Rayleigh[4] theoretically found the critical conditions for the formation of fine trickle; in 1929, Kiyohiko[5] prepared silk fiber by high-voltage electrostatic method; in 1964, Taylor[6] found Taylor cone, and he thought that droplets in the electric field are mainly affected by the

---

Corresponding Author:
Binjie Xin
School of Fashion Technology, Shanghai University of Engineering Science, Shanghai 201620, China
E-mail: xinbj@sues.edu.cn

two forces of the joint action: electric field force and surface tension. In 1966, Simons[7] invented an electro-spinning device with specified shape. Using this device, nanofiber aggregates of different shapes could be prepared. Since then, many researchers[8-12] reformed the electro-spinning device with different spinnerets and receiving devices to prepare different kind polymer fibers, meanwhile, computational fluid dynamics (CFD) was used to simulate electro-spinning jets. Recently, Jiang et al[13] invented a stepped pyramid-shaped spinneret which can increase the output of electro-spinning fiber. Zhang et al[14] have prepared highly oriented nanofibers by using a needleless linear spiral electro-spinning device, in this device, the traditional spinning needles were replaced by spiral coils.

Nowadays, many researchers have investigated the influences of various spinning methods on the performances of fibers. Bonhomme et al[15] have carried out the theoretical analysis on the microfluidic wet-spinning alginate micro fiber formation, it showed that the selection of fibers against pieces-of-gel is controlled by the stress exerted on the jet at the level of the spinneret. Xie et al[16] simulated and studied the effect of electric field on multi-needle electro-spinning, this study showed a possibility that by designing the electric field distribution, finer fibers, and better oriented fiber bundles could be prepared at a high production rate. Demir[17] studied the effects of the two processing variables of electro-spinning and wet-spinning, including the drawing rate and the extrusion rate, on the elastic fiber diameter. He found that the rate of drawing was inversely proportional whereas the rate of extrusion was directly proportional to fiber diameter. Cho et al[18] studied the effect of molecular weight and storage time of the spinning solution on the properties of the regenerated silk fibroin fabricated by wet-spinning and electro-spinning, the fiber diameter could be controlled from 100 to 800 nm by changing the molecular weight and storage time. Although many scholars have done a lot of research on wet spinning and electro-spinning, and have drawn many important conclusions, there are few studies on the comparison of the flow field and electric field simulation in the above two spinning processes.

In this study, PSA fiber was prepared by wet-spinning and electro-spinning. Morphology, molecular structure and chemical composition, crystallinity, thermal stability and mechanical properties of the fiber prepared by the two different spinning methods were systematically investigated. The influences of the two methods on the properties of PSA fiber would be compared and studied. Moreover, corresponding models referring to the two technologies were established by COMSOL multi-physics software for the investigation.

6.1.1 Fluid Movement Parameter Simulation of Wet-spinning

Schematic diagram and simulation model of wet-spinning are respectively shown in Fig.6-1 (a) and (b). As illustrated in Fig.6-1 (a), a certain volume of PSA solution is ejected from the single-needle spinneret under an extrusion force of nitrogen during the wet-spinning process, and then the PSA fluid is solidified and molded in a coagulating bath (aqueous solution) to form a fiber. The spinning parameters (pressure: 1.5 Pa; spinneret diameter: 0.05 mm; receiving length: 1.2 m) can be set in COMSOL to realize the simulation during the processing of fiber extrusion, solidification and molding. As shown in Fig.6-1 (b), the left side of image is the high pressure area and the right side is the low pressure area during the spinning process.

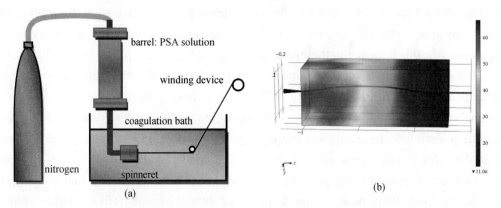

Fig.6-1　Images of wet-spinning: (a) Schematic diagram; (b) Simulation model

The models of wet-spinning under different spinneret pressures (a1, a2, a3) and spinneret diameters (b1, b2, b3) are illustrated in Fig.6-2. As can be observed from the images of series a, the diameter of the PSA fiber decreases with the increasing spinneret pressure from 0.5 to 1.5 Pa. A slight yellow area suggesting a very small pressure is observed in the left part of image (a1), and a faint red area is discovered in left part of image (a2) as the pressure increased from 0.5 to 1.0 Pa. A significant red area indicating a high pressure is found in left part of image (a3) as the pressure further increased from 1.0 to 1.5 Pa. As can be seen from the images of series b, the diameter of the PSA fiber increased with the increasing spinneret diameter from 0.05 to 0.15 mm. The color of the left part is yellow and does not change which indicates that the pressure is low and it does not change with the increasing diameter.

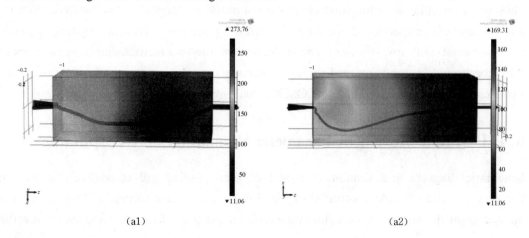

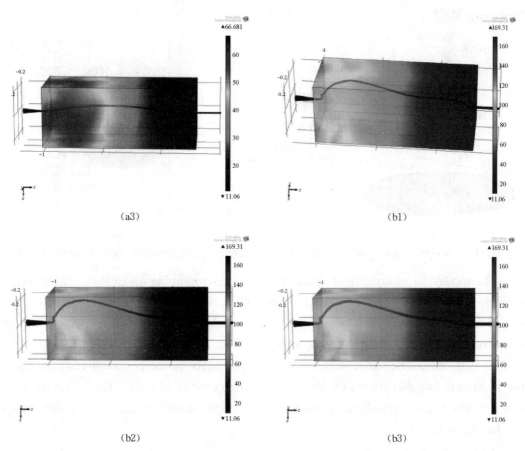

Fig. 6-2 Simulation models of wet-spinning with different spinneret pressures and diameters: (a1) 0.5 Pa; (a2) 1.0 Pa; (a3) 1.5 Pa; (b1) 0.05 mm; (b2) 0.10 mm; (b3) 0.15 mm

6.1.2 Electric Field Simulation of Electro-spinning

Schematic diagram and simulation model of electro-spinning process are respectively shown in Fig. 6-3 (a) and (b). The polymer solution/melt can be accelerated from the Taylor cone to the receiver under a high voltage static electricity and the fiber is stretched by the electrostatic field during the electro-spinning process. When the applied electrostatic field force is large enough, the polymer can overcome the surface tension to form a thin and long jet flow. During the electro-spinning, the solvent evaporates in the injection processing. Nonwoven fabric can be eventually obtained after the fiber depositing on the receiver. The electro-spinning parameters (spinneret diameter: 0.05 mm; voltage: 28 kV; electro-spinning distance: 15 cm) can be set in COMSOL to realize the simulation of electric field as exhibited in Fig. 6-3. The distribution of electric field is concentrated around the syringe needle but loosen away from the needle because a positive voltage is applied on the needles directly.

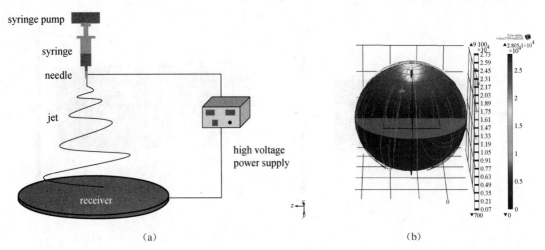

Fig. 6-3 Images of electro-spinning: (a) Schematic diagram; (b) Simulation model of electric field

The models of electric field simulation of electro-spinning under different spinneret diameters (c1, c2, c3) and voltages (d1, d2, d3) are shown in Fig. 6-4. As can be observed from the images of series c, the electric field lines are concentrated around the spinneret and the coverage area expands with the increasing spinneret diameter from 0.03 to 0.07 mm. The result indicates that the spinneret diameter not only affects the fiber diameter, but also affects the distribution of electric field force. The images of series d reveal that the electro-spinning voltage has no significant influence on the distribution of electric field.

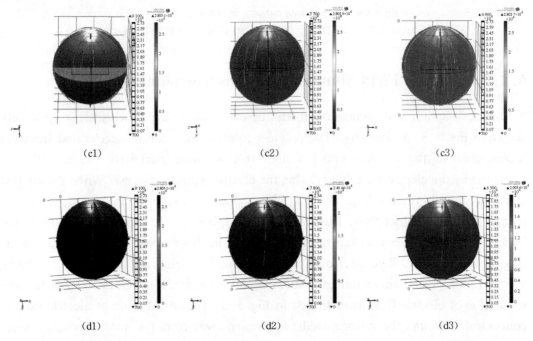

Fig. 6-4 Simulation models of electric field of electro-spinning under different spinneret diameters and voltages:
(c1) 0.03 mm; (c2) 0.05 mm; (c3) 0.07 mm; (d1) 20 kV; (d2) 24 kV; (d3) 28 kV

The models of electric field simulation of electro-spinning under multiple spinnerets and spinneret plate are shown in Fig. 6-5. The lines of electric field in image (e1) are uniformly distributed around the spinneret without interfering with each other. Although the lines of electric field generated by the spinnerets interact with each other when the number of spinneret increases from 1 to 3 and 5 [images (e2) and (e3)], the electric fields distribute uniformly around the overall unit. The different forms of electric fields indicate various spinning stress when the solution is ejected with different numbers of spinneret. As shown in image (e4), the electric field distribution of spinneret plate is similar with that of single-needle spinneret, and the electric field is uniformly distributed around the spinneret plate.

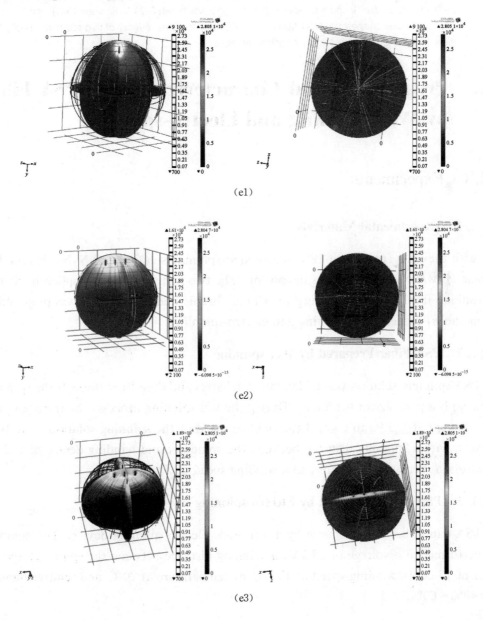

(e1)

(e2)

(e3)

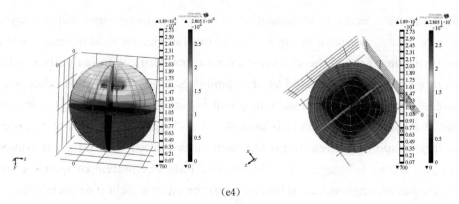

(e4)

Fig. 6-5 Simulation models of electric field of electro-spinning under multiple spinnerets and spinneret plate: (e1) Main view (left) and top view (right) of single spinneret; (e2) Main view (left) and top view (right) of three spinnerets; (e3) Main view (left) and top view (right) of five spinnerets; (e4) Main view (left) and top view (right) of spinneret plate

6.2 Preparation and Characterization of PSA Fiber by Wet-spinning and Electro-spinning

6.2.1 Experiments

6.2.1.1 Experimental Materials

PSA with the mass fraction of 12% was used as spinning solution and N,N,-dimethylacetamide (DMAc) was selected as dissolvent. The two materials were supplied by Shanghai Tanlon Fiber Co., Ltd., Shanghai, China. In this study, PSA fiber was prepared by two methods, namely wet-spinning and electro-spinning.

6.2.1.2 PSA Fiber Prepared by Wet-spinning

The PSA spinning solution was added into the barrel and then flow through the spinning nozzle with a pore size of 0.05 mm. During the wet-spinning process, the nitrogen pressure was kept at 1.5 Pa to control the flowing speed of the spinning solution. The fiber was spun with a spinning distance between the spinneret and winding device of 1.2 m, an extrusion speed of 15 m/min and a winding speed of 30 m/min.

6.2.1.3 PSA Fiber Prepared by Electro-spinning

The PSA nano-fiber was prepared by a self-made electro-spinning device. The spinning was performed at a voltage of 28 kV, a spinning distance between the spinneret and receiver of 15 cm, a winding speed of 45 r/min, temperature at 20 ℃ and relative humidity at 40%~60%.

6.2.1.4 Characterization

S-3400N Scanning Electron Microscope (SEM) with a resolution at nano-scale was used to analyze the morphological structure of the PSA fiber, the machine was operated at 10~15 kV. The diameter of the fiber is measured by Model 1.2.0 Nano Measurer Software.

American AVATAR 370 Fourier Transform Infrared Spectroscopy (FTIR) was used to investigate the molecular structure and chemical composition of the PSA fiber. The spectra data were recorded from 4000 to 500 cm^{-1} with a resolution of 4 cm^{-1} and the step size is about 1.929 cm^{-1}.

780 FirmV_06 X-Ray Diffraction (XRD) was used to characterize the crystalline structure of the PSA fiber with a CuKα radiation ($\lambda = 0.154$ nm) at a voltage of 40 kV and a current of 40 mA. The spectra were obtained at 2θ angles from 5° to 90° with a scanning speed of 0.8 sec/step.

YG006 Electronic Single Fiber Strength Tester was used to investigate the mechanical properties of the PSA fiber. The gauge length of the sample was 10 mm. The elongation speed was 20 mm/min. The measurements for each sample were conducted for 10 times and the average value was used for the result analysis.

Germany STA PT-1000 Thermal Gravimetric Analyzer (TGA) was used to investigate the thermal stability of the PSA fiber. The TG experiment was carried out under nitrogen atmosphere with gas flow of 80~100 mL/min. Samples were heated from room temperature to 700 °C at a heating rate of 20 °C/min.

6.2.2 Results and Discussion

6.2.2.1 Morphological Structure

SEM images of the fiber prepared by the two methods are shown in Fig.6-6. Significantly, the diameter of the fiber prepared by wet-spinning (fiber diameter: 230 μm) is larger than that (fiber diameter: 0.3~2.0 μm) prepared by electro-spinning. That is because there is die-swell effect[19-20] at the spinneret when the fiber is ejected by using wet-spinning, and then the diameter of the extruded fiber is larger than the spinneret diameter (50 μm). For the electro-spinning, the spinning solution is ejected from the needle under a high-voltage electric field which can overcome the surface tension of the fiber and quickly pull the fiber at a nano-scale diameter.

6.2.2.2 Fourier Transform Infrared Spectral

FTIR spectra of PSA fiber prepared by wet-spinning and electro-spinning are shown in Fig.6-7. As can be observed in Fig.6-7, the two preparation methods have no significant

Fig.6-6　SEM images of PSA fiber prepared by: (a) Wet-spinning; (b) Electro-spinning

effect on the position and shape of the characteristic peaks of PSA fiber. The absorption peaks exhibiting at about 3300 cm^{-1} reflect the stretching vibration of —N—H of amido bond, and they are presented in the form of a singlet. The absorption peaks at about 1660 cm^{-1} indicate the stretching vibration of —C=O— of amide bond. The absorption peaks at about 1590 and 1530 cm^{-1} are corresponding to the stretching vibration of —C=C— of benzene ring. The absorption peaks at 1500~1300 cm^{-1} are mainly caused by the in-plane bending vibration of C—H. The absorption peaks at 1300~1000 cm^{-1} correspond to the skeletal vibration of C—C and the peaks at 1000~650 cm^{-1} are the out-of-plane bending vibration of C—H. The absorption peaks at 1320 and 1140 cm^{-1} are caused by the symmetrical and anti-symmetric stretching vibrations of —SO$_2$—, respectively. The characteristic peaks mentioned above can confirm the presence of PSA[21-23].

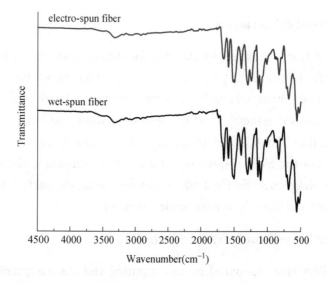

Fig.6-7　FTIR spectra of PSA fiber prepared by wet-spinning and electro-spinning

6.2.2.3 Crystal Structure

XRD patterns of PSA fiber prepared by wet-spinning and electro-spinning are shown in Fig. 6-8. Characteristic peak at 22° is observed in both samples.

Crystallinity of the fiber prepared by wet-spinning is not only influenced by the orientation effect of mechanical pressure, but also by the solidification conditions, including temperature, concentration and length of coagulating bath. In order to improve the crystallinity of the fiber, a winding and stretching system should be introduced during the wet-spinning process and then molecular rearrangements can be occurred[24-25].

Crystallinity of the fiber prepared by electro-spinning is influenced by the orientation effect of the applied electric field force and the crystallization time during the movement process of jet-flow. The crystallization time is affected by electro-spinning distance and voltage. The stretching effect is poor when the electro-spinning distance is too long or too short, in addition, the electric field intensity is low when the distance is too long and these will lead to a low crystallization[26-27]. Moreover, the higher voltage is, the faster jet velocity and the shorter crystallization time will be. It's difficult for the fiber to mold if the voltage is too low. Therefore, it is difficult to increase the crystallinity of fiber by adjusting the applied electric field force, the electro-spinning distance and voltage.

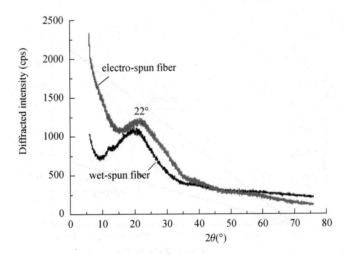

Fig. 6-8　XRD patterns of PSA fiber prepared by wet-spinning and electro-spinning

Comparing the mechanical pressure and the electric field force, supposing that the mechanical pressure of wet-spinning is applied to spinneret, and the electric field is uniform during electro-spinning, it can be concluded that:

$$F(M) = P \times S \tag{6-1}$$

$$F(M) = \frac{101\ 325}{2} \times 3.14 \times \left(\frac{0.000\ 18}{2}\right)^2 = 1.29 \times 10^{-3} \tag{6-2}$$

in where, $F(M)$ is mechanical pressure (N), P is intensity of pressure (Pa) and S is stressed area (m²).

$$F(E) = Eq = \frac{U}{d} \times q \tag{6-3}$$

$$F(E) = \frac{28\,000}{0.15} \times 1.6 \times 10^{-19} = 2.99 \times 10^{-15} \tag{6-4}$$

in where, $F(E)$ is electric field force (N), E is electric field intensity (N/C) q is electron charge (C), U is voltage (V) and d is spinning distance (m).

Significantly, the value of $F(M)$ is larger than that of $F(E)$, the result indicates that more mechanical pressure can be applied to the polymer solution when prepared by wet-spinning, which will lead to a more tight and regular molecular arrangement and an improved crystallinity of the prepared fiber.

The spinning pressure during the two preparation processes is illustrated in Fig. 6-9. For the electro-spinning, the pressure is generated by voltage, while for the wet-spinning, the pressure is generated by the mechanical pressure. As shown in the Fig., the spinning pressure increases as the raising intensity of mechanical pressure and voltage for wet-spinning and electro-spinning respectively.

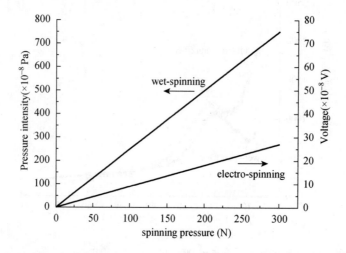

Fig. 6-9 Comparison of pressure during the wet-spinning and electro-spinning

6.2.2.4 Mechanical Properties

Physical parameters of mechanical properties of PSA fiber prepared by wet-spinning and electro-spinning are shown in Tab. 6-1. The value of breaking strength, elongation and initial modulus of the fiber prepared by wet-spinning are higher than that fabricated by electro-spinning.

Tab. 6-1 Parameters of mechanical properties of PSA fiber

Mode	Breaking strength (cN/dtex)	Elongation rate (%)	Initial modulus (cN/dtex)	Mean square deviation of breaking strength (cN/dtex)
Wet-spinning	0.411	29.70	0.098	0.091
Electro-spinning	0.399	28.40	0.086	0.104
Difference	3%	4.5%	13.95%	—

The breaking strength of the fiber is affected by initial thrust[28], internal structure of the molded fiber, molecule exchange[29] during solvent evaporation, fiber crystallization and subsequent drafting[30]. As discussed above, the value of $F(M)$ is larger than that of $F(E)$, which indicates that there is different stress between the extrusion molding of wet-spinning and drafting molding of electric field force. In this case, different tightness of the internal structure of the fiber can be generated, and then it can affect the breaking force. For the electro-spinning process, the fiber is directly extruded from spinneret into the air. During solvent evaporation, electrostatic force not only affects the morphologies of the fiber, but also affects the breaking force. It is observed from Tab. 6-1 that the electro-spinning parameters can significantly influence the mechanical properties of the fiber. In addition, the mechanical properties of the fiber prepared by wet-spinning are better than that by electro-spinning.

6.2.2.5 Thermal Stability

TG and DTG curves of PSA fiber prepared by wet-spinning and electro-spinning are shown in Fig. 6-10. Physical parameters of TG curves are illustrated in Tab. 6-2. As shown in Fig. 6-10, both the thermal decomposition behaviors of PSA fiber prepared by the two technologies are presented in the form of decomposition process of a weight loss period. The thermal decomposition process of PSA fiber can be divided into 3 intervals[31].

The first interval is the trace weight loss period (room temperature~400 ℃). The weight loss is mainly caused by the volatilization of bound water and various auxiliaries among high polymer molecules when the temperature rises from room temperature to 100 ℃. The weight loss is mainly caused by rupture of weak bond of high polymer and small molecule decomposition when the temperature continuously rises from 100 to 400 ℃.

The second interval is the thermal decomposition period (400~600 ℃). The weight loss in this high-temperature period may be caused by the reason that movement rate of high polymer macromolecule chain continuously increases and becomes increasingly severe until it breaks. In this case, the high polymer releases in the form of gas, thus causing weight loss. The two samples show a rapid decomposition at about 470 ℃. Corresponding to the rapid decomposition there is a weight loss peak in DTG curve and the

T_{max} (presented in Tab. 6-2) can be determined according to the value of the maximum peak[32]. As shown in the DTG curve, the peak point of the fiber prepared by wet-spinning is obviously higher than that by electro-spinning. The result indicates that the fiber fabricated by electro-spinning has a higher weight loss than that by wet-spinning when it is at a high temperature (470 ℃). That can be attributed to the poor crystallinity and high surface-to-volume ratio of the fiber fabricated by electro-spinning. In this case, the fiber can absorb more heat in unit time and leads to the thermal decomposition.

The third interval is the carbon stabilization period (600~700 ℃). As shown in Fig. 6-10, most of the polymers are carbonized in this period, and the raising temperature has little influence on the weight loss of the residual substance, and the two TG curves maintain stable. According to the Tab. 6-2, the initial decomposition temperature and the residual weight rate of the fiber prepared by wet-spinning are higher, and it indicates that the thermal stability of PSA fiber prepared by wet-spinning is better than that by electro-spinning.

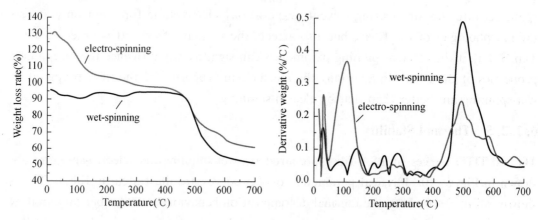

Fig. 6-10 TG curve (left) and DTG curve (right) of PSA fiber

Tab. 6-2 Parameters of thermal decomposition process of PSA fibers

Spinning method	T_0(℃)	$d\alpha/dt$	T_{max}(℃)	α(%)
Wet-spinning	460.90	0.471 5	495.41	41.10
Electro-spinning	392.57	0.240 1	491.22	36.48

T_0—The initial decomposition temperature;
$d\alpha/dt$—The maximum decomposition rate;
T_{max}—The temperature at the maximum decomposition rate;
α—The residual weight rate at 700 ℃.

6.2.3 Summary

In this part, PSA fiber was prepared by wet-spinning and electro-spinning. The prepared PSA fiber by the two methods was systematically investigated by SEM, FTIR, XRD and TG. The mechanical properties of the fiber were also analyzed. The experimental results can be summarized as follows. The spinning solution is extruded by mechanical pressure when the fiber is prepared by wet-spinning, while the solution is extruded by electrostatic force when the fiber is prepared by electro-spinning. The former has a larger acting force and higher intensity than that of the latter. PSA fiber prepared by wet-spinning has a larger diameter, better mechanical properties and thermal stability when compared with the fiber prepared by electro-spinning. The PSA nanofiber prepared by electro-spinning can be used for filtration and medical materials. The potential application of the PSA nanofiber will be wide when it is made into yarn. Because of the good mechanical properties and thermal stability of the PSA fiber fabricated by wet-spinning, it has bright application in the preparation of protective clothes.

6.3 Preparation and Characterization of Coaxial Electrostatic Spinning of PSA/PU Composite Fiber

Co-electro-spinning is one of the new expanded methods based on the previously known electro-spinning technologies[33-34], which can fabricate the core-shell fibers at a nanoscale. To develop new functional nanofibers, some researchers have presented the concept of co-electro-spinning (or coaxial electro-spinning) to produce the core-shell composite nanofibers[35-38]. It was reported that the polymer solutions could be supplied from a spinneret consisting of two coaxial capillaries. A core-shell droplet emerges and spurts from the spinneret, and the jet could be sprayed fast away from its tip when the electric field is sufficiently strong[39]. The sprayed jet could be solidified after evaporation of the solvent existing inside the jet, after that, the as-spun fibers with the core-shell structure could be deposited on the electrode receiver[40].

In recent years, many researchers have investigated the optimized co-electro-spinning methods and the related equipment, such as increasing the numbers of spinneret to improve the distribution performance of the electric field[41-42], or designing a new flat core-shell spinneret to expand the distribution of the electric field[43]. Zhou et al[44] reported a new needleless emulsion electro-spinning method for the scale-up fabrication of ultrathin core-shell polyacrylonitrile/isophorone diisocyanate (PAN/IPDI) fibers, which can be incorporated at the interfaces of polymer composites for interfacial toughening and self-repairing due to the polymerization of IPDI which can be triggered by environmental moisture. Zheng et al[45] designed a new multi-hole spinneret with equilateral hexagon hole array to meet a high production requirement and high quality of nanofibers. Wu et al[46] developed an approach for the scale-up of co-electro-spinning via a flat core-shell structure spinneret which has a flat surface involving shell-holes and core-needles to fabricate composite nanofibers with special morphologies. Meanwhile, some scientists have developed new functional materials at a nanoscale by using these co-electro-spinning methods. Lin et al[47] reported a polystyrene/polyurethane (PS/PU) fiber prepared by coaxial electro-spinning as a sorbent for oil soak-up. They showed that the resultant composite PS/PU fibers orientated randomly in a form of nonwoven mats with nanoscale structure. Zhu et al[48] synthesized p-CuO/n-TiO$_2$ composite nanofibers by using side-by-side electro-spinning combined with sol-gel process. They found that the configuration of spinneret had obvious effect on the preparation efficiency of the composite fibers.

For obtaining multifunctional fibers, the location and spinning direction of the spinnerets could be changed to prepare bi-component fibers. Tijing et al[49] used side-by-side angled two-nozzle electro-spinning to improve the quality of the electro-spun nanofibers. In addition, some researchers have developed special functional fibers by blending with different polymer materials for complementary performances. For example, Chen et al[50] prepared nanoscale fibers by using effective coaxial co-electro-spinning systems via combining the flexible thermoplastic elastomer polyurethane (TPU) component with a rigid thermoplastic component. Kessick et al[51] reported a formation of core-shell structure stemming from a two-component solution by a conventional electro-spinning, and the composite solution includes a conducting polymer poly(aniline sulfonic acid) and a non-conducting polymer poly(ethylene oxide).

Although many researches have been conducted for the development of core-shell nanofiber, very few works have been reported about the preparation of PSA/PU composite fiber. It is valuable to present a new method and the related spinning system which can be used to develop the PSA/PU core-shell nanofiber. It is well known that PSA fiber has excellent heat resistance, flame retardant and thermal stability, and it can be applied to develop those protective products used in aerospace, high-temperature environments and civil fields to satisfy the flame retardant requirements. Moreover, PU consists of hard and soft segments. The hard segments of PU can improve its mechanical properties due to a relatively high transition temperature, whereas the soft segments of PU have a better elasticity. What's more, it is reported that PU has a good blood compatibility, so it can bring a considerable value in medical devices[52]. However, the thermal stability of PU is not as good as PSA, and the mechanical properties of the PSA are not good enough for the designing of functional textile in some cases. Therefore, it is possible to combine these two polymers to spin the composite fiber, which endows both advantages of these two kinds of polymers.

In this work, the composite core-shell nanofiber with an elastomeric component PU and a rigid component PSA was prepared by co-electro-spinning technique. It is well known that PU has perfect mechanical properties, so it can be selected as the core component of the composite yarn, and PSA has good flame retardant property, so it can be selected as the shell component to improve the thermal stability of the composite yarn. It is feasible to improve the overall performance of the core-shell composite fiber by the combination of PU and PSA. Meanwhile, three-dimensional (3D) electric field simulation is used to investigate the formation of the core-shell structure, three sets of co-spinnerets are designed and fabricated for the comparison experiments, and three kinds of electric field distributions with different shaped needles have been studied systematically.

6.3.1 Experiments and Simulation

6.3.1.1 Material Preparation

PSA was used as spinning solution with an intrinsic viscosity of 2.3 dL/g and N,N-dimethylacetamide (DMAc) was selected as dissolvent in this study. They were supplied by Shanghai Tanlon Fiber Co., Ltd., Shanghai, China. PU (5526) was supplied by DuPont, USA. All of these materials were used without further purification.

A certain amount of PSA was dissolved in the DMAc solution using ultrasonic vibration for 60 min at 60 ℃. PU solution was prepared by dissolving the PU chips in DMAc, and the mixture solution was stirred for 4 h at the room temperature. The solid concentration (mass fraction) of the PSA solution was set to be 12% and that of the PU solution was 20% to satisfy the viscosity requirement for the electro-spinning process. The spinning was performed at 20 ℃ and RH of 40%~60%.

6.3.1.2 Experimental Setup

The schematic of the co-electro-spinning system is shown in Fig.6-11. Three blunt-type stainless steel core-shell spinnerets with different diameters (Tab.6-3) were made by positioning the core needle inside the center of a coaxial spinneret, as depicted in Fig.6-12. The solutions for the core and shell materials were separately injected into the spinneret via corresponding syringes and pumps. The composite fibers were prepared with the use of co-electro-spinning system by using PSA (mass fraction: 12%) as shell component and PU (mass fraction: 20%) as core component. A high-voltage supply was ap-

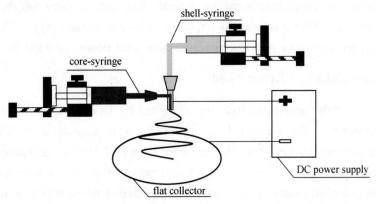

Fig.6-11 Schematic of the self-designed co-electro-spinning system

plied to the spinneret and the metal collector used in this experiment was flat-shaped. The experimental parameters of the coaxial electrostatic spinning process were set as follows:

(1) the extrusion speed was set to be 0.000 1 mm/s;

(2) the voltage and distance were 25 kV and 15 cm, respectively;

(3) all spinning processes were conducted at 45 ℃.

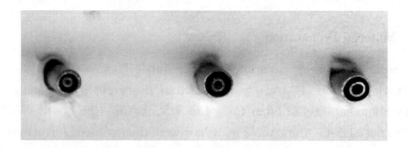

Fig. 6-12　Coaxial spinnerets (minimum sized on the left, maximum sized on the right)

Tab. 6-3　Parameters of the coaxial spinnerets

Mode of core-shell spinneret	Diameter of core needle (mm)		Diameter of shell needle (mm)
	Inner	Outer	
Minimum	0.34	0.64	1.25
Medium	0.51	0.82	1.36
Maximum	0.75	1.00	1.55

6.3.1.3　Imaging of Jet Path

A high-speed camera (HG-100K, Redlake Inc., San Diego, USA) equipped with a Nikon 24~85 mm, f 2.8 zoom lens was employed to digitalize the image sequences of the jet motion during the co-electro-spinning process. The used camera has the capability of recording images at a frame rate up to 100 000 frames per second (f/s). The light source used in the experiments was two fluorescent lamps with power of 2500 W.

6.3.1.4　Simulation of Electric Field

The 3D electric fields were simulated and analyzed by COMSOL Multi-physics (COMSOL Inc., Swedish). The electric field intensities were calculated by COMSOL software. Before the calculation, the physical geometries of the co-electro-spinning setups (e.g., electrode, spinneret and collector) and polymer solutions were established according to the practical dimensions, locations and relative permittivity used in the experiments.

6.3.2 Results and Discussion

6.3.2.1 Jet Path

Fig. 6-13 shows the composite droplets formed at the tip of three different core-shell spinnerets with working distance and applied voltage of 15 cm and 25 kV, respectively. The PU-core solution was color-dyed to enhance the contrast of the imaging object and the background. The red shell solution and white core solution can be identified clearly from the image.

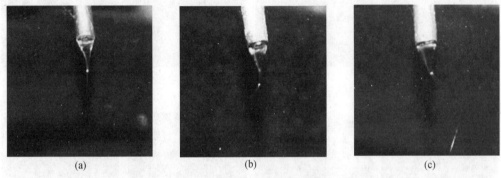

Fig. 6-13 Composite droplet formed at the tip of the core-shell spinneret: (a) Minimum core-shell spinneret; (b) Medium core-shell spinneret; (c) Maximum core-shell spinneret

The forming process of the Taylor cone with an increasing voltage from 10 to 25 kV is shown in Fig. 6-14. The magnified image of spinneret is also illustrated in Fig. 6-14. The composite droplets show a half ellipse shape when the voltage is set to be 10 kV [Fig. 6-14 (a-1), (b-1) and (c-1)]. In addition, it's difficult to draw the droplet to form a fiber when the voltage is low of 10 kV. The droplet becomes increasingly acute as the raising voltage from 10 to 25 kV. The Taylor cone can be formed and becomes sharp and stable when the voltage is raise to 25 kV. In this case, the fibers can be stretched and identified clearly. As observed in Fig. 6-14(a), (b) and (c), the different diameters of the coaxial spinneret have no significant influence on the forming process of the Taylor cone. The angle of jet spraying becomes larger with the increasing size of the needle diameter as marked by the yellow lines in Fig. 6-14(a-4), (b-4) and (c-4). It indicates that the distribution of electric field is associated and affected by the diameter of needles. The jet path at the stable stage of the spinning process is similar with the jet motion in a single-fluid electro-spinning process[53], however, it is different with the electrospun processing of off-centered core-shell spinneret which has many coils at the beginning of jet spraying[54].

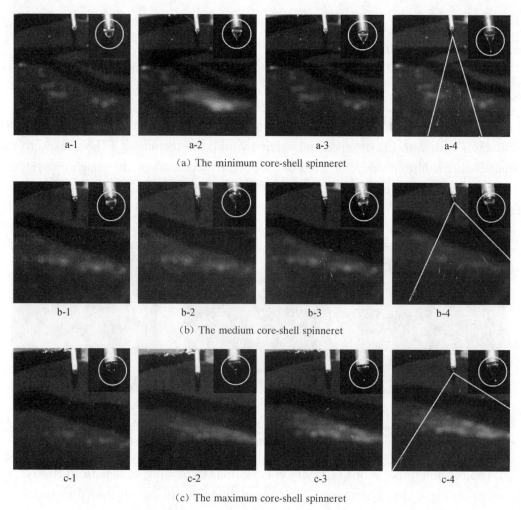

a-1　　　　　a-2　　　　　a-3　　　　　a-4
(a) The minimum core-shell spinneret

b-1　　　　　b-2　　　　　b-3　　　　　b-4
(b) The medium core-shell spinneret

c-1　　　　　c-2　　　　　c-3　　　　　c-4
(c) The maximum core-shell spinneret

Fig.6-14　The forming process of the Taylor cone with different voltages.
(1—10 kV; 2—15 kV; 3—20 kV; 4—25 kV)

6.3.2.2　Simulation and Distribution of Electric Field

The 3D electric fields formed by the three co-electro-spinning systems with different diameters of needles were simulated by COMSOL Metaphysics® Finite Elemental Analysis software. The simulations were conducted with a working distance and an applied voltage of 15 cm and 25 kV, respectively. The two conditions have been proved as the suitable parameters in the previous experiments for a stable spinning process. The lines of the simulated electric fields formed by the three types of spinnerets are demonstrated in Fig.6-15.

The distribution of the electric field of the electro-spinning is illustrated in Fig.6-15 (a-1), (b-1) and (c-1). The direction and length of the arrow represent the orientation and the intensity of the electric field lines, respectively. It seems that the electric field lines generated in the minimum sized needle are more concentrated around the central line of the spinneret, while the electric field lines are scattered and distributed outwards the central line of the spinneret for the maximum sized needle, which corresponds to the distribution angle of jet-spraying. That means the distribution of the electric field determines the size and range of the jet-whipping. It is obvious that the length of arrow is elongated with the increasing diameter of needle, which can be used to describe the change of its intensity.

Images of the top view [(a-2), (b-2) and (c-2)] and main view [(a-3), (b-3) and (c-3)] of the cross-sectional surface of electric field distribution for different co-spinnerets are also shown in Fig. 6-15. The electric field distribution of these images is similar with each other, belonging to an inhomogeneous electric field with an extremely high electric field concentrated in the surrounding area of the needle. The points of A, B and C in Fig. 6-15 (a-3), (b-3) and (c-3) are at the same position, however, they have various electric field intensities with different values. The calculated data demonstrates that the electric field intensity increases as the increasing size of co-needles, as shown in Fig. 6-15 (a-3), (b-3) and (c-3).

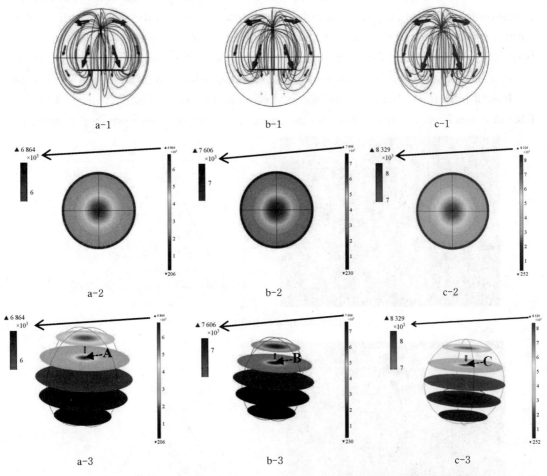

Fig. 6-15 Simulation of electric field of three co-spinnerets with different diameters: (a) The minimum core-shell spinneret; (b) The medium core-shell spinneret; (c) The maximum core-shell spinneret (a-1, b-1 and c-1 are simulation of electric field lines; a-2, b-2 and c-2 are simulation of top view of electric field distribution; a-3, b-3 and c-3 are simulation of main view of electric field distribution)

6.3.2.3 Morphological Structure

SEM images of PSA/PU composite fibers prepared by using the co-electro-spinning system with different diameters are shown in Fig. 6-16. The distributions of fiber diameter are also shown in Fig. 6-16. As shown in the SEM images, the shape of the electro-spun fiber is changed from straight to curved with the increasing diameter of the core-shell spinneret. In addition, the fiber presents helical structure when the diameter of the spinneret is large [Fig. 6-16 (c)] and it can be ascribed to a wide range of spraying and whipping. It is well known that the uniformity of the core-shell structure is one of the most important factors for the application of the core-shell nanofiber. The uniformity of the electro-spun fibers in this experiment is in the desired range because the proper crimp can increase the fiber strength which will be discussed in the following sections. Wu et al[55] mentioned that helical structures of the nanofibers are expected to improve the mechanical properties of the fibers, such as the better resiliency and flexibility. Wu et al also reported that helical conformation can be observed in biological systems, such as the plant tendril, and the electro-spun helical nanofiber resembles the tendrils in shapes. The results indicate that it is better to choose smaller diameter of the co-needles to spin the smooth and superfine fibers; however, the larger diameter of the co-spinnerets should be used to spin the fibers with better mechanical properties. The diameter distributions show that the diameter of the electro-spun fiber increases from 175 to 197 nm with the increasing diameter of the core-shell spinneret.

 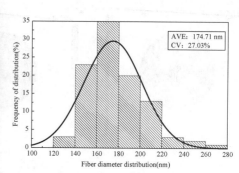

(a) The minimum core-shell spinneret

 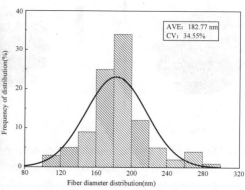

(b) The medium core-shell spinneret

 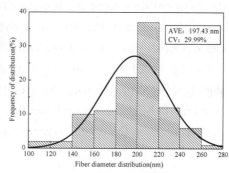

(c) The maximum core-shell spinneret

Fig. 6-16 SEM images (left) and diameter distributions (right) of fiber prepared with different core-shell spinnerets

TEM images of PSA/PU composite fibers prepared by using the co-electro-spinning system with different diameters are shown in Fig. 6-17. It demonstrates the impact of different diameters of the core-shell spinneret on the coaxiality of PSA/PU fiber. As shown in the figure, there is a significant difference among the internal structure of the three kinds of fibers. The minimum sized coaxial spinneret can be used to spin the fiber with superfine structure, however, the boundary between the core and shell part is not clear, as depicted in image of Fig. 6-17(a). The boundary between the core and shell part becomes obvious as the increasing diameter of the core-shell spinneret [Fig. 6-17 (b) and (c)]. Based on the simulation results, higher electric field can be generated by the bigger needle, and the fiber is fully stretched due to the strong electric force. The mechanical behavior of PSA droplets and PU droplets seems to be different during the spinning process; therefore, the overall morphology of the core-shell fibers should be investigated in the consideration of both the internal and external structure.

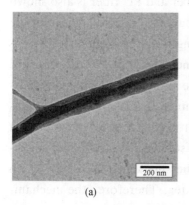 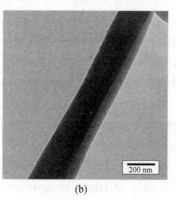 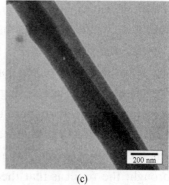

(a) (b) (c)

Fig. 6-17 TEM images of the electro-spun fibers by different core-shell spinnerets: (a) Minimum core-shell spinneret; (b) Medium core-shell spinneret; (c) Maximum core-shell spinneret

6.3.2.4 Crystal Structure

XRD curves of PSA/PU composite fibers prepared by using the co-electro-spinning system with different diameters are illustrated in Fig. 6-18, XRD curves of the pure PSA and PU are also shown in the Fig. for comparison. As demonstrated in Fig. 6-18, PSA nanofiber demonstrates a broad peak at 24° due to the amorphous nature, while the characteristic peak of PU is observed at 21°. The diffraction peak of the composite fibers slightly shifts to the left when compared with that of the pure PSA and it is more close to that of the pure PU. The results indicate that the diameter of the core-shell spinneret has no significant influence on the crystal structure of the composite nanofiber.

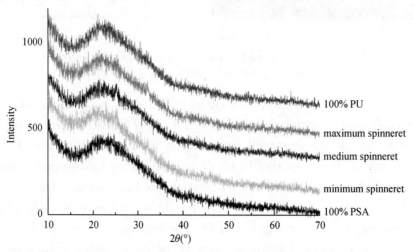

Fig. 6-18　XRD curves of the electro-spun PU fiber and PSA fiber by single-nozzle and the composite fiber by coaxial-nozzle

6.3.2.5　Mechanical Properties

The average tensile strength of the composite nanofibers with different spinnerets is depicted in Fig. 6-19. The average tensile strength of PSA fiber and PU fiber is also shown in this figure for comparison. As shown in Fig. 6-19, the pure PU has the highest tensile strength, while PSA fiber has the lowest value. The strength of the composite fiber is between the strength of PU fiber and PSA fiber, and it indicates that PU can improve the tensile strength of the composite fiber. In addition, the strength of the composite fiber increases with the raising diameter of the core-shell spinneret. This is possibly attributed to the helical structure of the composite fiber when it is prepared by the spinneret with the maximum diameter. Moreover, Sambaer et al[56] studied that the larger jet diameter covers a greater area for the fibers than that of the smaller jet diameter which brought the result is that the density of fibers became greater. Therefore, the mechanical strength of the composite fiber obtained from the co-nozzle system is greater than the pristine PSA fiber obtained from the single-nozzle system. It's a good way to improve the density of PSA without damaging the PSA.

　　The mechanical properties of PU fiber and PSA fiber as well as the PSA/PU composite fiber are illustrated in Tab. 6-4. The changing trend of the breaking elongation

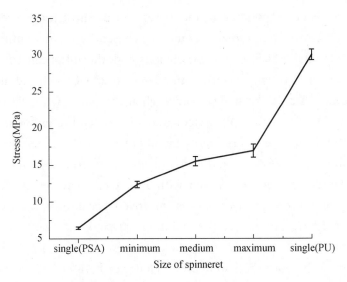

Fig. 6-19 Tensile strength curves of the electro-spun PU fiber and PSA fiber by single-nozzle and the composite fiber by coaxial-nozzle

and strain of the samples is similar with that of the tensile strength. It can be concluded that the mechanical properties of the composite fibers can be improved by the reinforcing phase of PU.

Tab. 6-4 Data of mechanical properties of the electro-spun PU fiber and PSA fiber by single-nozzle and composite fiber by coaxial-nozzle

Sample	Tensile strength (cN)	Breaking elongation (mm)	Strain (%)
PSA fiber	6.45	3.252	21.68
PSA/PU composite fiber prepared by the minimum core-shell spinneret	12.36	3.912	26.08
PSA/PU composite fiber prepared by the medium core-shell spinneret	15.52	4.99	28.01
PSA/PU composite fiber prepared by the maximum core-shell spinneret	16.95	4.70	31.43
PU fiber	30.008	6.69	76.174

6.3.2.6 Thermal Stability

DSC curves of PSA fiber and PU fiber as well as the composite fiber are illustrated in Fig. 6-20. Physical parameters of TG curves are illustrated in Tab. 6-5. DSC curve of the pure PSA has a diffuse peak from 30 to 130 ℃ due to the evaporation of water. The endothermic peak at around 250 ℃ may be resulted from the volatilization of water between polymer molecules and various additives[57]. DSC curve of the pure PU associated with the major thermal weight loss can be divided into three sections. The endotherm around 30 to 75 ℃ is corresponding to the disruption of short-range order of hard segments. The region between 150 to 200 ℃ may be caused by the disruption of long-range order of hard segments. PU fiber has an obvious endothermic peak corresponding to a high weight loss when the temperature is increased to 250 ℃. The results indicate that PSA fiber has a more stable thermal property when compared with PU fiber.

Based on the DSC curves of PSA fiber and PU fiber, the endothermic peak at around 80 ℃ of the composite fibers can be mainly attributed to the evaporation of water in the shell component of PSA. In addition, the enthalpy from 250 to 300 ℃ of the composite fiber is decreased when compared with PU fiber. The result indicates that the thermal stability of the composite fibers is improved because of the excellent thermal properties of PSA fiber. Moreover, the endothermic peak at about 255 ℃ of the composite fiber prepared with minimum core-shell spinneret is increasingly raised as the increased diameter of the core-shell spinneret. That can be ascribed to the large specific surface area of the small diameter of the composite fiber when it is fabricated by the minimum core-shell spinneret, and then more heat can be absorbed by the fiber in unit time. The result indicates that the composite fiber prepared with maximum core-shell spinneret has an effective thermal stability when compared with the fibers electro-spun by minimum and medium core-shell spinneret.

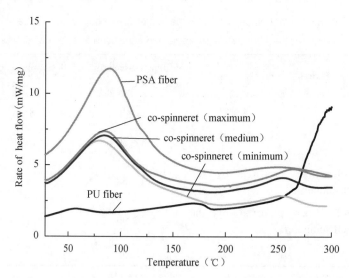

Fig. 6-20　DSC curves of the electro-spun PU fiber and PSA fiber by single-nozzle and the composite fiber by coaxial-nozzle

Both the TG and DTG curves of PSA fiber, PU fiber and PSA/PU composite fiber are illustrated in Fig. 6-21. There are two thermal decomposition peaks of PU fiber can be obviously observed between 200 and 500 ℃. Corresponding to the two rapid decompositions there is a corresponding weight loss peak in DTG curve and the T_{max} (presented in Tab. 6-5) can be determined according to the value of the maximum peak. The first peak at 300 ℃ is because of the decomposition of polyurethane molecule made up of long chains (soft chain segment), and the other at about 450 ℃ is attributed to the decomposition of short chains (rigid chain segments). As illustrated in Fig. 6-21, there is a thermal decomposition of PSA fiber at about 100 ℃ due to the evaporation of water, which is consistent with the result discussed in Fig. 6-20. A slight thermal weight loss is also observed in PSA fiber when the temperature is increased at 490 ℃. These results also prove that PSA has excellent thermal properties and it is assumed that PSA can improve the thermal properties of PSA/PU composite fiber when it is used as the shell component. The changing trend of the core-shell composite fiber is similar with PSA fiber, and it can be observed that the thermal decomposition peak at 485~490 ℃ is postponed when compared with PU fiber. What's more, the decomposition peak at about 485 ℃ of the composite fiber prepared with minimum core-shell spinneret is increasingly raised as the increased diameter of the core-shell spinneret. The results reveal that the composite fiber prepared with large core-shell spinneret has a good thermal stability when compared with the fibers electro-spun by small core-shell spinneret.

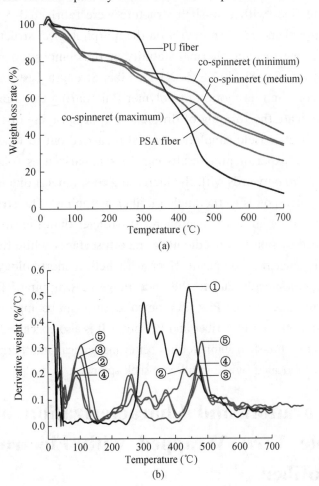

Fig. 6-21 DSC curves of electro-spun PU fiber and PSA fiber by single-nozzle and the composite fiber by coaxial-nozzle: (a) TG; (b) DTG. (1—PU fiber; 2—PSA fiber; 3—maximum co-spinneret; 4—medium co-spinneret; 5—minimum co-spinneret)

Tab. 6-5 Parameters of PSA fiber and PU fiber as well as PSA/PU composite fiber during the thermal decomposition

Sample	T_0(℃)	$d\alpha/dt$	T_{max}(℃)	α(%)
PU fiber	260	0.533 3	443.01	7.85
Minimum core-shell spinneret	—	0.239 4	470.88	33.94
Medium core-shell spinneret	—	0.262 8	476.34	34.72
Maximum core-shell spinneret	—	0.320 9	481.14	35.63
PSA fiber	390	0.407 6	483.30	43.04

T_0—The initial decomposition temperature;
$d\alpha/dt$—The maximum decomposition rate;
T_{max}—The temperature at the maximum decomposition rate;
α—The residual weight rate at 700 ℃.

6.3.3 Summary

PSA/PU composite fiber with core-shell structure were fabricated by electro-spinning. The influence of the diameter of spinneret on the morphological structure, crystal structure, mechanical properties and thermal stability of the composite fiber was characterized systematically by using SEM, TEM, XRD, Fiber Strength Tester and DSC. The images of the tailor cone and jet motion of polymer fluid during the spinning process were digitalized to investigate the forming of droplets and the spraying of the jet. The three-dimensional (3D) electric field simulation was also carried out to model the differences of electric field. The experimental results can be summarized as follows: The electric field lines are scattered outwards with the increasing co-spinneret diameter, and the field distribution angle corresponds to the angle of fiber whipping. The strength/intensity of electric field produced by large-sized spinneret is stronger than the small one. The core-shell fibers produced by small co-needle have smooth surface, while the core-shell structure of the fiber is unclear. Composite fiber with helical morphology can be prepared when used the large co-needle due to the wide range of whipping. The mechanical and thermal properties of the core-shell PSA/PU composite fiber can be improved due to the complementary properties between PSA fiber and PU fiber. It is also concluded that the composite fiber prepared by large core-shell spinneret has better mechanical and thermal properties when compared with that fabricated by the small core-shell spinneret.

6.4 Preparation and Characterization of Polyester Staple Yarn Wrapped with Electro-spun PSA Nanofiber

It has been reported that a large number of yarn spinning methods can be used to prepare different kinds of yarns, such as ring spinning[58], air spinning[59] and jet spinning[60]. The prepared yarns can satisfy diverse requirements for human activities by using high performance materials or changing the structure of yarn as well as spinning parameters[61]. However, the diameter of the yarn prepared by the traditional spinning technologies discussed above still remains at a micron scale. Very little research has been conducted on the fabrication of yarn at a nanoscale with the use of the traditional spinning methods. Because of the high surface-to-volume ratio, special hygroscopicity, permeability, and filterability of nanofiber, it can be used for many kinds of functional application purposes[62-63]. There are various methods can be used for the preparation of nanofiber, such as stretching method, template synthesis, and electric spinning. Electro-spinning is considered as the one of the most important and convenient methods for manufacturing fibers at a nanoscale when compared with the traditional methods (such as extrusion, phase separation, or drawing).

Although there are some new matrix electro-spinning methods, it is still difficult for the production of nanofiber to satisfy the requirements for industrial applications. In addition, it is hard for the nanofiber to prepare fabric because of its nanoscale diameter, low strength and poor scalability. In this case, one possible solution is to produce nanowrapped yarn, nanofiber as the shell component and microfiber as the core component, by electrostatic spinning and traditional spinning technologies. Therefore, the mechanical properties of the nanofiber can be guaranteed. Ravandi et al[64] prepared nylon 66 filament coated with nylon 66 nanofibers to study the wicking phenomena. The results showed that coating with nanofibers increases the equilibrium wicking height. Ding et al[65] reported a new approach for fabricating a superhydrophobic nanofibrous zinc oxide (ZnO) film through a simple electro-spinning coating technology. Lee et al[66] obtained titanium dioxide-coated nanofibers filters by electro-spinning polyamide nanofibers onto the surface of a conventional filter followed by electro-spraying a suspension of nanocrystalline titanium dioxide onto the electro-spinning nanofibers. Dabirian et al[67] explored the feasibility of electro-spinning nylon nanofibers onto nylon yarn, as well as poly (L-lactic acid) (PLLA) onto copper. Zhou et al[68] used single-needle electro-spinning to coat nanofibers onto monofilament arrays and then twisted them into hybrid yarns to improve cohesion.

Although much research has been conducted for the development of core-yarn nanofibers, very few works have been reported on the spinning of nanowrapped yarns composed of PSA and polyester (PET) fibers. It is well known that PET has good tenacity but poor hygroscopicity, while PSA has excellent thermal performance. It is valuable to present a new electro-spinning and wrapping method and a related spinning system that can be used to manufacture PSA/PET core-shell yarns. In this work, a new type of multi-nozzle jet electro-spinning was proposed to fabricate continuous PSA nanofiber wrapped and twisted on the surface of the core PET yarn. The effects of the electro-spinning voltage, rotation speed of funnel collector, and number of spinnerets on the PSA/PET core-shell yarn performances were investigated systematically.

6.4.1 Experiments and Simulation

6.4.1.1 Material Preparation

PSA with an intrinsic viscosity of 2.3 dL/g was used as spinning solution, and N,N-dimethylacetamide (DMAc) was selected as solvent. The two materials were supplied by Shanghai Tanlon Fiber Co., Ltd., Shanghai, China. PET yarn (32^s, wound package) was supplied by Xiangshui Xinmao Textile Co., Ltd, Xiangshui County, China. All of the materials were used without further purification.

A certain amount of PSA was dissolved in DMAc by using ultrasonic vibration for 60 min at 60 ℃. The solid concentration (mass fraction) of the PSA solution was set to be 10% to satisfy the viscosity requirements for the electro-spinning process. Spinning was performed at 20 ℃ and relative humidity of 40%~60%.

6.4.1.2 Experimental Setup and Electro-spinning Process

The electro-spinning and wrapping system for the preparation of PSA/PET core-shell yarn is illustrated in Fig.6-22. The system consists of a grounded purpose-made funnel collector (iron, diameter = 100 mm, angle = 30°, length = 70 mm), a high-voltage power supply [ES60P-20W/DDPM, Kansai Electronics (Suzhou) Co., Ltd., Jiangsu, China], a syringe pump system (stainless steel spinnerets, inner diameter = 1.25 mm, length = 3 mm), and a step motor used to wind up the coated yarns. There are two electro-spinning nozzles placed ahead of either side of the funnel collector, which has a square-shaped tunnel (5 mm) in the middle for PET staple yarns to run through, and the winder system is located between the nozzles with a distance of 200 mm. The two needles of the syringes are connected to the positive electrodes of a DC power supply. The funnel collector is placed opposite to the two positive nozzles, so that the electrostatic induction is presented as negative charges. Under these circumstances, nanofibers electro-spun from the nozzles are deposited onto the funnel collector, which also functions as a twister to wind PSA nanofibers on the surface of polyester staple yarns.

The preparation parameters, including the number of spinnerets, the electro-spinning voltage, and the rotation speed of the funnel collector, are listed in Tab.6-6. In addition, three different spinnerets with various numbers of tubes are shown in Fig.6-23. The effects of the fiber diameter and quantity of nanofibers on the yarn performance were investigated by changing the electro-spinning voltage and number of spinnerets. Variations in the fiber diameter could be implemented using the different settings of electro-spinning voltage; usually, fibers with small diameters could be spun at a high

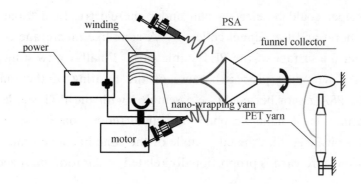

Fig. 6-22 Schematic of the preparation system for nano-wrapping yarn

voltage because of the high electric field force. The quantity of nanofiber which is proportionally related to the spinneret could be adjusted by the numbers of tubes in the spinneret.

Tab. 6-6 Experimental parameters for the preparation of core-shell yarn

Sample No.	Parameters		
	P_1	P_2	P_3
S_1	20	1	40
S_2	25	1	40
S_3	30	1	40
X_1	25	1	40
X_2	25	2	40
X_3	25	3	40
Y_1	25	3	20
Y_2	25	3	40
Y_3	25	3	60

P_1—Voltage (kV);
P_2—Spinneret number;
P_3—Rotating speed (r/min).

One-tube needle   Two-tube needle   Three-tube needle

Fig. 6-23 Three different spinnerets

PSA nanofiber could be electro-spun under the electric field force generated between the spinneret and the funnel collector. A cone-like membrane could be formed and wrapped on the surface of the PET staple yarn. Finally, a twisting process is executed during the wrapping process because of the rotating of the funnel collector. Therefore, the PSA nanofiber could be assembled with the PET staple yarn continuously. The formation of nano-wrapped yarn at different rotating speeds is shown in Fig. 6-24. It is observed that the edge angle of the cone-like membrane relative to the central line of the core yarn is proportionally related to the rotation speed of the funnel collector.

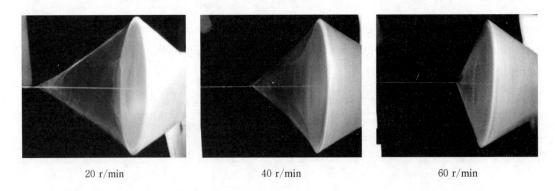

20 r/min   40 r/min   60 r/min

Fig. 6-24 Formation of nano-wrapped yarn at different rotating speeds

6.4.1.3 Characterization

The surface morphologies of the prepared core-shell yarns were investigated by scanning electron microscopy (SEM) (S-3400N, Hitachi, Tokyo, Japan) after the process of gold coating (coating time = 60 s). The mechanical behavior of the electrospun core-shell yarns was evaluated with a single-yarn tensile tester (YGB021DL) at a crosshead speed of 20 mm/min at room temperature. The thermal properties of the coaxial fibers were determined by thermogravimetry (TG) and differential thermogravimetry (DTG).

6.4.1.4 Electric Field Simulation

The three-dimensional (3D) electric fields were simulated and analyzed with COMSOL Multiphysics software (COMSOL Inc., Stockholm, Sweden). The electric field intensities were calculated with COMSOL software. The physical geometries of the electrospinning setups (e.g., electrode, spinneret and collector) and polymer solutions used for the simulation were consistent with the practical dimensions, locations, and relative permittivity used in our experiments.

6.4.2 Results and Discussion

6.4.2.1 Morphological Structure

Morphological structures of the core-shell yarns prepared by different electro-spinning voltages are shown in Fig. 6-25. As can be observed in Fig. 6-25 (a) that the polyester staple yarns were closely wrapped by a layer of electrospun PSA nanofiber. The magnified images of the cross section of the wrapped yarn can be discovered in Fig. 6-25 (b). Because the diameter of the nanofibers is so small, it is difficult to characterize the cross section of the single PSA nanofiber. As depicted in Fig. 6-25 (c), the core-shell composite yarns prepared with different electro-spinning voltages had similar twists.

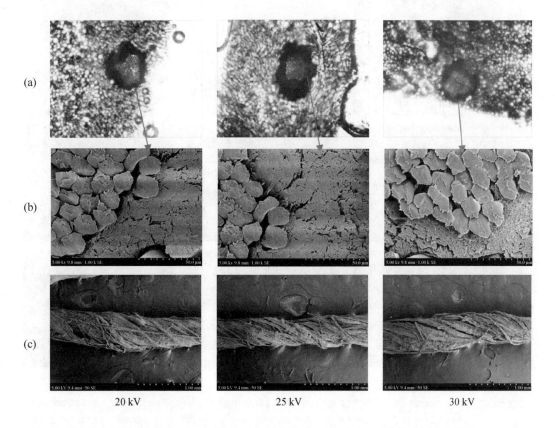

Fig. 6-25 Morphological structures of the core-shell yarns prepared at different funnel rotation speeds: (a) Optical micrographs; (b) SEM of cross section; (c) SEM of vertical structure

Morphological structures of the core-shell yarns prepared at different funnel rotation speeds are shown in Fig. 6-26. As observed in Fig. 6-26 (a), the amounts of the external-layer of the nanofibers corresponding to the three types of yarn are different although the preparation is the same. There are more nanofibers could be wrapped on the yarn when the funnel rotation speed is high. As shown in Fig. 6-26 (b), the structure of the nanowrapped yarn is loose when the rotation speed of the funnel is low (20 r/min) and there are few PSA nanofibers can be wrapped on the yarn. The structure of the nanowrapped yarn became dense with the rotation speed of the funnel increased from 20 to 60 r/min. The twisting degree of the core-shell yarn wrapped with PSA nanofiber is dependent on the funnel rotation speed. It can be seen from the angles of the nanowrapped nanofibers [Fig. 6-26 (c)] that an increase in the rotation speed of the twisting unit can cause a reduction in the nanofiber axial alignment. No twist could be achieved when the rotation speed of the funnel is 20 r/min, whereas the twisting was obvious when the rotation speed was 60 r/min. The results indicate that the twisting degree of the core-spun yarns increases with the increasing funnel rotation speed.

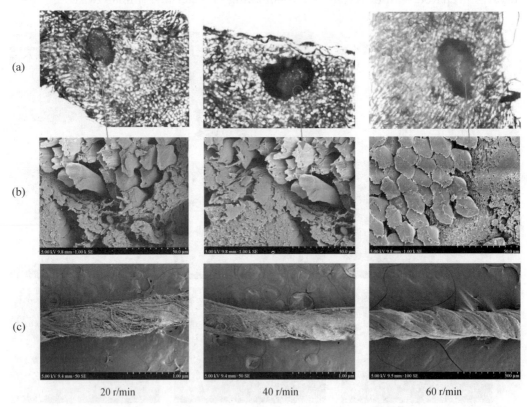

Fig. 6-26 Morphological structures of the core-shell yarns prepared by different funnel rotation speeds: (a) Optical microscope images of cross section of yarn; (b) SEM images of cross section of yarn; (c) SEM images of longitudinal section of yarn

Morphological structures of the core-shell yarns prepared with different numbers of spinnerets are shown in Fig. 6-27. The degree of compactness of the core-shell yarns is different because different amounts of PSA nanofibers can be wrapped on the yarn with various numbers of spinnerets. As can be observed in Fig. 6-27, more satisfied morphology of the core-shell yarn can be obtained when the three-tube needle is applied. A loose structure of PSA nanofibers was wrapped on the polyester staple yarns when the nanofibers were prepared with a one-tube needle.

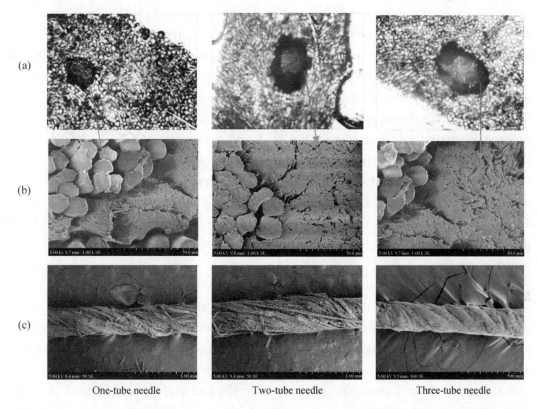

Fig. 6-27 Morphological structures of the core-shell yarns prepared by different numbers of spinnerets: (a) Optical microscope images of cross section of yarn; (b) SEM images of cross section of yarn; (c) SEM images of longitudinal section of yarn

6.4.2.2 Mechanical Properties

Mechanical properties are one of the most important properties of textile materials. In this work, polyester yarns nanowrapped with PSA fibers were prepared by electro-spinning with different parameters for the purpose of investigating the fracture mechanism and models which are suitable for the preparation of the nanowrapped yarns.

It is well-known that the electro-spun fibers collected on a traditional plate receiver are distributed randomly and that is corresponding to poor alignment or orientation of the fiber. To improve the mechanical strength of the nanowrapped assemblies, the bundles of PSA nanofibers were collected using a funnel receiver, so that large amounts of PSA nanofibers were clustered into one bundle or string, which can be twisted and wrapped on the surface of the polyester staple yarn. Therefore, it's possible to improve the mechanical properties of the yarn through twisting and wrapping the PSA nanofibers which are around the polyester staple yarn. Detailed experimental data are discussed in the following sections.

Usually, the fiber diameter of traditional polyester yarn is in the micron range. When polyester yarn is wrapped with PSA nanofibers, the tenacity of the composite yarns can be greatly improved because of these two reasons: (1) The combination between the polyester staple yarns and PSA nanofibers contributes to the prevention of breakage, and the overall tenacity should be equal or close to the sum of the tenacities of the two components. Compared with micron-scale fibers, the diameter of the electrospun PSA nanofibers is much smaller; the nanoscale PSA fibers can thus fill the micropores on the surface of polyester staple yarns, as illustrated by points A and B in Fig. 6-28; (2) The quantity and degree of twisting of the PSA nanofibers have obvious effects on the tenacity of the nanowrapped yarns. Large quantities and high degrees of twisting can increase the overall tenacity of the fiber, and this is also correct regarding to the reverse situation.

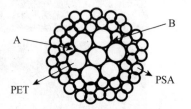

Fig. 6-28 Model of the cross section of the wrapped yarn

The influence of the electro-spinning voltage on the mechanical properties of the core-shell yarns is investigated. It is well-known that the diameter of nanofiber can be adjusted by changing the electro-spinning voltage; specifically, fine fiber can be electro-spun at high voltages. If the electro-spinning voltage has no obvious effects on the tenacity of the nanowrapped yarn, then the diameter variation of the nanofibers is not significant enough to affect its tenacity. The tensile curves and average tenacities of the core-shell yarns prepared at different voltages are shown in Fig. 6-29. As can be observed in Fig. 6-29 (a), the tenacity of PET yarns can be improved after wrapping with PSA nanofibers when compared with PSA yarn. Similar results can be obtained in Fig. 6-29 (b).

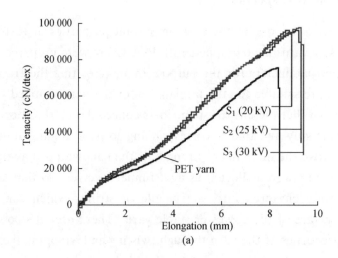

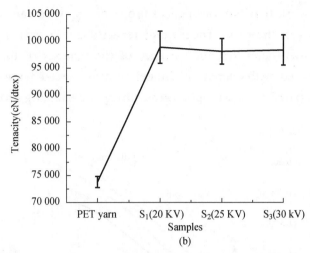

Fig. 6-29 Mechanical properties of the core-shell yarns prepared at different voltages: (a) Tensile curves; (b) Average tenacities

The data of tenacity, breaking elongation, initial modulus, and breaking time of the composite yarns are listed in Tab. 6-7. The data results indicate that the mechanical properties of the composite yarns can be obviously enhanced after cladding with PSA nanofibers when compared with PSA yarn. It can be also observed that there is no significant influence of the electro-spinning voltage on the mechanical properties of the composite yarns.

Tab. 6-7 Mechanical properties of the core-shell yarns prepared at different voltages

Samples	Tenacity (cN/dtex)	Breaking elongation (mm)	Initial modulus (cN/dtex)	Breaking time (s)
PET yarn	73 790 ± 2005	9.40 ± 0.71	2980 ± 125	1.12 ± 0.13
S_1 (20 kV)	100 240 ± 5909	9.55 ± 0.72	3740 ± 136	1.15 ± 0.13
S_2 (25 kV)	98 130 ± 4663	9.05 ± 0.69	3520 ± 135	1.09 ± 0.12
S_3 (30 kV)	97 370 ± 5525	10.25 ± 0.74	3820 ± 136	1.23 ± 0.15

In this work, a funnel-shaped collector was used to collect PSA nanofibers and wrapped them on the surface of the polyester. Therefore, the rotation speed of the funnel collector has an influence on the twisting of PSA nanofiber. It is well known that the twisting of fiber can improve the mechanical properties of the yarn. In this section, the influence of the funnel rotation speed on the mechanical properties of the core-shell yarns is investigated. The tensile curves and average tenacities of the nanowrapped yarns and PET yarn are presented in Fig. 6-30(a). The tenacity and average tenacity of the nanowrapped yarn increases with increasing number of twists at the same elongation because of the raising rotation speed of the funnel. The mark of "A" indicates the yielding points of the nanowrapped yarns spun at different rotation speeds. It can be concluded that the stress and strain can be improved at a high number of twists. Wu et al reported[69] in a similar work that the mechanical properties of yarn can be improved by increasing the twisting of the yarn. The average tenacity and its variance are illustrated in Fig. 6-30 (b). It was found that the variance of the tenacity of the prepared nanowrapped yarns increased with increasing funnel rotation speed because of the instable structure of the prepared nano-wrapped yarns during the spinning process.

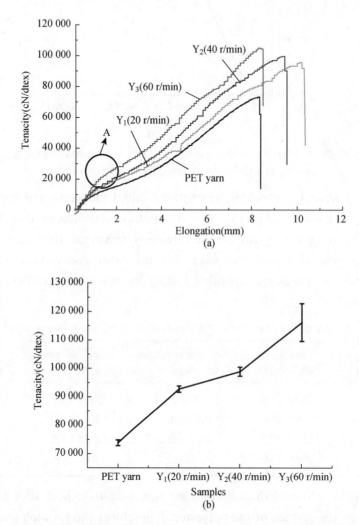

Fig. 6-30 Mechanical properties of the core-shell yarns prepared with different rotation speeds: (a) Tensile curves; (b) Average tenacities

The data of tenacity, breaking elongation, initial modulus, and breaking time of the composite yarns are listed in Tab. 6-8. It can be observed that the tenacity and initial modulus of PET staple yarns can be improved after wrapping with PSA nanofibers; however, further research should be conducted to investigate the breaking elongation of the composite fibers because the variation in the breaking elongation is not obvious with increasing tenacity. The elongation was neither the longest nor the shortest for the nanowrapped yarn with the maximum tenacity. The breaking elongation was found to have no positive or negative correlation with the tenacity.

Tab. 6-8 Mechanical properties of the core-shell yarns prepared with different rotation speeds

Samples	Tenacity (cN/dtex)	Breaking elongation (mm)	Initial modulus (cN/dtex)	Breaking time (s)
PET yarn	73 790 ± 2005	9.40 ± 0.71	2980 ± 125	1.12 ± 0.13
Y_1 (20 r/min)	92 710 ± 2123	9.70 ± 0.71	3620 ± 85	1.24 ± 0.16
Y_2 (40 r/min)	98 850 ± 3124	10.25 ± 0.74	3980 ± 127	1.23 ± 0.16
Y_3 (60 r/min)	116 140 ± 13 038	9.40 ± 0.70	4580 ± 148	1.13 ± 0.15

Spinneret is very important for the electro-spinning of nanofiber because spinneret can affect the physical properties of nanofiber. Needles are usually used as spinneret in laboratory. In this section, the influence of the number of spinneret tube on the mechanical properties of the core-shell yarns is investigated, including one-tube, two-tube and three-tube needles. All tubes inside the needles had the same diameter.

The tensile curves and average tenacities of the nanowrapped yarns prepared with different numbers of tubes are presented in Fig. 6-31. It is discovered that the variance of the tenacity increase as the increasing number of tube. Compared with the Fig. 29 and 30, it can be observed in Fig. 6-31 that there is a significant effect of the spinneret on the mechanical properties of the core-shell yarns. One reason is that a spinneret containing more tubes has the ability to produce more nanofibers; another is that more nanofibers can be wrapped onto the surface of the polyester staple yarn tightly. It can be concluded that the number of tubes inside the needle (spinneret) has a positive correlation with the tenacity of the nanowrapped yarn.

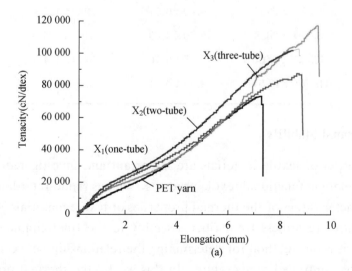

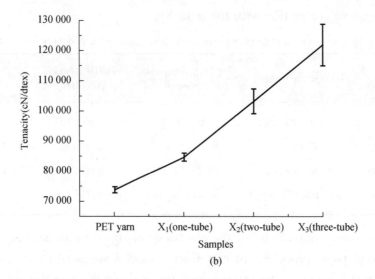

Fig. 6-31 Mechanical properties of the core-shell yarns prepared with different numbers of tubes: (a) Tensile curves; (b) Average tenacities

The data of tenacity, breaking elongation, initial modulus, and breaking time of the composite yarns are listed in Tab. 6-9. The tenacity of PET yarn can be enhanced after wrapping with PSA nanofiber and the tenacity of the core-shell yarns improves with the raising number of tubes during spinning process. In addition, the number of tube has no significant influence on breaking elongation, initial modulus and breaking time of the composite fibers.

Tab. 6-9 Mechanical properties of the core-shell yarns prepared with different numbers of tubes

Samples	Tenacity (cN/dtex)	Breaking elongation (mm)	Initial modulus (cN/dtex)	Breaking time (s)
PET yarn	73 790 ± 2005	9.40 ± 0.71	2980 ± 125	1.12 ± 0.13
X_1 (one-tube)	84 680 ± 2638	8.50 ± 0.69	3160 ± 126	1.02 ± 0.12
X_2 (two-tube)	103 240 ± 8030	9.15 ± 0.70	3740 ± 134	1.15 ± 0.13
X_3 (three-tube)	115 920 ± 13 541	9.40 ± 0.71	3720 ± 134	1.13 ± 0.13

6.4.2.3 Thermal Stability

Thermal properties of textile materials are important and investigated widely for the purpose of developing functional textile products, such as flame-retardant yarns or fabrics. The characterization of the thermal properties of textile materials is also necessary for quality control regardless from fiber assembly to the final commercial products. Thermal analysis is one method for determining the relationship between the weight of material and the controlled temperature. In this work, the thermal properties of the nanowrapped yarns (PET as core yarn, PSA as wrapped fibers) were investigated using the typical thermal analysis method; PET staple yarn was also characterized for the comparison with the newly developed yarns. It was previously reported that PSA has excellent thermal properties and that it can be used as the outer wrapping nanofibers to protect the core polyester yarns from decomposition at a high temperature.

DSC curves of the core-shell yarns prepared with different voltages, funnel rotation speeds, and numbers of spinneret are shown in Figs. 6-32, 6-33 and 6-34, respectively. As illustrated in TG curves, the weight loss rate of all the nanowrapped yarns are larger than that of PET yarn. The results indicate that the thermal resistance of PET yarns can be improved after wrapping with PSA nanofibers. As illustrated in Fig. 6-32, there is no significant difference of the curves among the nanowrapped yarns electro-spun at different voltages, and it indicates that the electro-spinning voltage has no obvious effect on the thermal properties of the nanowrapped yarns. As demonstrated in Fig. 6-33, the highest decomposition temperature of the core-shell yarns prepared with a high funnel rotation speed (40 and 60 r/min) postpones when compared with the composite yarn fabricated with a low rotation speed (20 r/min) as well as PET yarn. The result reveals that the thermal stability of the core-shell yarns can be enhanced with the increasing funnel rotation speed from 20 and 60 r/min. The changing trend of the curves in Fig. 6-34 is similar with that in Fig. 6-33, and the result also indicates that the thermal stability of the core-shell yarns can be improved with the increasing numbers of spinnerets from one-tube to three-tube. The increasingly improved thermal stability of the core-shell yarns can be attributed to the excellent thermal properties of PSA, what's more, an increased amount of PSA nanofibers can be wrapped on the surface of PET with increasing funnel rotation speed and numbers of spinnerets.

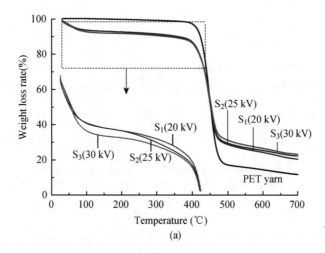

(a)

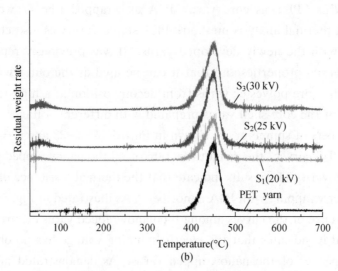

Fig. 6-32 DSC curves of the core-shell yarns prepared at different voltages: (a) TG; (b) DTG

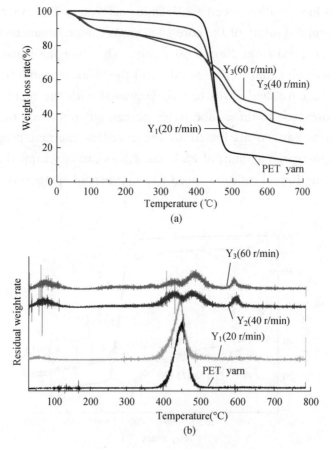

Fig. 6-33 DSC curves of the core-shell yarns prepared at different funnel rotation speeds: (a) TG; (b) DTG

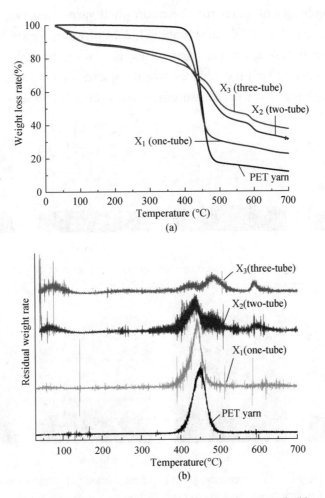

Fig. 6-34 DSC curves of the core-shell yarns prepared with different numbers of spinnerets: (a) TG; (b) DTG

6.4.2.4 Hygroscopicity

Hygroscopicity is an important physical property which can allow materials to be water absorbable, breathable, and antistatic. Nanofiber has excellent hygroscopicity because of its nanoscaled structure. The hygroscopicity of polyester staple yarn is poor, so it can improve or its hygroscopicity by blending or wrapping with functional materials with good hygroscopicity. The climbing height of water is a parameter used to evaluate the hygroscopic properties of textile materials.

The images of the climbing height of water for the core-shell yarns prepared with different voltages, funnel rotation speeds, and numbers of spinneret are shown in Figs. 6-35, 6-36 and 6-37, respectively. It can be observed that the climbing height of water increase with increasing voltage from 20 to 30 kV as well as increasing number of needles (spinnerets) from one-tube to three-tube (Figs. 6-35 and 6-37, respectively). In addition, the climbing height of water for these core-shell yarns increases as the raising testing time. As shown in Fig. 6-36, the climbing height of water for the core-shell yarn increases with increasing twists, but it starts to decrease with the further increased twist. It reveals that excessive twisting reduces hygroscopicity of the core-shell yarns. This changing trend is similar with the measurement with different testing time.

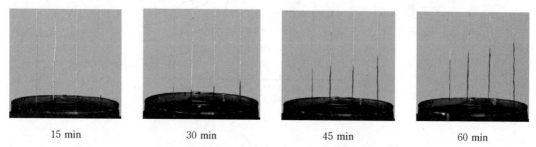

Fig. 6-35 Height of the hygroscopic climb for yarns prepared at different voltages
(in each panel, from left to right: PET yarn, S_1, S_2 and S_3)

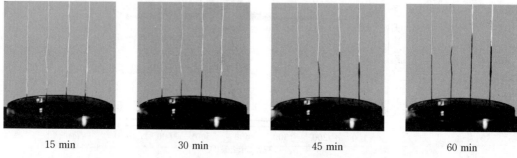

Fig. 6-36 Height of the hygroscopic climb for yarns prepared at different rotation speeds
(in each panel, from left to right: PET yarn, Y_1, Y_2 and Y_3)

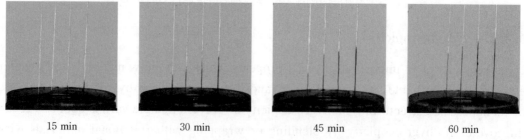

Fig. 6-37 Height of the hygroscopic climb for yarns prepared with different numbers of needles
(in each panel, from left to right: PET yarn, X_1, X_2 and X_3)

6.4.3 Summary

In this part, continuous PSA nanofiber wrapped and twisted on the surface of PET yarn was prepared by using a new type of multi-nozzle jet electro-spinning. The effects of the electro-spinning voltage, rotation speed of funnel collector, and number of spinnerets on the PSA/PET core-shell yarn performances were investigated systematically. Polyester yarns wrapped with PSA nanofibers provide an obvious improvement in terms of mechanical, thermal and hygroscopic properties. The improved thermal properties of the core-shell yarns because of the excellent thermal stability of PSA and the PSA nanofibers can protect the inside polyester staple yarn from decomposing at a high temperature. PSA nanofibers can also be used to improve the hygroscopicity of nanowrapped yarns. The hygroscopicity of the core-shell yarns improved with an increased voltage and an increased number of needles. The hygroscopicity of the core-shell yarns enhanced with the raising rotation speed (increased twisting) from 20 to 40 r/min, and then decreased with the further increased raising rotation speed (increased twisting) from 40 to 60 r/min.

References

[1] Xin B J, Chen Z M, Wu X J. Preparation and characterisation of PSA/CNT/TiO$_2$ nano composites and fibres. Fibres & Textiles in Eastern Europe, 2013, 21(1): 24-28.

[2] Chen W J, Xin B J, Wu X J. Fabrication and characterization of PSA nanofibers via electro-spinning. Journal of Industrial Textiles, 2013, 44(1): 159-179.

[3] Chen Z M, Xin B J, Wu X J, et al. Reparation and characterisation of PSA/CNT composites and fibres. Fibres & Textiles in Eastern Europe, 2012, 20(5): 21-25.

[4] Rayleigh L. On the equilibrium of liquid conducting masses charged with electricity. London, Edinburgh, and Dublin Philosophical Magazine and Journal of Science, 1882, 14(87): 184-186.

[5] Kiyohiko H. Process for manufacturing artificial silk and other filaments by applying electric current. US 1699615, 1929.

[6] Taylor G. Disintegration of water drops in an electric field. Proceedings of the Royal Society of London. Series A, Mathematical and Physical Sciences, 1964, 280(1382): 283-297.

[7] Simons H L. Process and apparatus for producing patterned non-woven fabrics. US 3280229, 1996.

[8] Chu B, Hsiao B S, Fang D, et. al. Apparatus and methods for electro-spinning polymeric fibers and membranes. US 6713011, 2004.

[9] Smith D J R, Darrell H M, Albert T, et al. Electrospun fibers and an apparatus therefor. US 6753454 B1, 2004.

[10] Fathona I W, Yabuki A. Mapping the influence of electro-spinning parameters on the morphology transition of short and continuous nanofibers. Fibers and Polymers, 2016, 17(8): 1238-1244.

[11] Purwar R, Sai G K, Srivastava C M. Electrospun sericin/PVA/clay nanofibrous mats for antimicrobial air filtration mask. Fibers and Polymers, 2016, 17(8): 1206-1216.

[12] Zernetsch H, Repanas A, Rittinghaus T, et al. Electro-spinning and mechanical properties of polymeric fibers using a novel gap-spinning collector. Fibers and Polymers, 2016, 17(7): 1025-1032.

[13] Jiang G J, Zhang S, Qin X H. High throughput of quality nanofibers via one stepped pyramid-shaped spinneret. Materials Letters, 2013, 106: 56-58.

[14] Zhang J N, Song M Y, Li D W, et al. Preparation of self-clustering highly oriented nanofibers by needleless electro-spinning methods. Fibers and Polymers, 2016, 17(9): 1414-1420.

[15] Bonhomme O, Leng J, Colin A. Microfluidic wet-spinning of alginate microfibers—A theoretical analysis of fiber formation. Soft Matter, 2012, 8(41): 10641-10649.

[16] Xie S, Zeng Y C. Effects of electric field on multineedle electro-spinning: Experiment and simulation study. Industrial & Engineering Chemistry Research, 2012, 51(14): 5336-5345.

[17] Demir M M. Electro-spinning and wet-spinning of elastic fibers. Sabanci University, 2001.

[18] Cho H J, Yoo Y J, Kim J W, et al. Effect of molecular weight and storage time on the wet-and electro-spinningelectro-spinning of regenerated silk fibroin. Polymer Degradation & Stability, 2012, 97(6): 1060-1066.

[19] Tanner R I. A theory of die-swell revisited. Journal of Non-Newtonian Fluid Mechanics, 2005, 129(2): 85-87.

[20] Hu C C, Cheng L, Gu R X, et al. Die-swell behavior of heterocyclic aromatic polyamide spinning solution. China Synthetic Fiber Industry. 2012, 35: 16-19.

[21] Li Y, Wei M Y, Lu W M, et al. Comparison of physical and chemical properties between polysulfonamide and aramid. Melliand China, 2004, 1: 14-16, 22-24.

[22] Liu J X, Tang Z Y, Zhang D R, et al. Preparation and performance of polysulfonamide nanocomposites and it's fiber. Technical Textiles, 2007, 25(2): 14-20.

[23] Tiwari A, Terada D, Yoshikawa C, et al. An enzyme-free highly glucose-specific assay using self-assembled aminobenzene boronic acid upon polyelectrolytes electrospun nanofibers-mat. Talanta, 2010, 82(5): 1725-1732.

[24] Lee H M, Mahapatra S K, Anthony J K, et al. Effect of coagulation condition on structure and properties of as-spun feather keratin-PVA fiber. Synthetic Materials Aging & Application, 2009, 44(14): 3721-3735.

[25] Dong X G. The caggulation formation and spinning process of polyacrylonitrile fiber. Shandong University, 2009.

[26] Fornes T D, Paul D R. Crystallization behavior of nylon6 nanocomposites. Polymer Degradation and Stability, 2003, 44: 3945-3961.

[27] Theron S A, Zussman E, Yarin A L. Experimental investigation of the governing parameters in the electrosp inning of polymer solutions. Polymer, 2004, 45: 2017-2030.

[28] Cramariuc B, Cramariuc R, Scarlet R. Fiber diameter in electro-spinning process. Journal of Electrostatics, 2013, 71(3): 189-198.

[29] Dong J, Yin C Q, Lu H X. Morphology control of polyimide fibers by phase separation under different coagulation bath conditions. Materials Science Forum, 2014, 789: 142-147.

[30] Tascan M. Optimization of process parameters of wet-Spun solid PVDF fibers for maximizing the tensile strength and applied force at break and minimizing the elongation at break using the taguchi method. Journal of Engineered Fibers & Fabrics, 2014, 9: 165-173.

[31] Wang H T, Wang K, Li C Q, et al. Study of the process of phase separation in polyacrylonitrile/dimethyl sulfoxide solutions using an optical microscope. Journal of Beijing University of Chemical Technology(Natural Science Edition), 2010, 37(1): 73-77.

[32] Yang W T. Characterization and testing of polymer materials. China Light Industry Press, 2008: 144.

[33] Loscertales I G, Barrero A, Larsen G, et al. Electrically forced coaxial nanojets for one-step hollow nanofiber design. Journal of the American Chemical Society, 2004, 126(17): 5376-5377.

[34] Huang Z, Zhang Y, Kotaki M, et al. A review on polymer nanofibers by electro-spinning and their applications in nanocomposites. Composites Science & Technology, 2003, 63(15): 2223-2253.

[35] Agarwal S, Greiner A, Wendorff J H. Functional materials by electro-spinning of polymers. Pro-

gress in Polymer Science, 2013, 38(6): 963-991.

[36] Qu H, Wei S, Guo Z. Coaxial electrospun nanostructures and their applications. J. Mater. Chem. A, 2013, 1(38): 11513-11528.

[37] Wang M, Jin H, Kaplan D L, et al. Mechanical properties of electrospun silk fibers. Macromolecules, 2004, 37(18): 6856-6864.

[38] Sun Z, Zussman E, Yarin A L, et al. Compound core-shell polymer nanofibers by co-electro-spinning. Advanced Materials, 2003, 15:1929.

[39] Reneker D H, Yarin A L, Hao F, et al. Bending instability of electrically charged liquid jets of polymer solutions in electro-spinning. Journal of Applied Physics, 2000, 87(9): 4531-4547.

[40] Hohman M M, Shin M, Rutledge G, et al. Electro-spinning and electrically forced jets. II. Applications. Physics of Fluids, 2001, 13(8): 2221-2236.

[41] Thoppey N M, Bochinski J R, Clarke L I, et al. Edge electro-spinning for high throughput production of quality nanofibers. Nanotechnology, 2011, 22(34): 486-498.

[42] Zheng Y S, Zeng Y C. Jet repulsion in multi-jet electro-spinning systems: From needle to needleless. Advanced Materials Research, 2014:852.

[43] Zheng Y, Zeng Y. Electric field analysis of spinneret design for multihole electro-spinning system. Journal of Materials Science, 2014, 49(5): 1964-1972.

[44] Zhou Z, Wu X, Ding Y, et al. Needleless emulsion electro-spinning for scalable fabrication of core-shell nanofibers. Journal of Applied Polymer Science, 2014, 131(20): 1366-1373.

[45] Zheng Y S, Xie S, Zeng Y C. Effects of structural parameters on nanofiber from a multihole electro-spinning setup. Advanced Materials Research, 2013:690-693.

[46] Wu H, Zheng Y, Zeng Y A. Method for scale-up of co-electrospun nanofibers via flat core-shell structure spinneret. Journal of Applied Polymer Science, 2014, 131(21): 8558-8572.

[47] Lin J, Tian F, Shang Y. Co-axial electrospun polystyrene/polyurethane fibres for oil collection from water surface. Nanoscale, 2013, 5(2): 745.

[48] Zhu C, Deng W, Pan J. Structure effect of dual-spinneret on the preparation of electrospun composite nanofibers with side-by-side heterojunctions. Journal of Materials Science: Materials in Electronics, 2013, 24:2287.

[49] Tijing L D, Choi W, Jiang Z, et al. Two-nozzle electro-spinning of (MWNT/PU)/PU nanofibrous composite mat with improved mechanical and thermal properties. Current Applied Physics, 2013, 13(7): 1247-1255.

[50] Chen S, Hou H, Hu P, et al. Effect of different bicomponent electro-spinning techniques on the formation of polymeric nanosprings. Macromolecular Materials and Engineering, 2009, 294(11): 781-786.

[51] Kessick R, Tepper G. Microscale polymeric helical structures produced by electro-spinning. Applied Physics Letters, 2004, 84(23):4807-4809.

[52] Cha D, Kim K W, Hee Chu G, et al. Mechanical behaviors and characterization of electrospun polysulfone/polyurethane blend nonwovens. Macromolecular Research, 2006, 14(3): 331-337.

[53] Wan Y Q, Wei J, Qiang J, et al. Modeling and simulation of the electro-spinning jet with archimedean spira. Advanced Science Letters, 2012, 10(1): 590-592.

[54] Shariatpanahi S P, Zad A I, Abdollahzadeh I, et al. Micro helical polymeric structures produced by variable voltage direct electrospinning. Soft Matter, 2011, 7(22):10548-10551.

[55] Wu H, Zheng Y, Zeng Y. Fabrication of helical nanofibers via co-electro-spinning. Ind. Eng.

Chem. Res. , 2015, 54(3): 987-993.

[56] Sambaer W, Zatloukal M, Kimmer D. The use of novel digital image analysis technique and rheological tools to characterize nanofiber nonwovens. Polymer Testing, 2010, 29: 82-94.

[57] Ye J Q, Zhang Y H, Ren J R, et al. Thermal stability analysis of PSA fiber. Journal of Donghua University (Natural Science), 2005,31:13-16.

[58] Hasanuzzaman, Dan P K, Basu S. Optimization of ring-spinning process parameters using response surface methodology. Journal of the Textile Institute, 2015, 106(5): 510-522.

[59] Xing M. Study on the characteristics of air jet vortex spinning. Cotton Textile Technology, 2009, 37(8): 15-18.

[60] Xing M. The latest development of open-end jet spinning technology. China Textile Leader, 2005: 28-32.

[61] Wang J. The method of increasing the air jet spinning's yarn strength in the high speed. Journal of Tianjin Institute of Textile Science and Technology, 1996,15:50-55.

[62] Ignatova M, Manolova N, Rashkov I. Electro-spinning of poly(vinyl pyrrolidone)-iodine complex and poly(ethylene oxide)/poly(vinyl pyrrolidone)-iodine complex a prospective route to antimicrobial wound dressing materials. Eur. Polym. , 2007, 43: 1609-1623.

[63] Dalton P D, Klee D, Moller M. Electro-spinning with dual collection rings. Polymer, 2005, 46(3): 611-614.

[64] Ravandi S A H, Sanatgar R H, Dabirian F. Wicking phenomenon in nanofiber-coated filament yarns. Journal of Engineered Fabrics & Fibers, 2013, 8(3): 10-18.

[65] Ding B, Ogawa T, Kim J, et al. Fabrication of a super-hydrophobic nanofibrous zinc oxide film surface by electro-spinning. Thin Solid Films, 2008, 516(9): 2495-2501.

[66] Lee B Y, Behler K, Kurtoglu M E, et al. Titanium dioxide-coated nanofibers for advanced filters. Journal of Nanoparticle Research, 2009, 12(7): 2511-2519.

[67] Dabirian F, Ravandi S A H, Hinestroza J P, et al. Conformal coating of yarns and wires with electrospun nanofibers. Polymer Engineering & Science, 2012, 52(8): 1724-1732.

[68] Zhou F L, Gong R H. Nano-coated hybrid yarns using electro-spinning. Surface & Coating Technology, 2010(21-22): 3459-3463.

[69] Wu S H, Qin X H. Uniaxially aligned polyacrylonitrile nanofiber yarns prepared by a novel modified electro-spinning method. Materials Letters, 2013, 106(9): 204-207.

Chapter 7 Development of Conductive Polymer Materials

Xiangqin Wang, Binjie Xin*, Zhuoming Chen and Na Meng

7.1 Introduction

Most of non-natural organic polymers (insulation materials), whose status is self-evident, have been utilized in our daily life. In recent years, conductive polymers have attracted wide attention and quite a lot of researches have been done for development of these polymers, with the ever-increasing requirements on science and technology. So far, there has been a steady increase in special demands on conductivity of polymers in numerous fields, such as industry, aviation, electronics and materials, etc.

It was proposed by Ferraris in the early 1970s that charge transfer reaction and large-scale superconduction phenomenon could occur to the compound of tetrathiafulvalene-tetracyanoquinodimethane (TTF-TCNQ). The high cis-polyacetylene was characterized by typical conjugated structure and polymerized by Hideki Shirakawa (Japan) based on appropriate high-concentration catalyst. By the late 1970s, it was proposed that the conductivity of polyacetylene with doped inorganic material (e. g. , borontrifluoride, iodine or arsenic fluoride) increased 10 orders of magnitude with a maximum value of 10^3 S/cm, furthermore, apparent color change occurred to the doped polyacetylene membrane which was presented in golden yellow (i. e. , gloss of metallic conductor)[1].

Corresponding Author:
Binjie Xin
School of Fashion Technology, Shanghai University of Engineering Science, Shanghai 201620, China
E-mail: xinbj@sues. edu. cn

Recently, many researchers have attempted to investigate on the conjugated π bond polymers, such as polypyrrole (PPy), polythiophene (PTs), polyaniline (PANI, being served as typical conductive polymer). It has been found that the conductivity of these materials increases orders of magnitude because of the doping, even reaches the level of metallic conductor with rich colors. The electrochemical properties of these materials are improved and optimized due to the special structure, in this case, these materials have become the primary choice in such fields as electromagnetic shielding, metal anticorrosion, sensor[2] and stealth technology[3-4]. It has captured more attention that heavy load of inorganic stealth material is overcome by the conductive polymers with features of light-weight and excellent film-formation properties.

The development of PANI starts rather late, in comparison with other conductive polymers. The polymerization mechanism, solubility and conduction mechanism of PANI are still in research, therefore, the investigation on the conduction mechanism, thermal stability and electrochromic properties of PANI has been still a research hotpot.

In this chapter, emulsion polymerization is proposed to prepare PANI/polyvinyl alcohol (PVA) emulsion and PANI/PVA ultrafine composite fiber is fabricated by electro-spinning. Morphological structure, conductive properties, mechanical properties, material composition and crystal structure, thermal properties and electrochromic properties of PANI/PVA ultrafine composite fiber are systematically characterized by optical microscope, scanning electron microscope (SEM), four-point probe resistivity tester, electronic single fiber strength tester, fourier transform infrared spectroscopy (FTIR), X-ray diffraction (XRD), thermal analyzer (TG) and electrochemical workstation.

7.1.1 Synthesis Method of PANI

According to the latest research on PANI served as conductive polymer, the synthesis methods can be classified into five categories, including chemical polymerization[5], electrochemical polymerization[6], radiation polymerization, enzyme catalyst polymerization and nano-PANI polymerization[7] (as illustrated in Fig. 7-1), which have caused wide attention due to their unique advantages.

7.1.1.1 Chemical Polymerization

The chemical polymerization of PANI[8-9] is conducted by using oxidant to polymerize aniline (AN) monomer. It mainly includes solution polymerization and emulsion polymerization, where the conductivity of PANI could be enhanced after treatment by these two methods. In addition, the solubility of PANI could be reinforced as well after the emulsion polymerization.

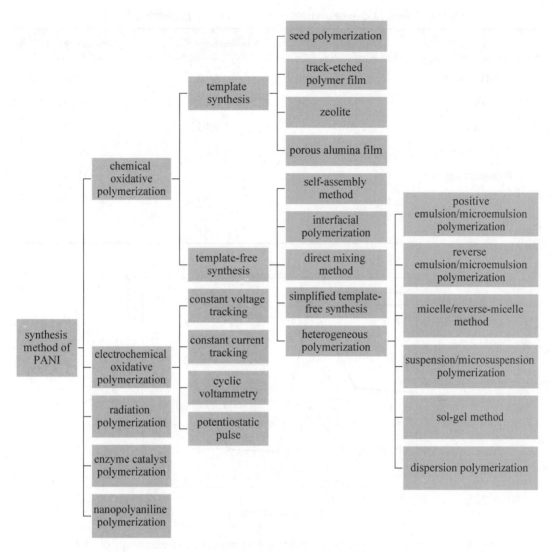

Fig. 7-1 Synthesis methods of PANI

Solution polymerization[10-12] is considered as a representative of chemical polymerization, the PANI with diverse electrical activities could be achieved by applying different mass fraction of oxidant. Usually, the chemical oxidative polymerization of AN is achieved in the three-phase system composed of aniline, acid, oxidant and water. The commonly used inorganic oxidant includes peroxides, such as $(NH_4)_2S_2O_8$ and H_2O_2, $K_2Cr_2O_7$, $KMnO_4$, Na_3VO_4, and ferric chloride, etc. The influence of diverse oxidants on conductivity and productivity of PANI are listed in Tab. 7-1 and the effects are clearly illustrated by the curves in Fig. 7-2[13-14]. The conductivity of PANI increases with increasing potential of oxidant and the maximum conductivity of PANI could be achieved under the potential of $K_2Cr_2O_7$.

Tab. 7-1 Effects of diverse oxidants on conductivity and productivity of PANI[1]

Oxidant	Potential(V)	Conductivity(S/cm)	Productivity(%)
$(NH_4)_2S_2O_8$	2.01	2.50	78.50
H_2O_2	1.78	0.07	33.90
$KMnO_4$	1.51	0.03	6.00
$K_2Cr_2O_7$	1.33	4.30	37.50
Potassium iodate	1.09	1.82	11.70
Na_3VO_4	1.00	1.80	23.30
Iron trichloride	0.77	1.73	2.30

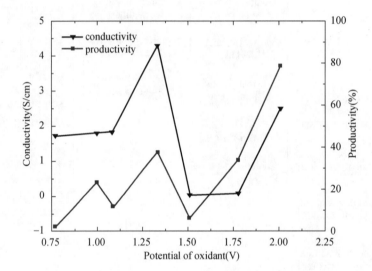

Fig. 7-2 Relationship of potential of oxidant with conductivity and productivity of PANI[9]

Emulsion polymerization[15] is achieved through following processes: the monomer is added into the emulsifier with strong mechanical stirring and oscillating to achieve the goal of uniform dispersion of the emulsion-like monomer, and then the polymerization is initiated by the initiator. Emulsion polymerization exhibits several advantages: the first is that the emulsion product could be applied in finishing directly without secondary pollution; the second is that the elliptic-shaped and rod-shaped high polymer could be achieved through demulsification and washing; the third is that the polymer with high molecular weight is prone to be fabricated and its production process is easy to control. However, it has following disadvantages: the properties of the final product will be affected by the doping of emulsifier during the reaction process; meanwhile, it requires complicated finishing technology for the preparation of solid polymer.

The organic dodecylbenzenesulfonic acid is selected as the primary dopant for the preparation of PANI via emulsion polymerization, and it can be used as the surface active agent for the preparation of PANI product with high molecular weight, excellent productivity, conductivity larger than 1 S/cm, solubility of 86% in N-methyl pyrrolidone (NMP) reagent. The productivity and solubility strengthen remarkably when compared with the PANI particles prepared via solution polymerization.

7.1.1.2 Electrochemical Polymerization

Electrochemical polymerization of PANI[16-17] mainly includes constant potential/current method and potential polarization method. The principle of electrochemical polymerization is that AN monomer is fed into the electrolyte in advance and then dispersed uniformly, eventually, PANI powder could be achieved based on the oxidative polymerization of AN on the anode after connecting power supply. In 1980, PANI membrane was fabricated by Diaz et al[18] using this method.

7.1.1.3 Template Polymerization

Template polymerization is an integrated utilization of many methods (e.g., chemical method and physical method), which provides more freedom for design, assembly and development of nano-structure materials. For example, applying nano-PANI on the matrix of organic film with pores, in addition, the control of particle dimension and morphology even the degree of orientation can be achieved. The conductive PANI with conductivity of 8.3×10^{-4} S/cm was fabricated by Yuan et al[19] via this method.

7.1.1.4 Radiation Polymerization

Optical energy or energy from other rays rather than oxidant is used to initiate the aniline polymerization during radiation polymerization[20]. The wavelength of radiation source and radiation shape have effects on the morphology of nano-PANI synthesized via this method, and products with different morphology can be obtained by adopting diverse visible radiation, for example, spherical product could be generated by ultraviolet radiation.

7.1.1.5 Nano-PANI Polymerization

The PANI particle at micro-scale or nano-scale contributes to doping and facilitates to interaction in the chain or out of the chain as well as enhancement of crystallinity. Preparation of nano-PANI could be achieved via microemulsion polymerization, which is composed of positive/reverse microemulsion polymerization and ultrasonic-radiation microemulsion polymerization. Water is selected as the dispersed and continuous phase for positive microemulsion polymerization, whereas oil is utilized as the dispersed and continuous phase for reverse microemulsion polymerization, for the purpose of oxidative polymerization of aniline monomer in the emulsion.

There are high requirements on selection and proportioning of emulsifier in positive microemulsion polymerization, which serves as the critical factor for preparation of PANI nanometer latex particle. The conductive nano-PANI was achieved by Moulton et al[21] on the basis of molar ratio configuration of AN, APS and DBSA as well as the conductivity measured by four-point probe of 5 S/cm, with organic acid (DBSA) as the dopant and inorganic acid (APS) as the oxidant. The polymerization may be conducted outside the micelle in the positive microemulsion polymerization, which is caused by the poor emulsification of emulsifier.

The reverse microemulsion polymerization presents isotropy, thermal stability and nano-scale dispersed phase. The nano-PANI is with particle size of 10 nm, uniform particle distribution and relatively-high crystallinity. In addition, the molar ratio of water and emulsifier has effects on particle size and morphology in reverse microemulsion polymerization.

The ultrasonic-radiation microemulsion polymerization, which is characterized by high conversion ratio, narrow distribution of particle size and no induction by oxidant, has become a novel approach for preparation of polymer nanoparticles. It features on ultrasonic instead of conventional oxidant being used for the induction of polymerization, in addition, stirring and shattering contributes to the dispersion, which increases the velocity of polymerization to a certain degree[22]. In the early 21st century, the PANI emulsion with conductivity of 3.0×10^{-2} S/cm was prepared by Atobe et al[23], with polyethylene oxide as the emulsifier, potassium iodate as the oxidant, inorganic hydrochloric acid as the dopant, and ultrasonic polymerization as the method.

7.1.2 Development of Conductive PANI Fiber

In recent years, quite a lot of researches focused on the preparation of conductive PANI fiber or PANI composite[24-27], as well as the investigation on the molecular structure, synthesis method, conduction mechanism and application of PANI. In the following section, different methods for the fabrication of PANI fiber are introduced, including in-situ polymerization, dry-wet spinning, wet-spinning, melting spinning and electro-spinning.

7.1.2.1 In-situ Polymerization

In-situ polymerization is a simple method to prepare conductive fiber, whose principle is that a PANI conducting layer is adhered to the surface of fiber-matrix to achieve conductivity, the matrix fiber can absorb the polymeric PANI macromolecule effectively, meanwhile, the generated PANI layer exhibits excellent conductivity, therefore, it is also called in-situ-adsorption polymerization. The polymerization processes can be classified into two categories: the first is that the fiber-matrix is impregnated into the AN monomer/dopant mixture solution in advance, and then the oxidant is applied under specified reaction conditions; the second is that the fiber-matrix is impregnated into the oxidant/dopant mixture solution in advance, and then the AN monomer is fed under specified reaction conditions, the conducting layer is formed based on the adhesion of finally-synthesized PANI on the surface of the fiber-matrix, thus endowing the fiber with conductivity.

In 1997, the experimental results reported by Liu et al[28] showed that V_2O_5 can exhibit better oxidation effect than that of an ordinary strong oxidant which is facilitated to change the point of view that a strong oxidant exhibits excellent oxidation effect, based on the investigation on effects of oxidants such as $(NH_4)_2S_2O_8$, $K_2Cr_2O_7$, $KMnO_4$, H_2O_2, V_2O_5 on the conductivity of PET fiber-matrix. The speed of polymerization reaction is high when $(NH_4)_2S_2O_8$ was selected as the strong oxidant. There was no sufficient time for the oligomer to diffuse towards fiber interior and surface, and it was separated out when being polymerized into the high polymer. When the weak oxidant was selected, it could control the oxidative polymerization speed of AN effectively and provide the synthesized oligomer with sufficient time to diffuse towards fiber interior and surface, and the PANI with high molecular weight was generated on the fiber. Therefore, appropriate polymerization velocity is the critical factor for the high-content PANI to be generated on the PET fiber-matrix. Meanwhile, conduction of pre-adsorption of PET fiber-matrix is to achieve enough AN monomer to be absorbed on the surface of PET fiber, therefore, most of the PANI polymerization occurs on the surface of PET fiber, thus forming compact conducting layer on the surface of PET fiber.

In 2000, the effects of such factors as concentration of AN monomer, concentration of doped acid, concentration of oxidant and reaction temperature on conductivity of conductive fiber were investigated by Pan et al[29], and the experimental results showed that the above-mentioned factors could affect the conductivity of conductive fiber and the velocity of polymerization reaction had remarkable effects on the PANI absorption.

The actual application of adsorption polymerization is limited due to its complicated preparation. Moreover, the conductivity of PANI fiber is quite important, that is because the conductivity of conductive fiber and the time for continuous conduction are affected by the absorption effect. Therefore, it is difficult for this preparation method to satisfy certain special demands on textile material.

7.1.2.2 Dry-wet Spinning and Wet Spinning

Both dry-wet spinning and wet spinning[30-21] are selected as the popular methods for the preparation of conductive PANI fiber. Previous research demonstrated that it is possible for PANI to be dissolved in few organics such as NMP and DMPU, and in certain inorganic concentrated sulfuric acid. The above-mentioned solvents are essential for dry-wet spinning and wet-spinning and have been widely used.

In 1993, it was proposed by Andreatta et al[33] that PANI fiber could be fabricated directly from the concentrated-sulfuric-acid-contained PANI solution. In addition, it was certified that sulfonated PANI fiber could be spun from the sulfuric-acid-contained solution or the sodium-hydroxide-contained solution, however, the fabricated noumenal PANI fiber was stiffened and fragile with poor wear resistance. Therefore, it presented no practical value when it was used as textile material.

In 1995, the wet-spun conductive PANI fiber was fabricated by Tzou et al[34], however, the PANI solution in N-N-methyl pyrrolidone (NMP) was unstable and changed into gel quickly. The mechanical properties of blended-spun PANI fiber were strengthened remarkably and its conductivity can be enhanced sufficiently under the action of appropriate dopant, with improved solubility of doped PANI.

The wet-spun composite conductive fiber was fabricated by Pan et al[35] by dissolving the blended PANI and PAN in the dimethyl sulfoxide/chloroform ($DMSO/CHCl_3$) solvent. The experimental results showed that the dispersion of PANI in the PAN matrix had direct effects on the conductivity of conductive composite fiber and its conductivity could be enhanced by improving the dispersion uniformity of PANI in the PAN matrix. That is because the conductivity of PANI in the PAN matrix could be extended properly like a channel of the power grid.

It can be concluded from previous researches that the key factors for preparation of blended PANI conductive fiber are the component and the proportion of the blended spinning solution as well as the ratio of appropriate dopant to eigen-state PANI. The dopant not only reinforces the conductivity but also improves the solubility of PANI. The organic sulfonic acid, which is usually selected as the preferred reagent for doping, is characterized by high boiling point and melting point, excellent environmental stability, both non-polar group and polar group on the molecular chain.

7.1.2.3 Melt Spinning

Melt spinning is characterized by high production speed, simple equipment, easy technology, and no solvent and precipitant. In addition, it seems to be feasible to prepare conductive fiber for the development of industrial production. However, the melt spinning can be conducted only when the polymer is in molted or plasticized state, it is impossible for PAN to be molted in conducting state, in this case, quite a lot of investigations have been done to achieve PANI plastination via applying plasticizer.

7.1.2.4 Electro-spinning

Electro-spinning has been popularly used in diverse fields, such as natural polymeric nanofiber, polymer nanofiber, polymer/inorganic composite nanofiber and inorganic nanofiber, meanwhile, remarkable achievements in certain fields, such as filtration, personal protection, sensor, energy source and photo electromagnetism, biology and medical, have been obtained. So far, electro-spinning has been selected as the effective method for preparation of micromaterial even nanomaterial and has potential applications in the future.

The basic principle of electro-spinning[36-38] is that the fiber at micro-scale even nano-scale is fabricated based on stretching and solidification of polymeric solution with a high-voltage electrostatic field. Namely, the charged polymer droplet is changed into a Taylor cone at the tip of capillary under the action of thousands to tens of thousands of volts of high-voltage electric field force. It is possible for the polymer to overcome its surface tension to form droplet at the tip of capillary when the electric field force increases to a certain degree, the nonwoven-structure fiber could be spun due to the evaporation or solidification of solvent in the droplet during spraying.

In 1996, the eletrospun conductive PANI fiber at nano-scale was fabricated by Reneker et al[39]. The experimental result showed that the conductivity, discoloration and visco-elasticity of the prepared fiber could be influenced dramatically by the doping of ion or molecule.

In 2000, the electrospun blended conductive PANI/PEO fiber was fabricated by Norris et al[40]. The diameter of the prepared PANI/PEO fiber (2.1 μm) is larger than that of PEO fiber (950 nm). However, the electrospun ultrafine fiber provided the similar ultraviolet spectrum curve with the prepared membrane, that is because the electrospun fiber can exhibit high porosity, and the higher vaporization speed of the doped fiber could be achieved when compared with the membrane structure, namely, it contributed to the optimization of mechnical properties of the blended PANI material.

In 2002, investigation on PANI/PMMA blended fiber was conducted by Desai et al[41]. The experimental results indicated that the concentration of spinning solution had great effects on the beaded fiber, the fiber diameter decreased and most of the fibers were at nano-scale, meanwhile, the diameter of the beaded droplet (occupying 20% of total amount of the fibers) decreased and its length shortened with the increasing PANI in spinning solution, which had significant effects on the fiber diameter.

In 2004, the conductivity of AMPS-doped PANI can be enhanced remarkably and the conductivity of electrospun PANI ultrafine fiber and spin-coated membrane was a little lower, that were proposed by Pinto et al[42]. The fiber with nickle film being coated on its surface presented a smooth and unifrom surface and no overlap could be found in the SEM image, it was found that the substrate of the metal-coated fiber exhibited great superficial area, combining electro-spinning and chemical precipitation.

In 2008, the electro-spun conductive composite nano-fiber was fabricated by Shin et al[38]. Great variation of multi-layer carbon nanotube/PANI/PEO composite nano-fiber could be charaterized by cyclic voltammetry. The self-generated heat by the internal carbon nanotube had sufficiently-strong effects on its conductivity, which facilitated to the enhancement of conductivity of the conductive fiber remarkably.

In 2009, it was proved by Shie et al[43] that the biocompatible PANI nano-fiber was an appropriate biosensor with the electrical properties, physical properties and chemical properties of organic polymer and metal. In 2010, the electrospun PLA/PANI blended fiber was fabricated by Picciani et al[44] and it was found that the crystal structure of fiber tube had great effects on fiber properties during electro-spinning.

In 2010, PAN/PANI ultrafine composite fiberwas fabricated by Cao et al[37]. It was found that the mass fractions of PAN and AN and the electro-spinning voltage had significant effects on the fiber properties. PANI was distributed in the PAN matrix uniformly with fiber conductivity of 0.02 S/cm, which exhibited excellent conductivity.

In 2011, the photocatalytic properties of PANI/PEO/TiO_2 conductive composite fiber were investigated by Neubert et al[45]; the electrospun PVA/PANI composite nano-fiber and PVA-PANI-$AgNO_3$ composite nano-fiber were studied by Shahi et al[46]; the PANI/nylon 6 blended composite fiber was prepared by Bagheri et al[47]; the PANI biosensor was developed by Shin et al[48]; the electromagnetic shielding device made of PANI composite was investigated by Zhang et al[49]. Recently, PANI has attracted a wider attention.

7.1.3 Application of PANI

The conductivity of conductive PANI remains on metal-scale and could be used as the alternatives of certain metals according to its scope of application. PANI is characterized by unique doping mechanism, excellent stability and proper electrochemical behavior. Because it only requires low cost and easy technology, PANI has potential applications in future in many science and technology fields, such as battery shielding material, electrochromic device, display and anti-corrosion material.

7.1.3.1 Anti-corrosion Coating

Anti-corrosion coating is the most significant industrial application of PANI in initial stage. It is not allowed for PANI to be used as the coating directly because of the insoluble and infusible properties of PANI, the excellent properties of PANI could be embodied when combining with the commonly-used matrix resin. At present, the investigations on selection of matrix resin, interaction of PANI and matrix, anti-corrosion mechanism and anti-corrosion effect have become mutual, and two PANI anti-corrosion coating systems have been developed successfully.

7.1.3.2 Rechargeable Battery

The conducting-state PANI exhibits excellent electrochemical behavior and reversibility, and has been selected as the appropriate choice of cathode material of lithium battery. The oxidation-reduction reaction of conductive polymer is achieved through the reversible process of doping-dedoping during the electrode reaction process, thus realizing the battery charge/discharge. The $LiNi_{0.8}Co_{0.2}O_2$/PANI composite fabricated by Mosqueda via in-situ polymerization exhibited superior reversibility when compared with conventional lithium electrode material. PANI/carbon black (PANI/C) composites were synthesized by Zhu et al[50] with ferric chloride as oxidant by adsorption polymerization in hydrochloric acid aqueous solution, and then the PANI/C conductive composite material with extremely-high conductivity was fabricated in the nitrogen atmosphere. The secondary battery exhibits relatively high discharge capacity with PANI/C composite as the anode, lithium as the cathode and $LiPF_6$/EC-DMC-EMC as the electrolyte.

7.1.3.3 Electromagnetic Shielding

PANI was popular in anti-static coating and battery shielding[51-52] after the breakthrough in its processability. The conductive PANI could reflect or absorb electromagnetic wave with specified frequency which contributes to the electromagnetic shielding because of the advantages of PANI, such as light-weight, regulatable electromagnetic parameters, excellent environmental stability, regulatable in insulator, semiconductor and conductor. The experimental results indicated that the electromagnetic wave within the range of 10 MHz~1 GHz could be shielded by the high-conductivity PANI and the shielding effect is greater than 20 dB. To prepare diverse antistatic flooring, the organic-sulfonic-acid-contained conductive PANI was mixed with other matrix by UNIX (U.S.).

7.1.3.4 Electrochromic Device

Electro-spinning refers to the process of discoloration of the material under the action of applied electric field force. The conducting-state PANI membrane fabricated by Kaner[53] exhibited reversible electrochromic behavior, and the discoloration process of PANI membrane presented in bright yellow, blue, dark blue and finally black respectively with the electro-spinning voltage increasing from -0.7 to 0.6 V, which brings unmeasurable applications for the development of military camouflage in the future.

In addition, PANI has potential applications in other fields, such as chemical modification, molecular-level circuit and artificial muscle by robot.

7.2 Structure of Electrochromic Device

In 1969, it was the first discovery by Deb et al[54] that WO_3 membrane (serving as an inorganic substance) can exhibit electrochromic properties. In the 1970s, the research on electrochromism mechanism has been reported[55-58] and the electrochromic properties of transition metal oxides were found, which focused on response time and application for electronic display device.

Until the end of 1980s, it was found that the oxidative-reductive organic material could present electrochromic properties as well, which provided much more superior synthesis technology, material consumption, response time of color change and category of changed color. In addition, the commonly-used organic electrochromic material exhibited relatively-high stability, such as viologen[55], polypyrrole (PPy)[56], PANI[57-58], polythiophene (PTh)[59] and metal phthalocyanine[60-61].

In the early 1990s, intelligent and energy-saving smart window was successfully developed by Zeng et al[62], and the electrochromic film was used as the matrix. So far, it has been widely adopted by researchers of Germany, England, France and America. The electrochromic device is the combination of electrochromic element and control circuit. Its characteristic is that the element color changes under the control of applied potential, and it is with reversibility and memory effect. In addition, the incident doses of electromagnetic radiation with different wavelengths can be adjusted and controlled, therefore, the light filtering, dimming control and energy saving can be achieved[63].

The basic structure of electrochromic device is shown in Fig. 7-3. As depicted in the figure, the complete electrochromic device is designed with seven-layer sandwich structure, including glass substrate, transparent conducting layer, electrochromic layer, ion conducting layer, ion storage layer and complementary layer. The electrochromic layer is the core layer and it has attracted increasing attention in the science research. The ion conducting layer provides ions with transmission channel between it and the electrochromic layer, and the ion storage layer is used for ion storage and charge balance. When applying the forward DC voltage to the transparent conducting layer, ions in the ion storage layer are extracted, pass through the ion-conducting layer, and enter into the electrochromic layer. As a result, the electrochromism occurs and the no-energy-consumption memory function is realized. When applying the reverse voltage, ions in the electrochromic layer are extracted and then enter into the ion storage layer again. As a result, the overall device returns to the original transparent state.

As shown in Fig. 7-4, glass refers to the glass substrate, ITO refers to the conductive film, EC refers to the electrochromic material, PE refers to the electrolyte material, and IC refers to the ion conductor.

Chapter 7
Development of Conductive Polymer Materials

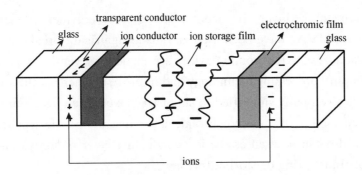

Fig. 7-3 Basic components of electrochromic device

7.2.1 Transparent Conductor (TC)

It is mainly used for conducting electricity, establishing electric field and realizing electrochromism. The most-commonly used material is the ITO glass and it has the advantages of good light transmittance, low resistivity and convenient for low-temperature preparation. To realize the bendable and folding performance of the device, new transparent electrode material is required to take the place of ITO. For example, the Pulis Company has prepared a new-type transparent conductive polymeric material which can be made into film through printing technology and can be bent. The research team of University of Yamanashi (Japan) has successfully prepared the organic transparent film made of PEDOT/PSS.

7.2.2 Electrochromic Layer (EC)

It is the core layer of overall electrochromic device and the occurrence layer of the electrochromic reaction. It is mainly presented in the form of electrochromic film which is used to prepare electrochromic material. The high polymer electrochromic material such as polythiophene, polypyrrole and PANI can be made into film through electrochemical deposition and organic vapor phase deposition, etc.

7.2.3 Ion Storage Layer (ISL)

It is also called the counter electrode layer and used for ion storage and charge balance. Material in this layer must have powerful ion storage ability. VZOs and TiO_2 film are the materials used currently and they have greater limitations on the modulation of spectral intensity and energy. In recent years, the scholars are dedicated in finding and adopting the electrochromic material with powerful ion storage ability as the ion storage layer.

7.2.4 Ion Conductor (IC)

It is used for transferring the ions and blocking the electrons between the electrochromic layer and the ion storage layer. In recent years, organic polymer gel-state electrolytes, such as PEO and PMMA, have been used and made significant breakthroughs.

Not all practical electrochromic devices are made of seven-layer materials. To decrease cost, simplify processing technology and improve efficiency, the structure of electrochromic device is usually simplified to six or five layers even less. For example, some materials can be used as the electrochromic layer due to the electrochromic performance, and they also can be used as the ion conducting layer or ion storage layer due to the function of transferring or storing theions[64].

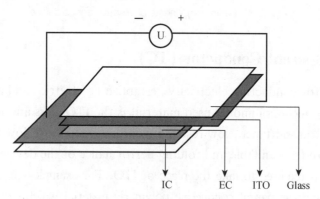

Fig. 7-4 Laminated structure of electrochromic device

7.2.5 Models of Electrochromic Reaction

Electrochromic is a phenomenon that the optical properties of materials are steady and reversibly change under electric field. According to this, dual implantation of electron and ion is considered as the starting point of all models used for explaining the electrochromic mechanism, and specific electrochromic mechanism is proposed according to the implantation process and the reaction after the implantation. So far, the proposed electrochromic mechanism models mainly includes electrochemical reaction model, color center model, inter valence transition model, polaron model and ligand field model, etc[65].

7.2.5.1 Electrochemical Reaction Model

It is considered that in the electrochemical reaction (also called tungsten bronze model), under the action of electric field, the positive ionhas single charge and small diameter (e.g., H^+ and Li^+, etc.) and the electron can be injected into the film from both ends of the film, the following reduction reactions can occur between them:

$$I^+ + e^- \longrightarrow I \qquad (7-1)$$

$$MeO_n + xI \longrightarrow I_x MeO_n \qquad (7-2)$$

In above Eqs., I^+ refers to the positive ion with single charge and small diameter, Me refers to the metal atom. If $n = 3$, it is the defect perovskite structure and corresponds to the WO_3 crystal structure. By using Eq. (7-1), the atom is obtained through the reaction between the positive ion and the electron, and then diffuses into WO_3 crystal lattice. As a result, the defect perovskite structure is changed into the tungsten bronze structure and the film color is changed into blue, as shown in Fig. 7-5. After the voltage reversal, the electron and positive ion move out of the electrochromic film and the film fades. Although the electrochemical reaction model provides no transmission channel for metal ion and electron, it is deemed that the Eq. (7-1) occurs in the crystal (grain) interface and the Eq. (7-2) completes via diffusion, this explanation complies with the following experiment result that the electrochromic effect decreases with the increasing density and order degree of the film.

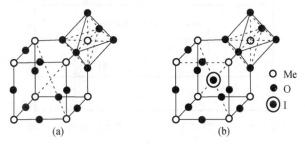

Fig. 7-5 Electrochemical reaction model: (a) Defect perovskite structure; (b) Tungsten bronze structure

However, Qian et al[63] studied the cyclic voltammetry curve and electrochromic phenomenon of WO_3 film in 1 mol/L lithium hexafluorophosphate organic electrolyte, it was considered that the ion implantation and extraction process fails to comply with the tungsten bronze model proposed by Crandall and Faughnan[66]: (1) after Li^+ implanted into the tungsten oxide film completely exiting, color fading fails to occur to the tungsten oxide film; (2) the 2.5 V electric potential which is lower than the 3.5 V electric potential (e.g., the electric potential for Li^+ gaining electrons in the aqueous solution) is adopted in the experiment. The results showed that the color change of film is not arising from the tungsten bronze since lithium tungsten bronze is not formed in the tungsten oxide film. If the film has certain catalytic mechanism, e.g., crystal (grain) interface, as the medium used for positive ion gaining or losing electrons, it can reduce the reduction potential of Li^+, and the electrochemical reaction model would be more perfect.

7.2.5.2 Deb Model

Deb model is also called the color center model. Zhang et al[67] proposed that amorphous WO_3 has an ionic crystal structure which is similar to the metal halide. It is reported that the amorphous WO_3 can be generated by vacuum evaporation. In this case, electropositive oxygen vacancy defect can be formed and electron injected from the cathode is captured by the oxygen vacancy and the color center is formed. The captured electron is unstable and it is easy to absorb the visible light photon and then be excited into the conduction band, thus making the WO_3 film color visible.

This original model explains this phenomenon: the colored-state WO_3 film fades after being heated in the high-temperature oxygen, as a result, the electrochromic ability disappears. But Crandall and Faughnan[66] thought that it is difficult for the WO_{3-y} film ($y = 0.5$) to generate a larger quantities of color centers when there is serious lack of oxygen.

7.2.5.3 Faughnan Model

Faughnan model is also called the dual implantation/extraction model or inter valence transition model as shown in Fig. 7-6.

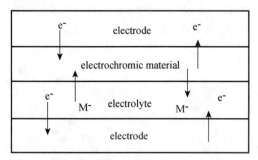

Fig. 7-6 Sketch of intervalence charge transfer model

Electrochromic mechanism of amorphous WO_3 proposed by Faughnan can be expressed in following formula:

$$xM^+ + xe^- + WO_3 = M_xWO_3 \qquad (7-3)$$

In this formula, M^+ refers to H^+ and Li^+, etc. When applying the electric field, the electron e^- and positive ion M^+ are simultaneously implanted into the defect positions among the atomic lattices of WO_3 film to form the tungsten bronze (M_xWO_3), and it is presented in blue. When applying the electric field in reverse direction, the electron e^- and positive ion M^+ in electrochromic layer are simultaneously detached; as a result, the blue disappears.

For the tungsten bronze, the transfer of electrons from different crystal lattices (position A and B) can be expressed as follows:

$$h\gamma + W^{5+}(A) + W^{6+}(B) = W^{6+}(A) + W^{5+}(B) \qquad (7-4)$$

7.2.5.4 Schirmer Model

Schirmer model is also called the polarization model[68]. After the electron injecting into the crystal, localization can be occurred at certain crystal lattice position due to the interaction between the electron and ambient crystal lattices. As a result, the small polaron is formed and the equilibrium configuration is destroyed. The photon is absorbed during the transition of small polaron among different crystal lattice positions. The polaron transition arising from light absorption is called the Franck-Condon Transition. During the transition process, electron transition energy is completely changed into the photon emission energy.

Both the Faughnan model and Schirmer model are established on the basis of dual implantation/extraction of ion and electron. They have the same physical essence, and the Faughnan mode can be considered as the semi-classical form of Schirmer model actually. For some time, these two models were widely accepted by the researchers, however, something contradictory with the experiment results was found.

7.2.5.5 Ligand Field Model

For the electrochromism, not only the electron transition of different-valence metal ions may occur, but also the d-d transition of metal ion and electron transition between the metal ion and coordination ion. Weng Jianxin, etc. named this electrochromic mechanism as the ligand field model. However, this model fails to provide the further explanation about the dual extraction mechanism of electron and ion; it is just a tentative idea[69].

7.2.6 Classification and Selection of Electrochromic Material

The electrochromic principle of electrochromic material mainly depends on the band structure of chemical composition and oxidation-reduction feature of the material. For example, modulate the absorption characteristics of film in the UV region and visible region or change the concentration of carrier in the film and plasma oscillation frequency by implantation and extraction of ion and electron, the IR reflection characteristics can be modulated.

Although the scholars have made a large quantity of researches on electrochromic material and its products, the uniform conclusion on electrochromic mechanism has been unclear. This is mostly because the electrochromic reasons are complicated and involved multiple aspects, such as chemical composition (doping) of material, fine structure of film, band structure, oxidation reduction feature, etc. Researching and mastering of electrochromic mechanism has great significance on the development of electrochromic device and product.

In general, the electrochromic material must have following performances:
(1) Good electrochemical redox reversibility;
(2) Short response time for color change;
(3) Reversible color change;
(4) High sensitivity to color change;
(5) Long cycle life;
(6) Certain memory storage function;
(7) Good chemical stability[70].

Among the electrochromic materials, the transition metal oxide especially the high-energy semiconductor tungsten oxide (WO_3) gained wide attention in the past 30 years. So far, multiple inorganic materials have been developed, e.g., Prussian blue, oxides of Mo, Nb and Ti (anodic coloring), oxides of Ni, Co and Ir (anodic coloring)[71]. The electrochromic material made of small organic molecules (such as viologen, its structure is shown in Fig. 7-7) is transparent under the stable dual-cation state. For the one-electron reduction, the state of radical cation is with special color. The electrochromic film made of polymeric viologen and the viologen replaced by N has been presented. In recent years, it is very important to improve the electrochromic property by increasing the absorption of organic molecule on the pore of metal oxide.

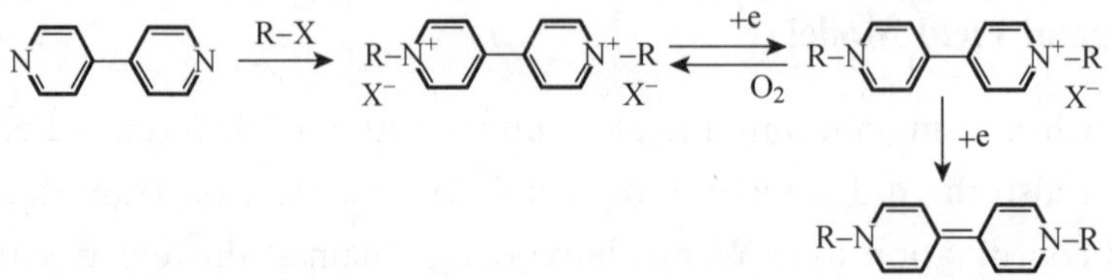

Fig. 7-7　Viologen molecular at dual-cation state and free radical cation state

The conjugate polymer is classified into three kinds of electrochromic materials. It has caused people's attention because of its advantages, including easy to be prepared, short reaction time, simple preparation technology, high optical contrast, easy to control molecular structure to make it with multi-color. The typical conjugate electrochromic polymer includes polythiophene (PTh), PANI and polypyrrole (PPy), whose structures are shown in Fig. 7-8.

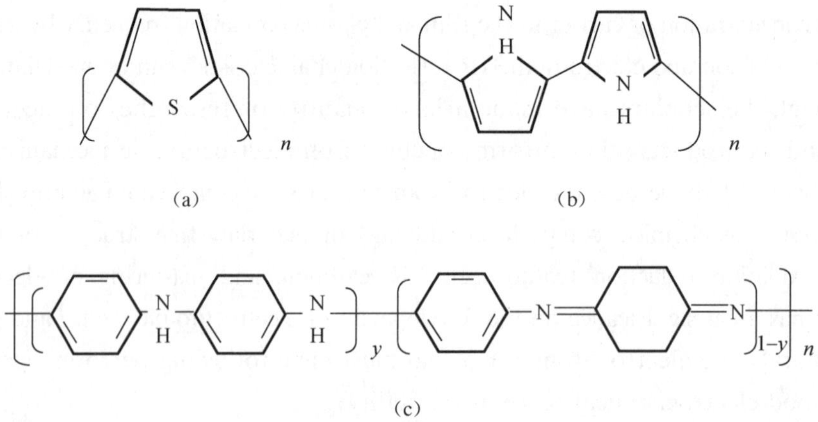

Fig. 7-8　Typical conjugate electrochromic polymer structure: (a) PAh; (b) PANI; (c) PPy

The electrical conductivity of electro-conductive polymer can be controlled with the help of reversible chemical or electrochemical doping process, and the conduction state of polymer can be improved from insulator to conductor. During the oxidation process, the backbone of electro-conductive polymer will be oxidized to form the positive charge due to the electron loss. For the purpose of maintaining the electric neutrality of the system, the positive charge will attract the negative ion in the solution to coordinate with it, thus maintaining the stable state; and the reverse process is called de-doping. The doping of electro-conductive polymer with positive-charge is called p-doping, while the doping of electro-conductive polymer with negative-charge is called n-doping.

7.2.6.1 Polythiophene and Its Derivatives

The thiophene has the feature of extremely-good environmental stability, easy to be prepared, good conductivity after doping. The poly (3,4-ethylene dioxy-thiophene) (e.g., PEDOT) is the most popular object among thiophene conducting polymers. PEDOT has good conductivity, stability, photoelectric characteristic and electrochromism and its film has good transparency, therefore, it is mainly used as the transparent electrode material and electrochromic active material.

When PEDOT is in oxidation state, π-electron is in high-energy state; the electron absorbs the spectrum and moves towards the near-infrared low-energy spectrum band, as a result, transparent blue appears in the visible spectrum. When PEDOT is in reduction state, conjugate interactive single-double bond structure makes the π-electron absorb the spectrum and move towards high-energy visible spectrum band; as a result, PEDOT is presented in dark blue. Some researchers compounded PEDOT with other material to increase the electrochromic performance. For example, Homas J used the layer-by-layer LBL technology (as shown in Fig. 7-9) to prepare PEDOT/PSS and BPEI composite membrane, added TiO_2 and carbon black to improve the anti-UV degradation ability and conductivity of the composite. The results showed that the conductivity is 250 times higher than before and the light transmittance increased 27%.

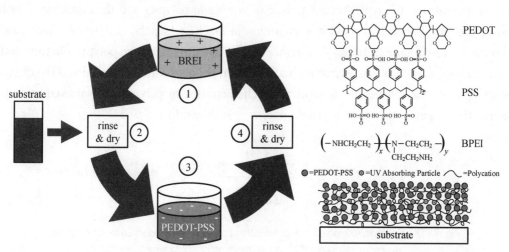

Fig. 7-9 Preparation process of PEDOT/PSS and BPEI composite membrane by LBL technology

Other scientists copolymerized the thiophene with other organic monomer to obtain the electrochromic material with multiple colors. Aubrey L Dyer copolymerized the thiophene derivative monomer with benzene to obtain the electrochromic material with multiple colors through controlling the proportions of two monomers and the substituent group of derivative. In addition, some scholars are researching on the performance of thiophenehalogeno-structure. Selmiye Alkan used the constant voltage to obtain the poly (3-bromine thiophene) and poly (3,4-two bromine thiophene) (e.g., PDBrTh) from the BFEE solution and prepared the high-performance electrochromic device whose cycle index reached up to 1000 (Fig. 7-10).

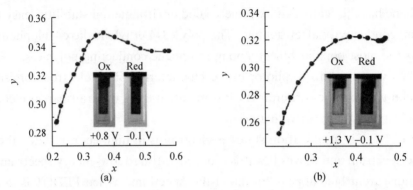

Fig. 7-10 Synthetic polymer at ITO glass surface: (a) poly (3-bromine thiophene); (b) poly (3, 4-two bromine thiophene). (Ox—Oxidizing substance; Red—Reducting substance)

Generally, electrochromic material made of polythiophene conducting polymer is the most-commonly researched, most-widely used and most-widely involved polymer, its huge market value is also obvious and it has a rather important influence on the color enrichment of electrochromic device.

7.2.6.2 Polypyrrole

The polypyrrole has the typical characteristic of conjugate polymer and the doping-state polypyrrole has good conductivity and it is presented in black, while the polypyrrole film is light yellow after de-doping. In 1979, Kanazawa et al[72] firstly prepared the polypyrrole film in the acetonitrile solution through electrochemical oxidation polymerization and observed the electrochromic phenomenon. Obvious electrochromic characteristic of polypyrrole is proved in proper electrolyte. The electrochromic reaction process[73] is shown in Fig. 7-11.

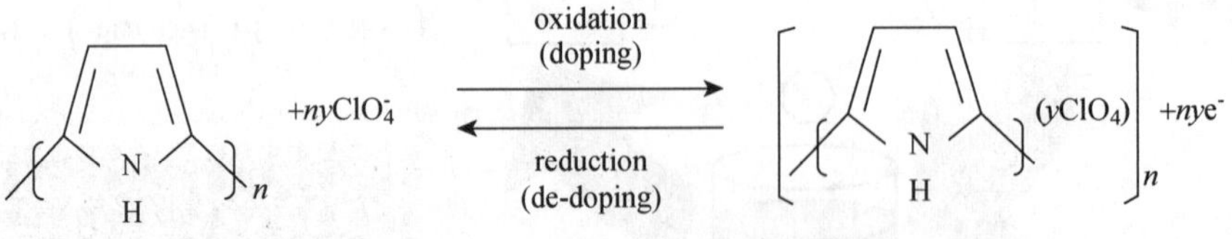

Fig. 7-11 Polypyrrole oxidation and reduction process. (n—polymerization degree; y—molar ration of $nClO_4^-$; e—electron of polypyrrole)

7.2.6.3 PANI

PANI is a kind of typical conductive polymer. Compared with other conductive polymers, PANI has following features: one is that it has the diversified structure, different redox states correspond to different structures; the other is its special doping mechanism, its property can be changed via doping and de-doping process. Due to its unique electrochemical activity and stronger chemical stability, PANI has become one of the most popular conductive polymers. At present, it has been widely applied in secondary battery, electrochromic device, sensor, electro-catalysis and anti-corrosion of metal, etc. However, PANI has extremely-poor solubility and it is difficult to be dissolved in any solvent due to the strong stiff of PANI chain and the interaction of inter-chain. This disadvantage greatly restricts the application of PANI[74]. The current research on PANI electrochromic fabric is still in initial stage, and a series of related problems have to be solved to fully develop this high and new technology.

7.2.7 Application of Electrochromic Technology

Electrochromic technology has unique advantages in high-tech fields, such as color-changeable sunglasses, high-resolution-ratio photoelectric photography equipment, photoelectric chemical-energy conversion, active optical filter, etc. The camera and laser can be used as the optoelectronic regulating valve for the purpose of the image recording, information processing, optical memory, optical switch, holography, decorating material, fully-protective material, etc[75]. Electrochromic technology has great application potential in many fields, and certain applications are introduced as follows.

7.2.7.1 Electrochromic Smart Window

The potential of electrochromic smart window (as shown in Fig. 7-12) in electrochromic devices is tremendous, and a few developed countries have applied it as the new-generation energy-saving material and the core research objective of electrochromic device. Through the applied voltage/current, the optical properties of smart window can be arbitrary changed according to actual demands. Dynamically adjust the input/output of solar energy and the spectrum in visible region to realize the continuous and reversible regulation of optical density. It only requires low power consumption and can be widely applied in the construction, automobile, airplane, spacecraft field, etc.

Fig. 7-12 Electrochromic smart window

7.2.7.2 Electrochromic Display Device

Electrochromic material has the bi-stable performance. The electrochromic display device (Fig. 7-13) made of electrochromic material requires no backlight, in addition, it requires no power consumption and the displayed contents are not changed after displaying the static image, in this case, energy saving can be achieved. Compared with other display devices, the electrochromic display device has the advantages of no blindness angle, high contrast, low manufacturing cost, wide working temperature range, low driving voltage, rich colors, etc. It has brighter application future in such fields as instrument display, outdoor advertising, static display, etc.

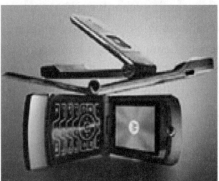

Fig. 7-13 Electrochromic display device

7.2.7.3 Electrochromic Storage Device

The scientific community has developed multiple electrochromic materials with different colors. Use the multi-color-changeable property of the electrochromic material and combine different colors (e.g., combining the tricolor material in different proportions) to record the continuous color information which is similar to thechromophotograph, in addition, it has the functions of erasure and rewrite. For example, the PANI derivative can display the tricolor, supposing that use the electrochromic material made of PANI derivative to record the continuous color information. Since it has the advantages of high contrast, rich colors, memory storage available, light weight, thin, easy to compatible with microelectronic circuit, all above-mentioned advantages facilitate it to play more important role in the electronic display and storage field. The application of electrochromic device in sensor display system can be observed in Fig. 7-14.

7.2.7.4 Electrochromic Fabric

Intelligent textile makes the high-tech sensor or sensitive element combine with the traditional structural material and functional material, the combinational design can be realized on the basis of modification, functionalization, and intelligent of current fiber

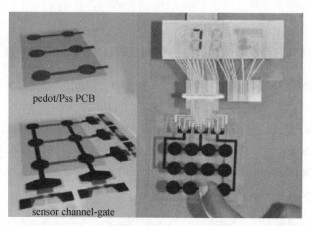

Fig. 7-14 Application of electrochromic device in sensor display system

fabric. Electrochromic fabric made of conducting polymers refers to the fabric whose color can change under the action of electric field. At present, research on it is still in initial stage.

It is expected that the intelligent electrochromic fabric will be widely applied in the military camouflage uniform. In the future, the solider can arbitrarily adjust the uniform colors according to different circumstances on the battlefield; as a result, it is difficult for the enemy to identify. It has the similar concept with the chameleon camouflage; its nature is to combine the electrochromic fabric with the detector used for monitoring the ambient color. With the data acquired by the color detector, the color, light and pattern of the uniform change immediately through the mode of fiber transferring the visual information, so as to achieve the goal of automatic control and automatic change of uniform color. Except for being used for the camouflage uniform, the electrochromic fabric can be used for the pattern-changeable dossal or canvas as well, and it can be used as the cloth-type display if it has enough brightness. The current difficulties mainly refer to the low electric conductivity of transparent conducting layer, poor transparency and low stability of chromophore. If the above-mentioned difficulties can be overcome, it is expected that the intelligent electrochromic fabric will bring considerable business opportunities. Several kinds of electrochromic uniforms can be observed in Fig. 7-15.

Fig. 7-15 Electrochromic uniforms

7.3 Preparation and Characterization of PANI Emulsion

PANI as a synthesized macromolecular compound is a special conductive polymermaterial[76-77]. The molecular structure of PANI is shown in Fig. 7-16. The monomer of PANI is aniline which contains amine, benzene ring and nitrogen atom as shown in the figure.

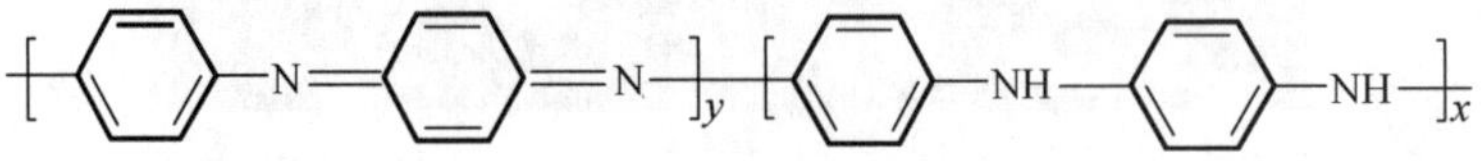

Fig. 7-16 Molecular structure of PANI

As can be seen in Fig. 7-16, PANI consists of oxidation unit (benzoquinone unit) and reduction unit (dianiline unit). The value of y ($0 \leqslant y \leqslant 1$) and x ($0 \leqslant x \leqslant 1$) indicates the degree of oxidation and reduction of PANI and it can be classified into three states based on the value of y: full-reduction state ($y=1$), medium-reduction state ($y=0.5$) and full-reduction state ($y=0$). The number of oxidation unit is equal to the number of reduction unit when $y=1/2$, it is called the medium-reduction state. PANI in medium-reduction state could be used as conductor after the doping with proton acid. The conducting-state PANI exhibits excellent electrochromic properties and the color will change under the action of applied voltage because of the oxidation-reduction reaction of PANI induced by electric field force. In this case, PANI presents rich colors with reversible change. PANI is characterized by broken circuit and memory function, which widens the application of PANI. The structure of PANI varies with the value of y (within the range of 0~1), and its color, molecular structure and conductivity vary correspondingly as illustrated in Tab. 7-2.

Tab. 7-2 Different chemical structures of PANI and corresponding colors

y	Name	Structure	Color	Properties
1.0	Colourless emeraldine	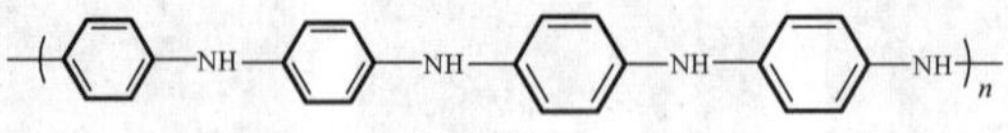	Colourless (light yellow)	Insulator
0.75	Primary-coloremeraldine	—	Light blue	Semiconductor
0.5	Emeraldine	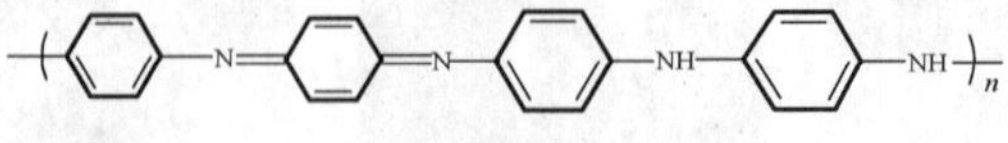	Dark green	Metallic state
0.25	Aniline black	—	Blue	Insulator
0	Full aniline black	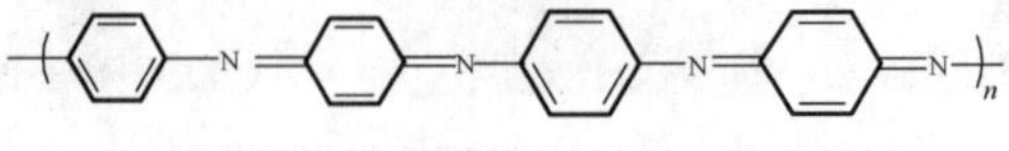	Blue purple	Insulator

The conductivity of PANI enhances under the doping action of proton acid, with water as the heat carrier and dodecylbenzenesulfonic acid (DBSA) as the surface active agent. The fabricated PANI emulsion rather than other solvent could be used for finishing directly, the film-formation properties, mechanical properties and conductivity[10] of PANI could be improved, the comprehensive properties of PANI could be optimized and its scope of application could be widened[80].

7.3.1 Experimental

7.3.1.1 Material and Equipment

The experimental reagents and equipment used in this study are listed in Tab. 7-3 and Tab. 7-4 respectively.

Tab. 7-3 Experimental raw material and reagents

Reagent	Specification
Aniline	Analytical reagent (AR)
Dodecylbenzenesulfonic acid	AR
$(NH_4)_2S_2O_8$	AR
Polyvinyl alcohol	AR
Acetone	AR
DMF	AR
Trichloromethane	AR
THF	AR
NMP	AR

Tab. 7-4 Experimental apparatus

Equipment	Specification
Digital-display top-mounted mechanical stirrer	RW 20 digital
Temperature controlled electric heating cover	DRT-TW
Low-speed desktop centrifuge	TDL 80-2D
Electronic balance	JT601N
Spin coater	KW-4A
Optical microscope	XS-213
Scanning electron microscope	S-3400N
FTIR infrared spectrometric analyzer	Nicolte AVATAR 380
UV-visible spectrophotometer	UV-1601PC
Vacuum drying oven	DZF-6050

7.3.1.2 Morphological Structure Analysis

To investigate on the dispersion of PANI particle in the PVA matrix, the Optical Microscope (XS-213, Jiangnan Optical Instrument Co., Ltd., Nanjing) was selected. The sample test was conducted based on the range of 10 μm (obtained from 3000X objective lens) because the image of the optical microscope was designed without the caliper scale. Then the caliper range of 10 μm was combined with the characterized sample image via photoshop software to obtain complete pattern, eventually, definition and analysis of the image was demonstrated and conducted, respectively.

Morphology of PANI particle on the PVA membrane surface was investigated by Scanning Electron Microscope (SEM, S-3400N, Japan). Before the SEM testing, gold was coating on the composite to obtain good conductivity requirements of SEM. The average diameter of PANI particle was measured by Image-Pro Plus 6.0 software.

7.3.1.3 Solubility Analysis

Acetone and water were selected for the demulsification of PANI/PVA emulsion, then PVA and residual DBSA reagent were filtered to obtain the PANI powders. 200 mg of PANI powders (4 in total) were dissolved in 10 mL DMF, trichloromethane, THF and NMP respectively with low-temperature ultrasonic treatment for 30 min. Subsequently, a low-speed stirring was carried out in ice-water bath for 6 h. PANI/organic solvent with mass sum of M_0 was then filtered and dried in oven at 50 ℃. The weight of filter material after solvent removal was M_1, and the weight of residual PANI was $M_1 - M_0$ and the solubility R (mg/mL) could be achieved as Eq.(7-5).

$$R = \frac{200 - (M_1 - M_0)}{10} = \frac{200 - M_1 + M_0}{10} \tag{7-5}$$

7.3.1.4 Determination of Infrared Spectrum

The molecular structure of PANI/PVA composite membrane was characterized by the Infrared Spectrometric Analyzer (AVATAR 380 FTIR) with spectral data within the range of 3800~750 cm^{-1} and step size of 1.929 cm^{-1}.

7.3.1.5 Determination of UV Absorption Spectrum

The determination of UV absorption spectrum was performed by the UV-visible Spectrophotometer (Shimadzu, Japan). Baseline correction for the UV-visible Spectrophotometer should be conducted firstly (only one baseline correction was required without cuvette replacement).

7.3.1.6 Investigation on DBSA Doping Mechanism

PANI/PVA composite membrane with uniform thickness and area of 4 cm^2 was fabricated by Spin Coater (KW-4A) and the conductivity of the sample was investigated by four-point probe resistivity tester. The testing data should be corrected according to the thickness correction factor of the sample.

7.3.2 Results and Discussion

7.3.2.1 Effects of Synthesis Parameters on Conductivity of PANI

Basing on previous researches[781-85], it was found that the content of monomer PANI is closely linked to the dosage of doped proton acid, polymerization temperature, content of oxidant even the reagent applying sequence. The dosage of oxidant, polymerization temperature and concentration of proton acid were characterized by $L_9(3^4)$ orthogonal table in this experiment. The experimental scheme was listed in Tab. 7-5 and the experimental data was demonstrated in Tab. 7-6.

Tab. 7-5 Level of form factors

Level	A Mass of proton acid(g)	B Polymerization temperature(℃)	C Mass of oxidant(g)
1	2.31	0	1.56
2	3.08	10	2.08
3	3.85	20	2.6

Tab. 7-6 Orthogonal $L_9(3^4)$

Experimental No.	A	B	C	Other	Conductivity(S/cm)
1	1	1	1	1	0.83
2	1	2	2	2	0.78
3	1	3	3	3	0.52
4	2	1	2	3	1.26
5	2	2	3	1	0.97
6	2	3	1	2	0.54
7	3	1	3	2	0.89
8	3	2	1	3	0.55
9	3	3	2	1	0.97
K_1	2.13	2.98	1.92	2.77	
K_2	2.77	2.30	3.01	2.21	
K_3	2.41	2.03	2.38	2.33	
k_1	0.71	0.99	0.64	0.92	
k_2	0.92	0.77	1.00	0.74	
k_3	0.80	0.68	0.79	0.78	

(continued)

Experimental No.	A	B	C	Other	Conductivity(S/cm)
Range (R)	0.21	0.32	0.36	0.19	
Influencing sequence (primary-secondary)			C B A		
Optimum solution			$C_2 B_1 A_2$		

K—the sum of corresponding experimental results at the same level;
k—the arithmetic mean value of K, where $k = K/3$ in this experiment;
R—the difference of maximum value and minimum value of k of each factor.

The optimized process was determined by the conductivity. The content of proton acid, oxidant and polymerization temperature were depended on the sequence of K_1, K_2, K_3 or k_1, k_2, k_3 and the higher value corresponding to K_i and k_i was selected. As can be judged from Fig.7-17, the perfect process was $C_2 B_1 A_2$ (mass of protonic acid is 3.08 g, mass of oxidant is 2.08 g, polymerization temperature is 0 ℃).

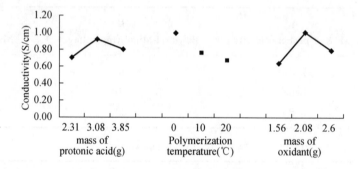

Fig. 7-17 Trend graph of the optimized PANI polymerization process

7.3.2.2 Solubility

PANI prepared via emulsion polymerization was dissolved in different solvents and its solubility was illustrated in Tab.7-7. DBSA was selected as the proton acid which was not only characterized by doping but also can be used as surface active agent, namely, the solubility of PANI could be enhanced under the induction of anion of organic sulfonic acid.

Tab.7-7 Solubility of PANI in different organic solvents

Solvent	DMF	Trichloromethane	THF	NMP
Solubility (mg/mL)	20.6	9.4	17.3	24.7

It could be clearly observed in the table that the solubility of PANI in NMP reaches 24.7 mg/mL, whereas its solubility in trichloromethane is only 9.4 mg/mL. By the classification of solvents by Tess[86], it could be found that trichloromethane is considered as a solvent of weak hydrogen bond, while DMF, THF and NMP are classified into the solvents of medium hydrogen bond. It could be concluded that the trichloromethane exhibits the poorest solubility, while NMP presents the strongest solubility. That is because the solubility parameters of PANI-doped solution provide similar polarity to NMP, PANI can exhibit the highest solubility in NMP and higher solubility in DMF according to the principle of "better compatibility with similar solubility parameters".

7.3.2.3 UV Absorption

The transition of electronic energy level in PANI macromolecule was investigated by UV-visible spectroscopy with ITO conductive membrane as the substrate. The UV-visible absorption spectrum of conducting-state PANI composite membrane fabricated via emulsion polymerization is demonstrated in Fig. 7-18.

As can be observed in Fig. 7-18, obvious characteristic peak of doped-state PANI are observed and the peak at 370 nm in the near ultraviolet region is the energy gap of $\pi \rightarrow \pi^*$ transition related to benzene structure; while the peak at 428 nm in the visible region is the transition of polaron band π^*. The peak at 572 nm is the $\pi b \rightarrow \pi q$ absorption related to quinines and the peak at 830 nm is the energy gap of transition of π polaron band. All the above-mentioned characteristic peaks are the basic peaks of the doped-state PANI, which directly indicates the band structure of PANI and structure change during the doping process[87].

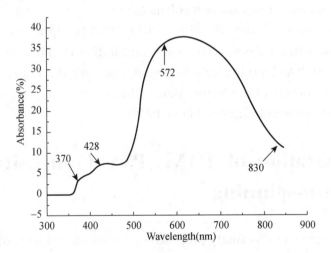

Fig. 7-18 UV-Visible spectrum of PANI

According to the previous research[88], the characteristic peak arising from $\pi \rightarrow \pi^*$ transition would move towards the long-wave direction because of the application of PVA. That is because the excited-state solvent can exhibit stronger polarity than that of ground-state solvent in most of $\pi \rightarrow \pi^*$ transitions, and π^*-orbital can present extremely-strong polarity and stronger action between it and PVA solvent, deducting the energy significantly. However, the π^*-orbital presents weaker action between it and PVA solvent if it exhibits lower polarity, therefore, the energy decreases a little and the energy difference between π and π^* decreases. Therefore, it can be concluded that the transition energy of $\pi \rightarrow \pi^*$ transition in the polar solvent is far lower than that in non-polar solvent, and the characteristic peak arising from the $\pi \rightarrow \pi^*$ transition in the polar solvent PVA moves towards the long-wave direction.

7.3.2.4 Doping Mechanism of PANI

(1) The PANI/PVA composite emulsion with improved conductivity and solubility was fabricated via emulsion polymerization in this study. It could be found from FTIR results that the prepared PANI composite emulsion presents the characteristic absorption peak of conducting-state PANI indicating that the synthesized substance is the conducting-state PANI.

(2) The experimental results indicate that the high-speed stirring contributes to the dispersion of PANI particles during the polymerization process, and it can be observed that the agglomeration disappears completely when the stirring speed is set to be 2200 r/min. The PANI particles are found to be disperses independently on the PVA matrix and presents in the form of compact mesh.

(3) The PANI particle with excellent conductivity is synthesized and its solubility has been improved and optimized in NMP solvent of 24.7 mg/mL according to the investigation on the mechanism of emulsion polymerization.

(4) The macromolecule organic sulfonic acid DBSA is doped in this experiment and the emulsion polymerization is selected as the preparation method, in addition, the doping mechanism of PANI proton acid is investigated and the entire process of PANI changing from eigen-state to conducting state is illustrated. This study can provide a further development and a wide application of PANI.

7.4 Preparation of PANI/PVA Composite Fiber by Electro-spinning

In 1934, electro-spinning was firstly proposed by Formhals and a set of electro-spinning device was developed. As a result, ultrafine fiber even nano-fiber could be spun by applying polymer with a certain visco-elasticity in a strong electric field[89-90]. In 1966, an electro-spinning device used for preparing light-weight and fine non-woven fabric was developed by Chi et al[91]. In 1972, an electro-spinning device used for fabricating acrylic acid fiber with diameter of 0.05~1100 μm was designed by Baumgarten, in addition, the relationship between fiber diameter and solution viscosity, electro-spinning distance and composition of environmental gas was investigated. In 1981, it was found by Larrondo and Manley that fiber diameter was closely linked to electro-spinning voltage, electro-spinning temperature and viscosity of melted material, whereas spinneret diameter had little effect on fiber diameter.

Quite a lot of theoretical and experimental investigations on electro-spinning has been conducted with the rapid development of nanotechnology. At present, diverse polymeric fibers with diameters from micrometer to nanometer have been fabricated and corresponding researches focused on investigation and preparation of micro/nano-fiber. If the fiber with linear structure and mesh structure could be fabricated via distributing the conductive polymer with poor solubility on the fiber uniformly, it facilitates to the development of related industries to a great extent[92].

The electro-spinning parameters for PANI/PVA are studied systematically in this section. In order to obtain controllable preparation of electrospun fiber, main electro-spinning parameters for the preparation of PANI/PVA composite fiber are investigated via changing certain factors, such as electro-spinning voltage, electro-spinning distance and mass fraction of PANI, etc.

7.4.1 Experimental

7.4.1.1 Material and Equipment

The experimental reagents in this study include Polyvinyl alcohol (PVA, AR) and PANI/PVA emulsion. The experimental equipments are listed in Tab.7-8.

Tab.7-8 Equipment list

Equipment	Specification
Spinneret	5 mL, 10 mL, 20 mL
Medical stainless steel needle	1.2 mm×40 mm (50 mm/60 mm)/0.79 mm (outer diameter×length/inner diameter)
PTFE tube	2.0 mm×2.4 mm (inner diameter×outer diameter)
Silicone tube	2.0 mm×4.0 mm (inner diameter×outer diameter)
Electro-spinningmachine	KH-2
Rotary viscometer	NDJ-1
Precision torsion balance	JN-A
Electronic single fiber strength tester	YG006
Scanning electron microscope	S-3400N
Thermogravimetric analyzer	ST PT-1000
X-ray diffraction	K780 FirmV_06

7.4.1.2 Electro-spinning Machine

A high-voltage electro-spinning device was selected in this experiment as shown in Fig. 7-19. The spinneret should be properly fixed on the propeller after the PANI/PVA spinning solution was extruded to the spinneret and then be clad by the aluminum foil paper or ITO conductive film (around the roller). In addition, the aluminum foil paper should be transversely straightened to guarantee its surface smoothness and the spinneret should be vertical to the rotary roller. Subsequently, the tip of the spinneret was connected to the anode of DC high-voltage power supply, while the collector was connected to the cathode of high-voltage power supply. To regulate the distance and angle between the tip of spinneret and collector, the position of spinneret pump was adjusted.

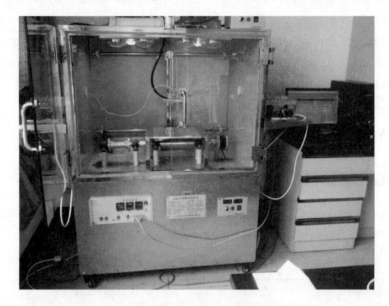

Fig. 7-19 Electro-spinning equipment

The electro-spinning parameters for preparation of PANI ultrafine composite fiber are listed in Tab. 7-9.

Tab. 7-9 Electro-spinning parameters for preparation of PANI/PVA

Sample No.	Electro-spinning voltage(kV)	Electro-spinning distance(cm)	Mass fraction of PANI(%)
F01	20	16	0.0
F02	20	16	0.2
F03	20	16	0.4
F04	20	16	0.6
F11	24	10	0.2
F12	24	12	0.2
F13	24	14	0.2

(continued)

Sample No.	Electro-spinning voltage(kV)	Electro-spinning distance(cm)	Mass fraction of PANI(%)
F14	24	16	0.2
F15	24	18	0.2
F16	24	20	0.2
F21	14	18	0.2
F22	17	18	0.2
F23	20	18	0.2
F24	23	18	0.2
F25	26	18	0.2
F26	29	18	0.2

7.4.1.3 Determination of Emulsion Viscosity

The determination of PANI/PVA emulsion viscosity was conducted by the rotary viscometer (NDJ-1, Shanghai Precision Science Instrument Co., Ltd.). The pure PVA emulsion and the PANI/PVA emulsion with different mass fraction of PANI, including 0.2%, 0.4% and 0.6%, were numbered as N1, N2, N3 and N4 respectively. Each sample was tested for 5 times to achieve the average value. The sample should be stewed for 1 min after the test to eliminate the effects caused by the inertia during the test process. Different stewing time was set for the repeated testes and it could be found that minimum error could be achieved when the testing time was set to be not less than 1 min. In this case, the stewing time of 1 min was selected for the test of emulsion viscosity in this study.

7.4.1.4 Test of Mechanical Properties

A long piece of fabric with length of 20 mm and width of 5 mm was selected (along the longitudinal direction of the fiber) as the sample. Weighing of each sample was performed in advance, the sample was placed in the constant-temperature laboratory for 24-hour treatment and its tensile mechanical properties were characterized by the electronic single-fiber strength tester (YG006, Shanghai). The relationship between the surface density and mechanical properties was investigated according to the standard "GB/T 1040.3—2006 *Plastics-Determination of Tensile Properties*[93] - *Part 3: Test Conditions for Films and Sheets*".

In addition, the test was conducted under specified circumstances (ambient temperature = (20 ± 2) ℃; ambient humidity = (50 ± 10)%; clamping length = 10 mm; drawing speed = 10 mm/min; initial tension = 0.1 cN; force measurement accuracy = 0.01 cN, 10 times of test for each sample; average value of each parameter analyzed eventually).

7.4.1.5 Characterization by Scanning Electron Microscope

The micro-morphology of ultrafine composite fiber was characterized by SEM (S-3400N, Hitachi, Japan; magnification = 5～300 000, continuously regulatable; working voltage = 10 kV～15 kV; probe current = 4 pA～20 nA, continuously regulatable) after a layer of metal film with thickness of 10 nm was coated on the surface of PVA/PANI ultrafine composite fiber.

The images of samples fabricated with different electro-spinning parameters were digitalized by the Image-Pro-Plus 6.0 software via selecting 100 fibers from each SEM photo and measuring their diameters to obtain the frequency distribution and the average diameter of fiber.

7.4.1.6 Characterization by X-ray Diffraction

The properties of the polymer are determined by its aggregated-state structure which is considered as the main aspect of aggregated-state structure. To investigate the effects of electro-spinning on the aggregated-state structure of PANI/PVA ultrafine composite fiber, XRD (k780FirmV_0) was utilized for the purpose of investigation on effects of electro-spinning on crystallinity of fiber. The dedicated JADE 5.0 software was selected to calculate the crystallinity of the polymer with diffraction angle 2θ within the range of $5°～70°$ and scanning speed of 0.8 sec/step. In addition, the relationship between the crystallinity and electro-spinning parameters was investigated based on the fitting curve.

7.4.1.7 Characterization by Thermogravimetric Analyzer

The thermal stability of the electrospun PANI/PVA ultrafine composite fiber was characterized by TGA (STA PT-1000, Germany). The sample should be tested under the protection of nitrogen with the testing temperature rising from room temperature to 800 ℃ and temperature rise ratio of 10 ℃/min.

Basic principle of TGA is described as follows. TG curve is used to indicate the relationship between the sample weight and the controllable temperature by the equipment, while DTG curve is used to record the first-order derivative of TG curve to the temperature, namely, the relationship curve between the weight change and temperature.

7.4.2 Results and Discussion

7.4.2.1 Viscosity of Spinning Solution

The interaction of molecular chain is characterized by the viscosity of polymer solution, and the curve of emulsion viscosity varying with the mass fraction of PANI is demonstrated in Fig. 7-20.

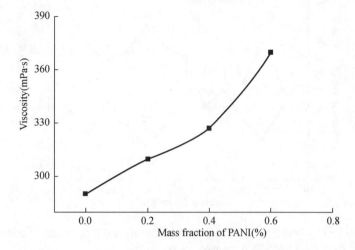

Fig. 7-20 Viscosity of emulsion with different Mass fraction of PANI

It can be seen in Fig. 7-20 that the viscosity of pure PVA emulsion reaches 290 MPa · s and the emulsion viscosity increases with the increasing mass fraction of PANI. That is because PVA molecular chain exhibits excellent flexibility and the entanglement could be found, in addition, strong hydrogen-bonding interaction would be generated between certain —NH— on PANI molecular chain and —OH on PVA molecular chain, which could improve and optimize the viscosity of PVA emulsion sufficiently. However, the emulsion viscoelasticity decreases remarkably although the emulsion concentration exceeds the critical value with the excessive increasing mass fraction of PANI. That is because PANI exhibits poor solubility, and entanglement could be found in the synthesized PANI molecular chain. In this case, the extensional viscosity and relaxation time of the emulsion can affect the electro-spinning process, and the relaxation time is shorter than the stretching time, namely, the solution viscosity decreases, thus taking effects on the electro-spinning.

7.4.2.2 Mechanical Properties

The mechanical properties of the PANI/PVA ultrafine composite fiber mainly depend on the mechanical properties of the fiber, entanglement state of the fibers and the surface density. The relationship between surface density and mechanical properties of the electrospun PANI/PVA ultrafine composite fiber is shown in Fig. 7-21. The prepared PANI/PVA composite is made of PVA emulsion with the mass fraction of PANI of 0.2%, electro-spinning voltage of 20 kV and electro-spinning distance of 16 cm.

As can be seen in Fig. 7-21, the breaking strength presents a trend of slow increasing and the variation range of breaking strength is within the range of 2.6~7.8 cN/dtex

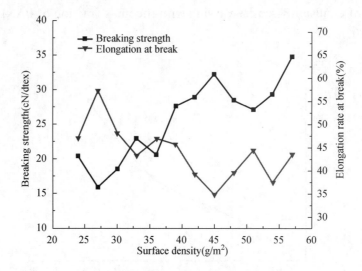

Fig. 7-21 Relationship between surface density and mechanical properties of PANI/PVA ultrafine composite fiber

with the increasing surface density; whereas the elongation at break exhibits a reverse variation trend of fluctuant relief. To achieve further investigation on the effects of surface density on the mechanical properties of PANI/PVA ultrafine composite fiber, single-factor variance analysis was conducted as listed in Tab. 7-10. The table depicted that the surface density of the PANI/PVA composite has remarkable effects on its mechanical properties although the variation curve is complicated. Therefore, the surface density is considered as one of the main contributing factors.

Tab. 7-10 Variance analysis of mechanical properties of PANI/PVA nonwoven fabric with different surface densities

Parameter	S_A	S_e	A	e	F	$F_{0.05}(A, e)$
Breaking strength(cN/dtex)	55.07	3.41	7	16	36.91	2.66
Elongation rate at break(%)	3 765.75	243.97	7	16	35.28	2.66

S_A—Interblock deviation square sum;
S_e—Block-interior deviation square sum;
A—Interblock freedom degree;
e—Block-interior freedom degree;
F—Ratio of interblock mean square and block-interior mean square;
$F_{0.05}(A, e)$—Critical value of F distribution.

7.4.2.3 Morphological Structure

(1) Effects of mass fraction of PANI in the spinning solution on morphological structure of PANI/PVA ultrafine composite fiber.

To study the effects of mass fraction of PANI on the morphology, fiber diameter and fiber distribution of the electrospun fiber, electro-spinning was conducted with different mass fractions of PANI. SEM images and histograms of fiber diameter distribution of ultrafine composite fiber with different mass fractions of PANI are depicted in Fig. 7-22. The composite was prepared with electro-spinning voltage of 20 kV and electro-spinning distance of 16 cm.

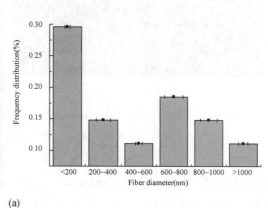

(a)

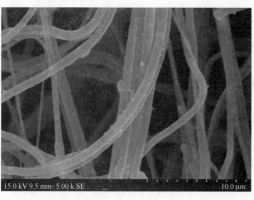
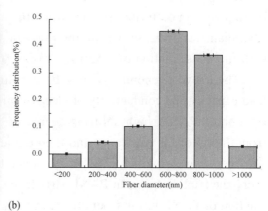

(b)

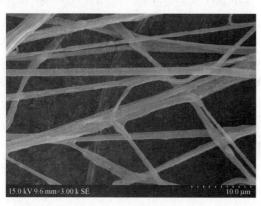
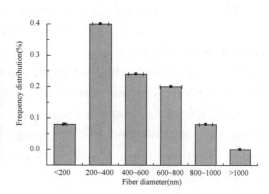

(c)

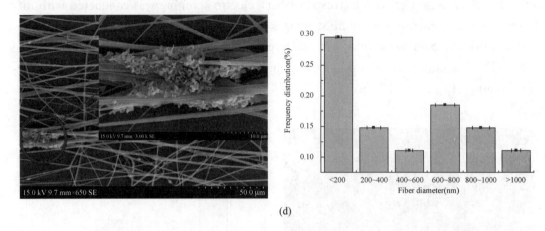

Fig. 7-22　SEM images and histograms of fiber diameter distribution of PANI/PVA composite with different mass fractions of PANI: (a) 0.0%; (b) 0.2%; (c) 0.4%; (d) 0.6%

It can be clearly observed in Fig. 7-22 that the fiber surface is smooth with high-speed electro-spinning process and the average fiber diameter is only 300 nm before the PANI particles adding into the spinning solution and the mass fraction of PVA is 4.5%. Sudden steep of fiber diameter could be found and the average fiber diameter reaches 730 nm with smooth fiber surface when the mass fraction of PANI is 0.2%. The average fiber diameter decreases within the range of 400~550 nm when the fiber surface is rough with the mass fraction of PANI increasing from 0.4% to 0.6%.

The above phenomenon may be explained by the following reasons. Because the load capacity and conductivity of electro-spinning solution increases with increasing mass fraction of PANI, high splitting-degree of jet could be found during the jet spraying from the tip of the capillary, and then the jet is stretched and whipped. Resultantly, finer fiber diameter could be achieved under the same electro-spinning conditions. However, the friction between PANI particle and PVA matrix increases with increasing mass fraction of PANI during fiber stretching. In this case, the resistance during fiber stretching is generated. As can be seen in Fig. 7-22 (d), agglomeration of PANI particles could be found in the fiber and it is affected by the electro-spinning conditions.

(2) Effects of electro-spinning distance on morphological structure of PANI/PVA ultrafine composite fiber.

The complicated effects of electro-spinning distance on morphological structure of PANI/PVA ultrafine composite fiber could be found during electro-spinning process. Therefore, electro-spinning distance is selected as a significant parameter influencing electro-spinning. SEM images and histograms of fiber diameter distribution of ultrafine composite fiber with different electro-spinning distance are demonstrated in Fig. 7-23.

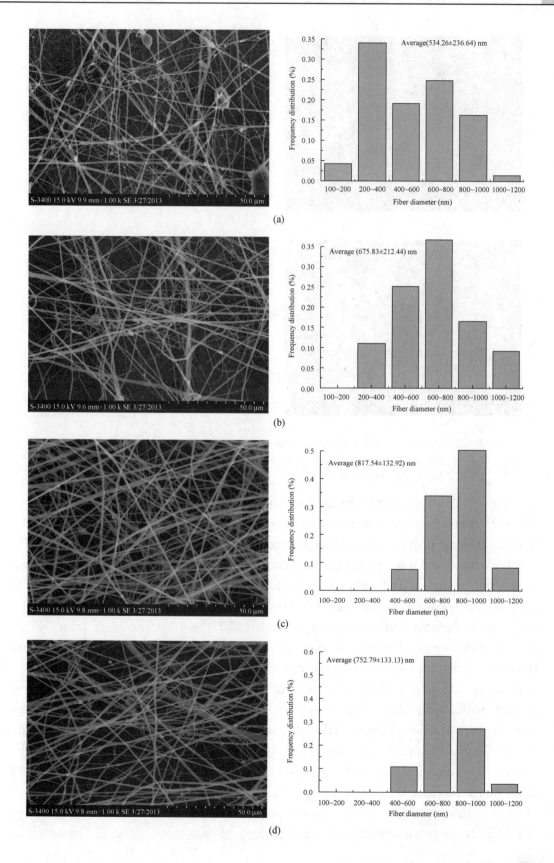

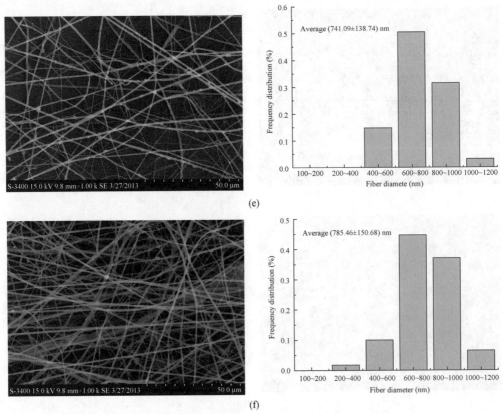

Fig. 7-23 SEM images and histograms of fiber diameter distribution of PANI/PVA composite with different electro-spinning distance: (a) 10 cm; (b) 12 cm; (c) 14 cm; (d) 16 cm; (e) 18 cm; (f) 20 cm

Fig. 7-23 shows SEM images and histograms of fiber diameter distribution of PANI/PVA composite with different electro-spinning distance. It can be observed in the figure that much PANI are distributed on the fiber uniformly because the spinning solution is extruded and it is difficult for the generated jet to be stretched and for the solvent to be volatilized completely when the electro-spinning distance is short. These result in fiber entanglement and large droplet as shown in Fig. 7-23 (a). The volume of droplet significantly decreases and the average diameter of the fiber increases from 543 to 817 nm as the increasing electro-spinning distance from 10 to 14 cm. It can be found that the fiber diameter decreases and becomes gentle with the increasing electro-spinning distance from 14 to 18 cm. Moreover, the fiber diameter decreases from 817 to 741 nm and the dispersion degree also decreases as the increased electro-spinning distance. In addition, the mass fraction of PANI in the fiber increases remarkably and becomes uniformly because an appropriate electro-spinning distance can eliminate the reaction force of the electric field force and the electric charge can transfer the electric field force from PANI to PVA. In this case, the electric charge can move from the emulsion-contained spinneret plate to the collector and the energy required for polymer acceleration could be achieved. It can be concluded from the experimental data that suitable electro-spinning effect are obtained when the electro-spinning distance is 18 cm.

(3) Effects of electro-spinning voltage on morphological structure of PANI/PVA ultrafine composite fiber.

Variation of electric field intensity has effects on the charge density on the sprayed surface during the electro-spinning process, thus affecting the stretching of electric field force to the jet. Therefore, electro-spinning voltage is selected as an important parameter influencing electro-spinning. SEM images and histograms of fiber diameter distribution of PANI/PVA ultrafine composite fibers with different electro-spinning voltage are depicted in Fig. 7-24.

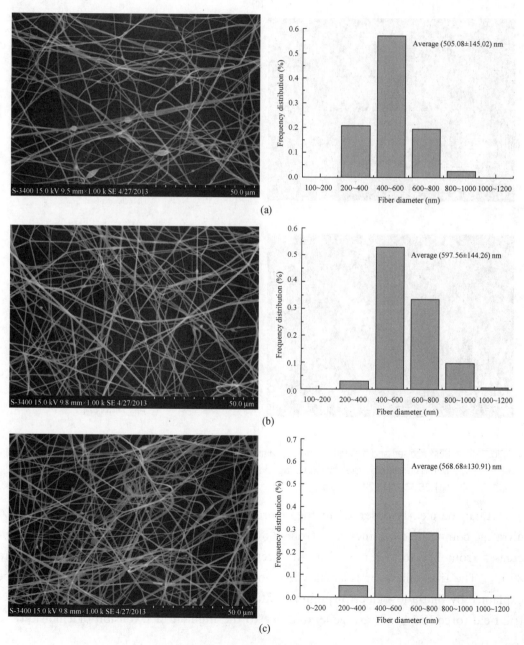

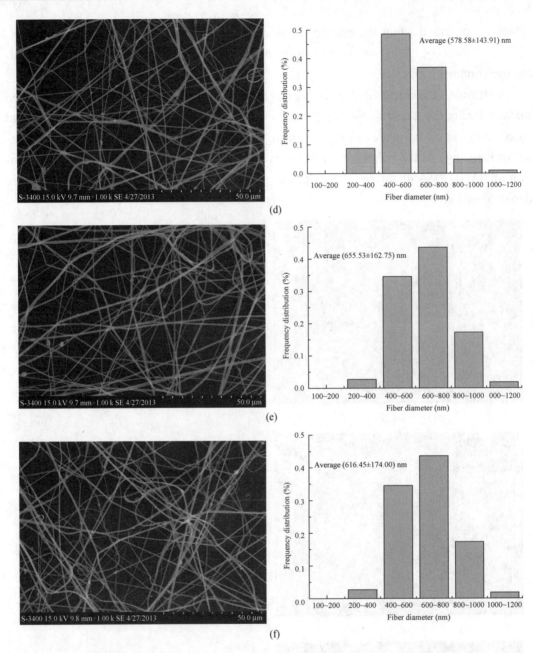

Fig. 7-24　SEM images and histograms of fiber diameter distribution of PANI/PVA composite with different electro-spinning voltage: (a) 14 kV; (b) 17 kV; (c) 20 kV; (d) 23 kV; (e) 26 kV; (f) 29 kV

It can be clearly observed in the figure that the average diameter of PANI/PVA ultrafine composite fiber increases from 505 to 656 nm and the diameter distribution increases from 131 to 163 nm with the electro-spinning voltage increasing from 14 to 26 kV. The fiber diameter is small when electrospun with a low voltage and this can be explained by that a low electro-spinning voltage, such as 14 kV, can result in a low electric field force and beads are generated due to the high surface tension of emulsion, so very few droplets on the surface are extruded to form finer fiber. In addition, the electric charge on the jet surface enhances gradually with the increasing electro-spinning voltage, and the conductive PANI particle can be sprayed with PVA emulsion. However, it is difficult for fiber to be stretched with a high spraying speed and then it leads to a large diameter of the prepared fiber. As the electro-spinning voltage increased to a relatively high value of 23 kV, stable fiber jet can be formed in such a strong electric field and the distribution of the fiber with a large diameter decreases to 144 nm. Therefore, it can be concluded that a good electro-spinning effect can be achieved when the electro-spinning voltage is 23 kV.

7.4.2.4 Crystal Structure

The aggregated-state structure of PANI/PVA ultrafine composite fiber was characterized by XRD, and it was found that the mass fraction of PANI, electro-spinning voltage and electro-spinning distance have effects on the crystallinity of PANI/PVA ultrafine composite fiber.

(1) Effects of mass fraction of PANI on the crystallinity of PANI/PVA ultrafine composite fiber.

XRD patterns of PANI/PVA ultrafine composite fibers fabricated with different mass fractions of PANI are illustrated in Fig. 7-25. Diffraction peaks at 19.4°, 32°, 39.5°, 46.1° are observed, which indicates the existence of PVA. In addition, peaks at 9.0° and 23.5° with slight difference in intensity can be discovered in sample N2, N3 and N4 which reveals the existence of PANI.

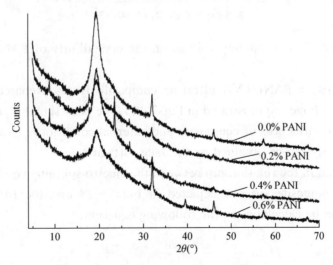

Fig. 7-25 XRD patterns of PANI/PVA ultra composite fibers fabricated with different mass fraction of PANI

The relationship between mass fractions of PANI and fiber crystallinity is depicted in Fig. 7-26 and it can be clearly observed that the mass fraction of PANI has great effects on the crystallinity of PANI/PVA ultrafine composite fiber. The crystallinity of PANI/PVA composite increases firstly as the mass fraction of PANI raised from 0.0% to 0.2%, and then the crystallinity decreases with the mass fraction of PANI further increasing from 0.2% to 0.6%. It was reported that the fiber surface becomes rougher and agglomerated with the mass fraction of PANI of 0.4%. Therefore, it can be concluded that the fiber with excellent morphology could be fabricated by an appropriate application of PANI, thus improving the crystallinity of the fiber. However, the increased viscosity of spinning solution can enhance the viscous resistance enhances leading to a decreasing action force used for guiding the orientation of molecular chain. In this case, the crystallinity of the fiber decreases because of the excessive solid content of spinning solution.

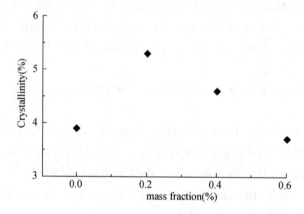

Fig. 7-26 Relationship between mass fraction of PANI and the crystallinity of PANI/PVA fiber

(2) Effects of electro-spinning voltage on the crystallinity of PANI/PVA ultrafine composite fiber.

XRD patterns of PANI/PVA ultrafine composite fibers prepared with different electro-spinning voltage are illustrated in Fig. 7-27. The peak signal is poor that is possibly because of the influences of equipment and external conditions. However, low crystallinity of the fiber can be obtained by the Jade software.

As can be seen in the relationship between the electro-spinning voltage and the crystallinity of the composite fiber is depicted in Fig. 7-28 and the fitting relationship between them can be calculated by the following equation.

$$y = ax^3 + bx^2 + cx + d \tag{7-6}$$

Where, y refers to crystallinity; x refers to electro-spinning voltage; a, b, c and d are equal to -0.0089, 0.5979, -13.1882 and 102.612, respectively.

Then, the relationship between them is as follows:

$$y = -0.0089x^3 + 0.5979x^2 - 13.1882x + 102.612 \tag{7-7}$$

The vertex coordinates of the function could be achieved based on Eq. (7-7) and the vertex coordinates of the function are (19.65, 6.80) and (25.14, 7.53), which indicates that the crystallinity of ultrafine composite fiber increases with the electro-spinning voltage increasing from 19.65 to 25.14 V.

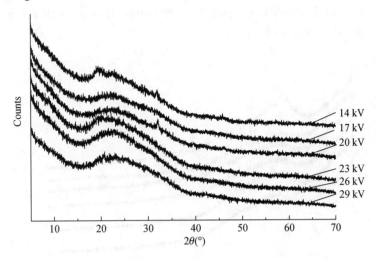

Fig. 7-27 X-ray patterns of PANI/PVA ultrafine composite fibers fabricated with different electro-spinning voltage

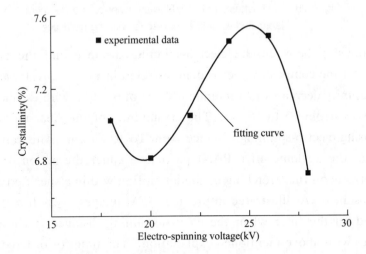

Fig. 7-28 Relationship between electro-spinning voltage and the crystallinity of PANI/PVA ultrafine composite fibers

The applied electric field force has effects on the crystallinity of the fiber. The macromolecule is orderly arranged under the action of external force, which contributed to improve the crystallinity of polymer. However, the excessive electrostatic field facilitates to shorten the spraying time of the jet and has effects on the orderly-arranged molecular structure, meanwhile; it destroys the ordered arrangement of the macromolecular chain, thus decreasing the crystallinity. The experimental results showed that the crystallinity of the fiber increases firstly and then sharp declines with the further increased electro-spinning voltage.

(3) Effects of electro-spinning distance on the crystallinity of PANI/PVA ultrafine composite fiber.

XRD patterns of PANI/PVA ultrafine composite fibers with different electro-spinning distances are depicted in Fig. 7-29.

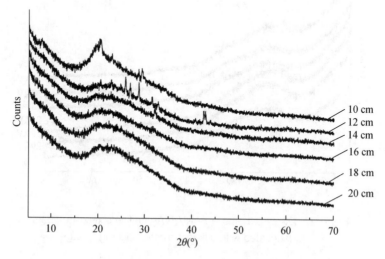

Fig. 7-29 XRD patterns of PANI/PVA ultrafiner composite fibers fabricated with different electro-spinning distances

The relationship between the electro-spinning distance and the crystallinity of PANI/PVA ultrafine composite fiber is demonstrated in Fig. 7-30. It can be observed that the crystallinity decreases to a minimum value of 6.89% with the electro-spinning distance increased from 10 to 16 cm. The crystallinity of the fiber increases with the further increasing electro-spinning distance from 16 to 20 cm. This result can be explained by that the agglomerative PANI particles (under the action of electric field force) have effects on the stretching of the jet motion within a specified range of electro-spinning distance. As illustrated in previous SEM images, very few PANI particles could be found on the fiber when the electro-spinning distance is set to be 16 cm, which complies with above-mentioned explanation. The time for ordered arrangement of macromolecule prolongs with the continuous increasing electro-spinning distance, and then the crystallinity of the fiber can be improved. However, the electric field intensity weakens because of the excessive electro-spinning distance and this will lead to a decreasing stretching action to the fiber. Therefore, the action force used for guiding orientation of molecular chain attenuates and the crystallinity of the fiber is poor. It is found that the fitting curve presents a low fitting coefficient when conducting the fitting of relationship between the crystallinity and electro-spinning distance, therefore, there is no need to conduct the fitting of cubic function.

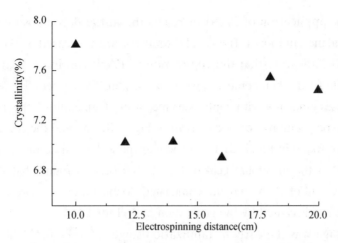

Fig. 7-30 Relation ship between electro-spinning distance and the crystallinity of PANT/PVA ultrafine composite fibers

7.4.2.5 Thermal Stability

DSC curves of PANI/PVA ultrafine composite fibers fabricated with different mass fraction of PANI are demonstrated in Fig. 7-31. The data referring to TG curves of PANI/PVA ultrafine composite fibers are illustrated in Tab. 7-11.

As can be observed from Fig. 7-31 (a), two obvious weight-loss stages of pure PVA fiber could be found at 212 and 400 ℃ due to the breaking of side chain and main chain of PVA, respectively. The thermal decomposition process of PANI/PVA ultrafine composite fibers could be divided into three decomposition stages: the first stage is at room temperature to 300 ℃; the second stage is at 300~500 ℃; the third stage is at 500~800 ℃.

The first stage was the slight weight-loss stage (room temperature ~300 ℃). The weight loss is mainly caused by the decomposition of low-weight oligomer when the temperature increases to 150 ℃. As depicted in Fig. 7-31, PANI/PVA ultrafine composite fibers fabricated with different mass fractions of PANI present similar weight-loss trends, the weight loss rate and thermal decomposition process of pure PVA ultrafine fiber are obviously different from other PANI-contained samples, it can be seen from the curves that diverse samples present similar TG curves when the temperature increases to 200 ℃. From data listed in Tab. 7-11, it can be found that the weight loss rates of different ultrafine fibers are lower than 12% when the temperature increases from room temperature to 150 ℃, and the weight loss is mainly caused by the decomposition of low-molecule polymer. However, the maximum weight loss rate of pure PVA can be achieved at 212 ℃, therefore, it can be concluded that the application of PANI results in the decreasing of 50~60 ℃ of thermal stability of PANI/PVA ultrafine composite fiber, however, it presents more stable variation of thermal stability and the high-porosity fiber could be fabricated in high-temperature environment, without sharp decomposition.

The second stage is the thermal decomposition stage (300~500 ℃). As can be seen in Tab. 7-11, the application of PANI increases the initial decomposition temperature of PANI/PVA ultrafine composite fiber. The temperature increased as the increasing mass fraction of PANI indicating that the application of PANI can improve the thermal stability of the composite fiber when compared with that of pure PVA fiber. The weight loss rate of the sample increases gradually with continuous increasing temperature, the TG curve indicates the high-speed decomposition process as shown in Fig. 7-31 (a) and the corresponding weight-loss peaks could be found in Fig. 7-31 (b). The maximum thermal decomposition rate could be determined based on the maximum peak value. It can be also found that the decomposition mainly occurs at C—C of PVA and the conjugated double-bond structure of PANI, furthermore, the weight loss is caused by the generation of NH_3 or CO_2.

The third stage was the carbon stabilization stage (500~700 ℃). In the carbon stabilization stage, the high-molecular polymer has been carbonized and the residual lessens with the increasing temperature. It can be clearly observed in Fig. 7-31 that stable weight loss rate of PANI/PVA ultrafine composite fiber could be achieved; whereas the weight loss rate of pure PVA decreases to 3.4%.

As can be clearly observed from Fig. 7-31, the sequence of residual rates of diverse ultrafine fibers with increasing temperature from room temperature to 800 ℃ are: 0.6% PANI>0.4% PANI>0.2% PANI>0.0% PANI (namely pure PVA). That is because the main chain of PANI contains B=Q=B benzoquinone structure with strong absorption electron and it is difficult for it to be split because of the conjugated bond generated by double bonds on the benzene ring. In addition, the conductivity of PANI contributes to the uniform transmission of the energy in ultrafine composite fiber from outer layer to inner layer, thus facilitates to relieve the thermal decomposition of PANI/PVA ultrafine composite fiber in high temperature of 400~500 ℃. Therefore, it can be concluded that the addition of PANI in PVA matrix can improve the thermal stability of ultrafine composite fiber.

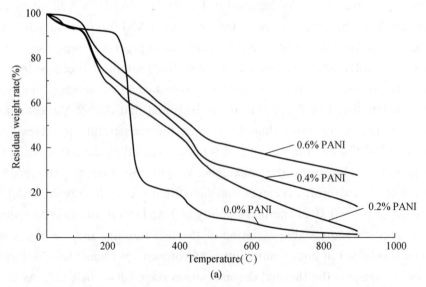

(a)

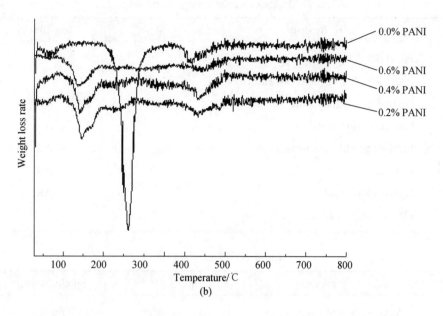

Fig. 7-31 DSC curves of PANI/PVA ultrafine composite fibers fabricated with different mass fractions of PANI：(a) TG; (b) DTG

Tab. 7-11 Parameters of PANI/PVA ultrafine composite fibers during TG

Mass fraction of PANI(%)	T_0(℃)	T_{max}(℃)	Residual weight rate at 700 ℃ (%)
0.0	212.0	264.0	3.4
0.2	398.2	423.0	13.1
0.4	407.6	440.0	23.8
0.6	426.0	446.0	34.8

T_0—The initial decomposition temperature;
T_{max}—The temperature at the maximum decomposition rate.

7.5 Study on Electrochemical Behavior of PANI Composite

7.5.1 Experimental

7.5.1.1 Material and Equipment

The experimental materials and reagents are listed in Tab. 7-12. The experimental equipments are listed in Tab. 7-13.

Tab. 7-12 Experimental materials and reagents

Reagent	Specification
PANI/PVA emulsion	—
PANI/PVA ultrafine composite fiber	—
Polymethyl methacrylate	AR
Propylene carbonate	AR
Lithium perchlorate trihydrate	AR
Ethanol	AR
Hydrochloric acid	AR
ITO conductive film	—

Tab. 7-13 Equipment list

Equipment	Specification
Vacuum drying oven	DHG-9075A
Spin coater	KW-4A
Electric blast drying oven	WD-5000
Stratometer	DMR-1C
UV-visible spectrophotometer	UV-3600
Electrochemical workstation	CHI 600D

7.5.1.2 Preparation of PANI/PVA Composite

The PANI/PVA composite membrane was prepared based on the preparation method illustrated in Section 7.4.1, after that, each sample was numbered and packed for storage after drying.

To investigate on the electrochemical behavior of the PANI/PVA ultrafine composite fiber, flexible ITO conductive film rather than aluminum foil paper was selected for electro-spinning for the purpose of guaranteeing the surface morphology of the fiber and enhancing the adhesion of fiber to the ITO conductive film. Under the circumstance of other unchanged technologies, the ultrafine composite fiber/ITO conductive film substrate was prepared for testing and investigation.

7.5.1.3 Preparation of Electrolyte

PMMA gel electrolyte was prepared based on the previous research. $LiClO_4 \cdot 3H_2O$ and PMMA were placed in the vacuum drying oven at 120 and 80 °C for 24 h, respectively. A proper amount of dry $LiClO_4$ was fed into the beaker filled with PC solution and then ultrasonic stirred for 10 min to achieve completely dissolved $LiClO_4$. Then PMMA powders were added into the solution and stirred at 120 °C until the complete dissolution of crystallized-state substance was achieved. Eventually, the viscous-state gel solution was prepared, cooled and sealed.

7.5.1.4 Preparation of Electrochromic Device

PANI/PVA composite emulsion can be coated on ITO conductive film directly. PMMA gel electrolyte can be coated on PANI/PVA membrane based on the above-mentioned method after the membrane was formed due to the emulsion drying. PMMA gel electrolyte can be coated onto PANI/PVA membrane following by coating PANI/PVA composite emulsion onto the ITO conductive film. Then mutual extrusion between the PANI/PVA coated ITO film and the PMMA gel electrolyte coated ITO film was carried out to remove air bubble. Subsequently, the electrochromic device (ECD) was prepared as shown in Fig. 7-32 after 4 h drying at 60 °C in the vacuum drying oven was conducted to the membrane.

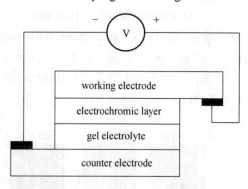

Fig. 7-32 Basic structure of ECD

As shown in the figure, the PANI/PVA-coated ITO conductive film (symmetric-type electrochromic device, SSECD), the copper-plated woven or the silver-fiber-woven knitted fabric (asymmetric electrochromic device, ASECD) could be selected as the counter electrode.

7.5.2 Results and Discussion

7.5.2.1 Electrochemical Behavior of PANI/PVA Membrane

The conducting-state PANI has been widely favored by researchers because of unique conductivity and discoloration, its electrochromic behavior is characterized by cyclic voltammetry (I-E curve) which indicates that symmetric triangular-wave scanning voltage (as illustrated in Fig. 7-33) was applied to the working electrode with corresponding material area and the response curve of voltage varying with current is recorded. Reduction reaction occurs to the electrode during cathode scanning; while oxidation reaction occurs to the electrode during anode scanning, where the characteristic peak of current of the former is the peak shape of the cathode; whereas the characteristic peak of current of the latter is the peak shape of the anode. Therefore, the essential process of cyclic voltammetry is the process of oxidation reaction and reduction reaction by triangular-wave scanning. It is selected as the effective means for the investigation on electrode mechanism, which could not only achieve the reversibility of electrode reaction but also analyze the peak height and peak value of cathode and anode precisely[94].

The conductive PANI was prepared with molar ratio of DBSA : AN : APS of 1.03 : 1.00 : 1.00 and the PANI/PVA composite membrane with thickness of 1 μm was fabricated via spin coating, after that, the membranes were dried for the test of electrochemical behavior. Color change of PANI membrane with scanning voltage of −0.3～1.3 V (protonic acid doped at 0 and 20 ℃) is illustrated in Fig. 7-34 (a) and (b).

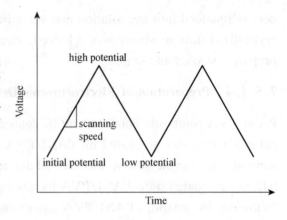

Fig. 7-33　Diagram of triangle scanning wave

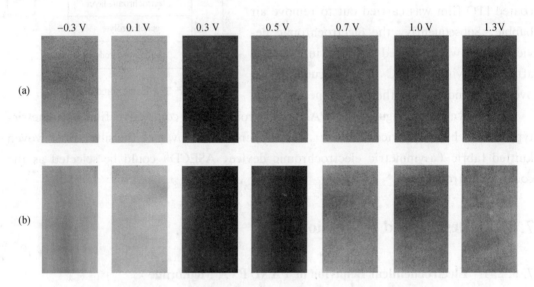

Fig. 7-34　Pictures of PANI/PVA composite membranes fabricated with different doping temperatures: (a) 0 ℃; (b) 20 ℃

The membranes prepared at different polymerization temperatures present a color variation trend: green→light blue→light yellow. It can be clearly observed in Fig. 7-34 that the color saturations (green, light yellow and light blue) of PANI/PVA membrane polymerized at 20 ℃ are higher than those of membrane polymerized at 0 ℃.

The experiment was conducted with specified parameters (1 mol/L hydrochloric acid solution as electrolyte, scanning voltage = −0.3～1.3 V, scanning speed = 50 mV/s), the PANI/PVA composite membrane was used as the working electrode, the saturated calomel electrode was used as the reference electrode, and the platinum electrode was used as the counter electrode. The curves of cyclic voltammetry (CV) of PANI/PVA composite membrane are depicted in Fig. 7-35.

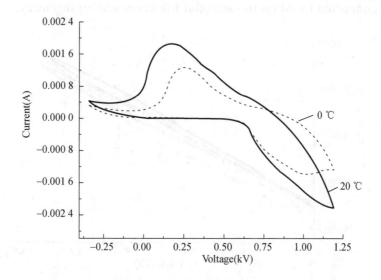

Fig. 7-35 CV curves of PANI/PVA composite membranes at 0 ℃ and 20 ℃)

CV curves of PANI/PVA composite membranes polymerized at 0 and 20 ℃ are demonstrated in Fig. 7-35 and a pair of redox peaks could be found from both curves. The characteristic peaks at 0.1~0.2 kV during positive scanning may be caused by cation; whereas a broadband peak appears at 0.8~1.1 kV during reverse scanning are caused by the proton in the PANI chain and two dual-polarized free radicals. During the scanning in cyclic voltammetry, the transition of electron in the PANI long-chain is the transmission of absorption peak moving towards long-wave direction and then transition to the visible region of the spectrum, presented in yellow, green and blue, etc. The unpaired lone electron pair in the amino-group dissociates in the conjugated system, thus increasing the delocalized degree characterized by increasing conductivity and current. In addition, the PANI/PVA composite membrane polymerized at room temperature exhibits superior electrochromic effect when compared with membrane polymerized at low temperature.

7.5.2.2 Electrochemical Behavior of PANI/PVA Ultrafine Composite Fiber

The prepared PANI/PVA ultrafine composite fiber was characterized by cyclic voltammetry with scanning speeds of 5, 20, 50 and 100 mV/s, hydrochloric acid solution of 1 mol/L as the electrolyte and scanning voltage within the range of −0.3~1.3 V. CV curves of PANI/PVA ultrafine composite fiber are illustrated in Fig. 7-36. As can be seen in Fig. 7-36, the redox peak of CV curve of PANI/PVA ultrafine composite fiber is not as apparent as that of PANI/PVA composite membrane. However, the PANI/PVA ultrafine composite fiber still presents the color change sequence of yellow, green and blue during the cycle of positive scanning and reverse scanning with relatively-light color intensity. Comparison of distribution of PANI particles on PVA membrane and PVA fiber could be conducted based on the material thickness and arrangement.

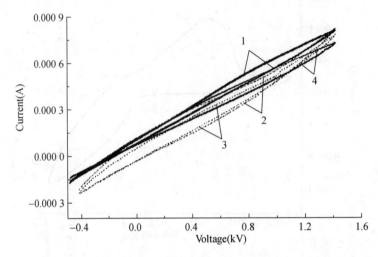

Fig. 7-36 CV curves of PANI/PVA ultrafine composite fibers at different scanning rate (1—5 mV/s, 2—20 mV/s, 3—50 mV/s, 4—100 mV/s)

The structure model of PANI particles distributing on PANI/PVA composite membrane is illustrated in Fig. 7-37. It could be observed that PANI particles are dispersed on PANI/PVA composite membrane randomly with isotropic conductivity transmission and equal probability. PANI/PVA ultrafine composite fiber exhibits different conductivity from that of PANI/PVA composite membrane because the fiber is composed of ultrafine fibers and a large number of gaps could be found between fibers, in addition, the degree of orientation of the fiber has effects on it, in this case, the conductivity transmission is with directionality.

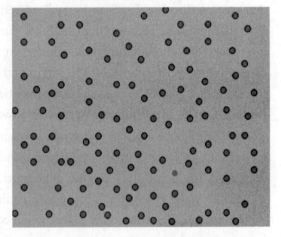

Fig. 7-37 Structure model of PANI particles distributing on PANI/PVA composite membrane

Structure model of PANI particles distributing on PANI/PVA ultrafine composite fiber is demonstrated in Fig. 7-38. As can be seen in Fig. 7-38, quite a lot of models of PANI particles distributing on PVA fiber could be achieved. The models in Fig. 7-38 (a), (c) and (e) exhibit superior conductivity when compared with other models. That is because the compact arrangement of PANI particle contributes to the transition of reaction force between particles under the action of low reaction force, thus resulting in charge transfer which is proved by electrochromic behavior. Similarly, the model of compact and dislocated type in Fig. 7-38 (d) presents superior electrochemical behavior than that of mode of dispersed type in Fig. 7-38 (b); the model of dispersed and heteromeral displacement type in Fig. 7-38 (h) provides better electrochemical behavior than that of modes in Fig. 7-38 (f) and (g). That is because the distance between PANI particles are different which has great effects on electrochemical behavior. Furthermore, thickness of fiber and arrangement of the fiber are also considered as the contributing factors of electrochemical properties, and only single-layer fiber and isodirectional distribution of the fiber are taken for examples.

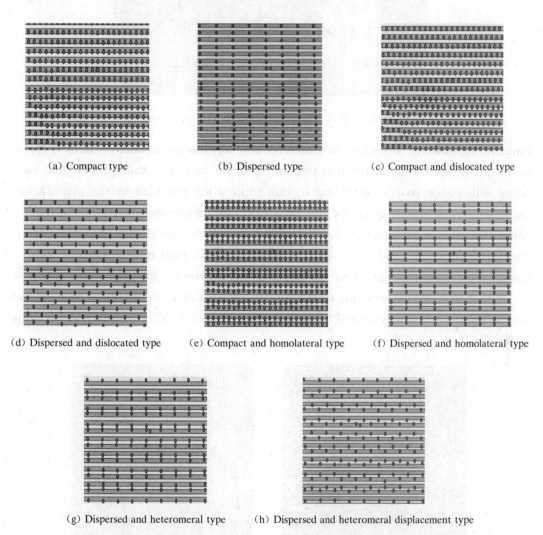

Fig. 7-38 Structure models of PANI particles distributing on PANI/PVA ultrafine composite fiber

7.5.2.3 Preparation and Characterization of Electrochromic Device Made of PANI/PVA

To observe the distribution of ITO particles on PET precisely, optical microscope (magnification of 3000) was selected in this section.

As can be seen in Fig. 7-39, the diameter of indium-tin oxide particle is within the

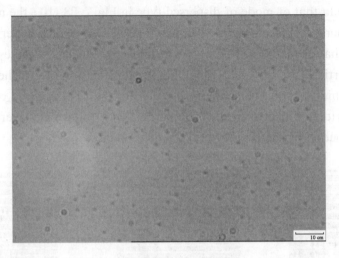

Fig. 7-39 Optical microscope image of ITO conductive film

range of nanometers to tens of nanometers, and continuous conductive mesh is formed based on the uniformly-distributed indium-tin oxide particles, thus proving the membrane with conductivity. According to the information provided by the manufacturer and examination report issued by the authority, the membrane presents in transparent color with specified parameters (surface resistance = 300~500 Ω/cm^2, thickness of ITO conductive film = (0.188 ± 0.019) mm, linearity <1.5%, light transmittance ≥ 86%), it has great effects on the discoloration effect of electrochromic material.

The morphological structure of the copper-plated conductive woven and silver-fiber-woven knitted fabric was studied by SEM (Epson Perfection V700 Photo) and illustrated in Fig. 7-40.

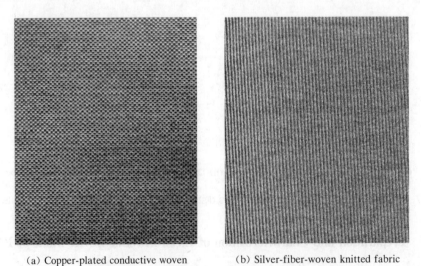

(a) Copper-plated conductive woven   (b) Silver-fiber-woven knitted fabric

Fig. 7-40 SEM images on the electrode material of ECD

The fabrics as depicted in Fig. 7-40 are with compact structure and excellent conductivity and can be selected as the material of electrochromic device, no matter woven or knitted fabric.

The electrochromic behavior of copper-plated conductive fabric with the voltage range of $-1.5 \sim 1.5$ V is demonstrated in Fig. 7-41. The color of fabric changes from dark green into black, and then it changes into golden yellow under an increasing voltage. The electrochromic behavior of the fabric is actually the electrochromism of PANI, and the copper-plated conductive fabric is used to provide PANI with ion for its oxidation-reduction reaction and achieve the conduction of internal ion rather than electrochromism.

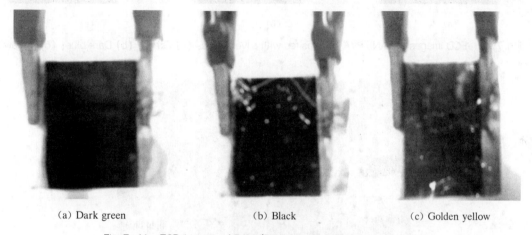

(a) Dark green　　　　(b) Black　　　　(c) Golden yellow

Fig. 7-41　ECD images of PANI/PVA composites with copper fabric

The electrochromic behaviors of the prepared electrochromic device with applied voltage of $-1.5 \sim 1.5$ V are depicted in Figs. 7-42 and 7-43. The asymmetric sandwich-structure electrochromic device made of ITO conductive film/(PANI/PMMA electrolyte)/conductive silver fabric is illustrated in Fig. 7-42 and obvious color change of the electrochromic device could be found. The symmetric sandwich-structure electrochromic device composed of ITO conductive film/(PANI/PVA ultrafine composite fiber)/PMMA electrolyte/(PANI/PVA ultrafine composite fiber)/ITO conductive film is demonstrated in Fig. 7-43 and the color changes from dark green to dark blue and then yellow. The electrochromic layer is made of fiber at nano-scale and the color depth of the fiber is light. Therefore, light yellow of the electrochromic layer fails to be recorded, however, it could be conjectured that the sequence of electrochromic behavior of PANI/PVA ultrafine composite fiber is light green, light blue and light yellow as the increased voltage based on the existing color change. In this case, the observable symmetric electrochromic material (ITO conductive film as the conductive substrate) was studied in this section when assembling the fabricated ultrafine composite fiber with the conductive fabric. That is because it's difficult for human to observe the extremely-light color change in electrochromic behavior. Meanwhile, further investigation should be conducted on the enhancement of thickness of ultrafine composite fiber and electrochromic behavior.

Fig. 7-42 ECD images of PANI/PVA composites with silver fabric: (a) Green; (b) Dark blue; (c) Yellow

Fig. 7-43 Color chart of symmetric ECD of ultrafine composite fiber mat

(1) The previous experiment indicates that the PANI/PVA composite membranes prepared at room temperature and low temperature present such electrochromic behavior with color change sequence of green, blue and yellow, however, it was found that obvious color change occurs to the membrane prepared at room temperature and its preparation time is 4~6 h longer than that of membrane prepared at low temperature, in addition, obvious redox peak could be found on the CV curve of the membrane prepared at room temperature, by comparing the images.

(2) Cyclic voltammetry was conducted to the PANI/PVA ultrafine composite fiber fabricated at scanning speeds of 5, 20, 50 and 100 mV/s, it could be found that the curves present similar trends and the redox peaks disappear nearly. It was found that the electrochromic behavior is linked to the distance between the PANI particles and the directionality of the fiber, based on the establishment of model used for indicating the distribution of PANI particle in the PVA fiber.

(3) The thickness of PANI/PVA ultrafine composite fiber, which is far less than the thickness of manually-coated membrane, has the significant effects on the properties

of electrochromic device made of PANI/PVA ultrafine composite fiber, therefore, the ultrafine composite fiber provides relatively-light color change.

7.6 Conclusions

In this chapter, the preparation method, research progress and application of PANI are introduced. The structure of electrochromic device is also provided. In addition, different kinds of PANI composites are fabricated and the effects of preparation parameters on the relevant properties are systematically investigated. The conclusions can be obtained as below.

(1) PANI/PVA composite emulsion with different conductivity is prepared via emulsion polymerization. The FTIR results reveal that the prepared PANI composite emulsion presents the characteristic absorption peak of conducting-state PANI, which indicates that typical conducting-state PANI has been synthesized.

(2) During the polymerization of PANI, high-speed stirring contributes to the dispersion of PANI particles, it is difficult for the agglomeration to be found when the stirring speed is set to be 2200 r/min and the PANI particles are dispersed in the PVA matrix independently with compact mesh. Based on the investigation on the mechanism of emulsion polymerization, the fabricated PANI composite exhibits good conductivity and its solubility is improved and optimized with the solubility in NMP of 24.7 mg/mL, which facilitates to widen application of PANI. In addition, the doping mechanism of PANI proton acid is investigated and the entire process of PANI changing from eigen state to conducting state is illustrated, which can provide a further development and a wide application of PANI.

(3) Concentration of electro-spinning solution has great effects on the fiber, the viscosity and viscoelasticity of the PVA emulsion increase gradually with the increasing mass fraction of PANI; however, a sharp decline occurs to the viscoelasticity of the PVA emulsion even fails to prepare the ultrafine fiber because of the excessive increasing mass fraction of PANI. The mechanical properties of the PANI/PVA ultrafine composite fiber becomes complicated with the increasing surface density, therefore, the single-factor analysis is conducted to investigate the effects of surface density on the mechanical properties of the fiber, and the experimental results show that surface density is one of the main contributing factors of affecting the mechanical properties of the PANI/PVA ultrafine composite fiber.

(4) Such parameters as electro-spinning voltage, electro-spinning distance and mass fraction have rather significant effects on the PANI/PVA ultrafine composite fiber. It can be clearly observed that the fiber surface is smooth and the average fiber diameter is only 300 nm when mass fraction of PVA is 4.5%; sudden steep of fiber diameter could

be found and the average fiber diameter reaches 730 nm with smooth fiber surface when the mass fraction of PANI is 0.2%; the average fiber diameter decreases within the range of 400~550 nm while roughness with varying degree could be found on the fiber surface when the mass fraction of PANI increases from 0.4% to 0.6%. Within the range of spinnable electro-spinning distance, average diameter of the fibers increases from 543 to 817 nm, however, the volumes of droplet and bead continuously decrease until they disappear when the electro-spinning distance is set to be 14 cm. It can be found that the fiber diameter decreases with the increasing of the electro-spinning distance, the dispersion degree decreases from 236 to 139 mg/m^3, meanwhile, it could be observed that PANI particles are distributed on the fiber in a uniform state. It can be clearly observed that the average diameter of PANI/PVA ultrafine composite fiber increases gradually, in addition, the average fiber diameter increases from 505 to 656 nm and the diameter distribution increases from 131 to 163 nm, with the electro-spinning voltage increasing from 14 to 26 kV. As can be seen in SEM images and histograms of fiber diameter distribution, the average fiber diameter increases gently within specified voltage range.

(5) Based on the XRD, the electrospun fiber presents a trend of first increasing and then decreasing with the increasing of mass fraction of PANI. It can be concluded that appropriate application of PANI contributes to preparation of fiber with excellent morphology, thus facilitating to the enhancement and improvement of fiber crystallinity. The ultrafine fibers with different mass fractions of PANI are characterized by TG and the experimental results show that the stability of PANI/PVA ultrafine composite fiber is improved and optimized.

(6) It can be concluded that the primary and secondary sequence of contributing factors during emulsion polymerization are respectively oxidant, concentration of proton acid and polymerization temperature, based on the orthogonal experimental results. The conductivity of PANI decreases with the increasing of concentration of DBSA when it has reached certain degree within specified range, and the optimum proportion of DBSA : AN : APS is 1.03 : 1.00 : 1.00.

(7) The PANI/PVA composite membranes prepared at room temperature and low temperature present such electrochromic behavior with color change sequence of green, blue and yellow, however, it can be found that obvious color change occurs to the membrane prepared at room temperature and its preparation time is 4~6 h longer than that of membrane prepared at low temperature, in addition, obvious redox peak could be found on the CV curve of the membrane prepared at room temperature, by comparing the images.

(8) Cyclic voltammetry is conducted to the PANI/PVA ultrafine composite fiber fabricated at scanning speeds of 5, 20, 50 and 100 mV/s, it could be found that the curves present similar trends and the redox peaks disappear nearly. It is found that the electrochromic behavior is linked to the distance between the PANI particles and the di-

rectionality of the fiber, based on the establishment of model used for indicating the distribution of PANI particle in the PVA fiber. The thickness of PANI/PVA ultrafine composite fiber, which is far less than the thickness of manually-coated membrane, has significant effects on the properties of electrochromic device made of PANI/PVA ultrafine composite fiber, therefore, the ultrafine composite fiber felt provides relatively-light color change.

Certain disadvantages still can be found in the study and investigation on following aspects could be conducted:

(1) Only effects of concentration of organic sulfonic acid (belonging to proton acid), polymerization temperature and concentration of oxidant on conductivity of PANI have been investigated for the emulsion polymerization; further study on effects of other parameters on emulsion polymerization and conductivity of PANI could be conducted as well for the purpose of complete PANI preparation technology, such as titration speed and titration time of the oxidant, concentration of emulsifier PVA.

(2) Only effects of electro-spinning voltage, electro-spinning distance and mass fraction of PANI have been researched for electro-spinning; furthermore, investigation on other electro-spinning parameters could also be performed to achieve the complete preparation technology of electrospun PANI composite, such as flow velocity of spinning solution, spinneret diameter, motion state (axial rotation and circular rotation, etc) of collector and environmental factor, etc.

(3) The PANI composite presents non-uniform thickness because of various reasons when the electrochemical behavior of PANI composite is characterized, therefore, the electrochromic behavior is characterized by non-uniformity, in this case, investigation on the thickness of PANI composite could be carried out to achieve more perfect characterization of electrochromic behavior.

References

[1] Huang Hui, Guo Zhongcheng. Preparation and Application of Conductive Polyaniline. Beijing: Science and Technology Press, 2010: 41-42.

[2] Kane M C, Lascola R J, Clark E A. Investigation on the effects of beta and gamma irradiation on conducting polymers for sensor applications. Radiation Physics and Chemistry, 2010, 79(12):1189-1195.

[3] Meng Zhaohe, Ren Penggang, Wang Xudong, et al. Application of conductive polymeric materials in stealth technology. Polymer Materials Science and Engineering, 2004 (5):29-32.

[4] Fang Kun, Wu Qiye, Liu Wenyan, et al. Application of conductive polymer electrochromic materials and their application in aircraft and military camouflage. Aerospace Materials Technology, 2004, (2):21-25.

[5] Li Xinguui, Li Bifeng, Huang Meirong. Emulsion polymerization of aniline and its application. Plastics, 2003, 32(6):32-39, 45.

[6] Yusoff A R B, Shuib S A. Metal-base transistor based on simple polyaniline electropolymerization.

Electrochimica Acta, 2011, 58:417-421.

[7] Li Cheng, Yang Xiaogang, Huang Wenjun, et al. Synthesis and application of polyaniline nanomaterials. Microelectronics Technology, 2011(2):92-97,117.

[8] He Ji. Preparation and application of conductive polyaniline Polyaniline. Science and Technology Information (Science and Technology), 2008(18):391-392.

[9] Ma Li, Tang Qi. Study on polyaniline of conductive polymeric materials. Journal of Chongqing University (Natural Science Edition), 2002(2):124-127.

[10] Zhou Zhentao, Yang Hongye, Wang Kejian, et al. Chemical synthesis, structure and conductivity of polyaniline. Journal of South China University of Technology (Natural Science Edition), 1996(7):72-77.

[11] Kan Jinqing, Mu Shaolin. Effects of oxidants on polyaniline properties. Acta Polymerica Sinica, 1989(4):466-471.

[12] Yu Huangzhong, Chen Mingguang, Huang He. Effects of different types of acid doping on structure and conductivity of polyaniline. Journal of South China University of Technology (Natural Science Edition), 2003(5):21-24.

[13] Yu Huajiang, Wu Kezhong, Wang Qingfei, et al. Determination of the effect of different protonic acid doping on the rate of aniline polymerization. Journal of Hebei Normal University (Natural Science Edition), 2008, 32(1):64-67, 84.

[14] Yang Chunming, Chen Dizhao, Fang Zheng, et al. In-site UV-visible spectroscopy and its factor analysis of polyaniline doped with protonic acid in organic solvents. Chemical Journal of Chinese Universities, 2002(6):1198-1201.

[15] Fu Heqing, Zhang Xinya, Huang Hong, et al. Preparation of polyaniline by emulsion polymerization and its conductivity. Journal of Chemical Industry and Engineering(China), 2005, 56(9):1790-1793.

[16] Bai Ying, Li Xin, Liu Jinquan. Effect of electrochemical polymerization time on the photoelectric properties of polyaniline nanofilm. Materials Research, 2012, 26(1):50-53.

[17] Wang Hongzhi, Liu Weihong, Li Jian, et al. Preparation of polyaniline film by constant potential pulse method and its characterization. Chemical Journal of Chinese Universities, 2012, 33(2):421-425.

[18] Diaz A F, Logan J A. Electroactive polyaniline films. Electroanal Chemical, 1980, 111(1):111-114.

[19] Yuan G L, Kuramoto N, Su S J. Template synthesis of polyaniline in the presence of phosphomannan. Synthetic Metals, 2002, 129(2):173-178.

[20] Huang Zhihui, Shi Lei, Zhou Junting, et al. Preparation of polystyrene-elastic silver complex by γ-radiation and its transport properties. Proceedings of the Seventh National Symposium on Polymer and Structural Characterization, Shanghai, 2010:168-169.

[21] Moulton S E, Innis P C, Kane M, et al. Polymerisation and characterisation of conducting polyaniline nanoparticle dispersions. Current Applied Physics, 2004, 4:402-406.

[22] Xia H S, Wang Q. Synthesis and characterization of conductive polyaniline nanoparticles through ultrasonic assisted inverse microemulsion polymerization. Journal of Nanoparticle Research, 2001, 3:399-409.

[23] Atobe M, Chowdhury A N, Fuchigami T, et al. Preparation of conducting polyaniline colloids under ultrasonication. Ultrason Sonochem, 2003, 10(2):77-80.

[24] Zhang Huiqin. Preparation of polyaniline conductive fibers. Journal of Zhongyuan Institute of Technology, 2005, 16(5):38-41.

[25] Liu Weijin, Wu Guoming. Preparation of polyaniline and polyaniline conductive fibers. Synthesized Fiber Industry, 1996, 19(4):47-51.

[26] Li Min. Preparation of polyaniline conductive fibers. Modern Silk Science and Technology, 2012, 27(2):79-82.

[27] Li Li, Yang Jiping, Chen Xiaochen, et al. Preparation and characterization of conductive polyaniline fibers. Polymer Materials Science and Engineering, 2011, 27(4):151-158.

[28] Liu Weijin, Wu Guoming, Ma Weihua, et al. Preparation of polyaniline/polyester conductive composite fibers. Polymer Materials Science and Engineering, 1997(S1):71-74.

[29] Pan Wei, Huang Suping, Jin Huifen. Preparation of polyaniline/polyester conductive fiber. Journal of China Textile University, 2000,26(2):96-99.

[30] Zhang F, Halverson P A, Lunt B, et al. Wet spinning of pre-doped polyaniline into an aqueous solution of a polyelectrolyte. Synthetic Metals, 2006, 156:932-937.

[31] Pomfret S J, Comfort N P, Monkman A P, et al. Electrical and mechanical properties of polyaniline fibres produced by a one-step wet spinning process. Polymer, 2000, 41(6):2265-2269.

[32] Jiang J M, Pan W, Yang S l, et al. Electrically conductive PANI-DBSA/Co-PAN composite fibers prepared by wet spinning. Synthetic Metals, 2005, 149(2):181-186.

[33] Andreatta A, Smith P. Processing of conductive polyaniline-UHMW polyethylene blends from solution sinnon-polar solvents. Synthetic Metals, 1993, 55:1017-1022.

[34] Tzou K T, Gregory R V. Improved solution stability and spinnability of concentrated polyaniline solutions using N,N-dimethyl propyleneureaasthe spin bath solvent. Synthetic Metals, 1995, 63:109-112.

[35] Pan Wei. Preparation of Soluble Polyaniline and its Composites. Shanghai: Donghua University, 2004.

[36] Wanna Y, Pratontep S, Wisitsoraat A, et al. Development of nanofibers composite polyaniine/CNT fabricated by electro spinning technique for CO gas sensor. The 5th IEEE Conference on Sensors, 2006: 342-345.

[37] Cao Tieping, Li Yuejun, Wang Ying, et al. Preparation and characterzation of PAN/PANI composite nanofibers by electro-spinning. Acta Polymerica Sinica, 2010 (12):1464-1469.

[38] Shin M K, Kim Y J, Kim S I, et al. Enhanced conductivity of aligned PANI/PEO/MWNT nano fibers by electro-spinning. Sensors and Actuators, 2008, 134:122-126.

[39] Repeker D H, Chun I. Nanometer diameter fibers of polymer produced by electro-spinning. Nanotechnology, 1996 (7):216-223.

[40] Norris I D, Shaker M M, Ko F K, et al. Electrostatic fabrication of ultrafine conducting fibers: Polyaniline/polyethylene oxide blends. Synthetic Metals, 2000, 114:109-114.

[41] Desai K, Sung C. Electro-spinning nanofibers of PANI/PMMA blends. Materials Research Society Symposium-Proceedings, 2002, 736:121-126.

[42] Pinto N J, Carrion P, Quinones J X. Electroless deposition of nickel on electrospun fibers of 2-acrylamido-2-methyl-1-propanesulfonic acid doped polyaniline. Materials Science and Engineering: A, 2004, 366(1):1-5.

[43] Shie M F, Li W T, Dai C F, et al. In vitro biocompatibility of electro-spinning polyaniline fibers. World Congress on Medical Physics and Biomedical Engineering. Munich: Springer Berlin Heidel-

berg, 2010.

[44] Picciani P H S, Medeiros E S, Pan Z L, et al. Structural, electrical, mechanical, and thermal properties of electrospun poly(lactic acid)/polyaniline blend fibers. Macromolecular Materials and Engineering, 2010, 295(7):618-627.

[45] Neubert S, Pliszka D, Thavasi V, et al. Conductive electrospun PANI-PEO/TiO_2 fibrous membrane for photo catalysis. Materials Science and Engineering B-Advanced Functional Solid-State Materials, 2011, 176(8):640-646.

[46] Shahi M, Moghimi A, Naderizadeh B, et al. Electrospun PVA-PANI and PVA-PANI-$AgNO_3$ composite nanofibers. ScientiaIranica, 2011,18(6):1327-1331.

[47] Bagheri H, Aghakhani A. Polyaniline-nylon-6 electrospun nanofibers for headspace adsorptive microextraction. Analytica Chimica Acta, 2012, 713:63-69.

[48] Shin Y J, Kameoka J. Amperometric cholesterol biosensor using layer-by-layer adsorption technique onto electrospun polyaniline nanofibers. Journal of Industrial and Engineering Chemistry, 2012, 18(1):193-197.

[49] Zhang Z, Jiang X, Liu Y, et al., Fabrication and EM shielding properties of electro-spinning PANi/MWCNT/PEO fibrous membrane and its composite. Proceedings of the SPIE, 2012, 8342:44.

[50] Zhu C E, Rem L, Wang L X, et al. Preparation and characterization of polyaniline/carbon black conducting composites as anode of lithium secondary battery. Polymer Materials Science and Engineering, 2005, 21(6):217-220.

[51] Wan M, Yang J. Mechanism of proton doping in polyaniline. Journal of Applied Polymer Science, 1995, 155(3):399-405.

[52] Sun G C, Yao K L, Liao H X, et al. Microwave absorption characteristics of chiral materials with Fe_3O_4-polyaniline composite matrix. International Journal of Electronics, 2000, 87(6):735-740.

[53] Kaner R B. Gas, liquid and enantiomeric separations using polyaniline. Synthetic Metals, 2002, 125(1):65-71.

[54] Deb S K. Electrochromic charaters of WO_3. Applied Optics, 1969, 58(3):192-194.

[55] Colton R J, Guzman A M, Rabalais W J. Photochromism and electrochromism in amorphous transition metal oxide films. Accounts of Chemical Research, 1978, 11(4):170-176.

[56] Jaeger C D, Bard A J. Electrochemical behavior of donor-tetracyanoquino-dimethane electrodes in aqueous media. Journal of the American Chemical Society, 1980, 102(17):5435-5442.

[57] He Y C, Xiong Y Q, Qiu J W, et al. Preparation and properties of all solid-state electrochromic thermal control thin film. Physics Procedia, 2011: 61-65.

[58] Grant B, Clecak N J, Oxsen M, et al. Study of the electrochromism of methoxy-fluorene compounds. Journal of Organic Chemistry, 1980, 45(4):702-705.

[59] Kaya I, Yildirim M, Aydin A. A new approach to the Schiff base-substituted oligophenols: The electrochromic application of 2-3-thienylmethylene aminophenol based co-polythiophenes. Organic Electronics, 2011, 12(1):210-218.

[60] Mortimer R J. Electrochromic materials. Chemical Society Reviews, 1997, 26(3):147-156.

[61] Mortimer R J. Electrochromic Materials. Chemical Society, 2011(41):241-268.

[62] Zeng Liqun, Li Tanping. Study on the development trend of deep processing of flat glass. New Materials Industry, 2004(4):35-41.

[63] Qian Jing, Fu Zhongyu, Li Xin. Research progress of electrochromic devices based on conducting

polymers. Chemical Research and Application, 2008, 20(11):1397-1404.

[64] Ling Ping, Zhao Xuequan, Guan Li, et al. Synthesis and electrochromic properties of novel oligothiophene derivatives. Materials Research and Application, 2010,4(4):321-324.

[65] Luo Wei, Fu Xiangkai, Zhou Jie. Electronic luminescence and electrochromic. Materials Review Journal, 2006 (1):1-6.

[66] Crandall R S, Faughnan B W. Comment on the cluster model of alkali-metal tungsten bronzes. Physical Review B Condensed Matter, 1977, 16(4): 1750-1752.

[67] Zhang J G, Benson D K, Deb S K, et al. Chromic mechanism in amorphous WO_3 films. Journal of The Electrochemical Society, 1997, 144(6): 2022-2026.

[68] Wittwer V, Schirmer, Schlotter P. Disorder dependence and optical detection of the Anderson transition in amorphous $HxWO_3$ bronzes. Solid State Commun, 1978, 25(12): 977.

[69] Niu Wei, Bi Xiaoguo, Sun Xudong. Research and development of electrochromic mechanism. Materials Review, 2011,25(2):107-110.

[70] Shen Qingyue, Lu Chunhua, Xu Zhongzi. Color-changing mechanism of electrochromic materials and their Research progress. Materials Review, 2007, 21:284-292.

[71] Argun A A, Aubert P H, Thompson B C, et al. Multicolored electrochromism in polymers: Structures and devices. Chemistry of Materials, 2004, 16(23):4401-4412.

[72] Kanazawa K K, Diaz A F, Geiss R H, et al. "Organic Metal" polypyrrole, a stable synthetic "metallic" polymer. Journal of the Chemical Society Chemical Communications, 1979, 19(19): 854-855.

[73] Kuwabata S, Yoneyama H, Tamura H. Redox behavior and electrochromic properties of polypyrrole films in aqueous solutions. Bulletin of The Chemical Society of Japan, 1984, 57(8): 2247-2253.

[74] Zhang Miao, Feng Huixia, Shao Liang, et al. Synthesis of water-soluble polyaniline and its application in corrosion protection. Applied Chemical Industry, 2008, 37(5):573-585.

[75] Ding Lei. Preparation and Properties of WO_3-TiO_2 Electrochromic Materials and Devices. Tianjin: Tianjin University, 2009.

[76] Tang Ying, Li Weiyi, Guo Hui, et al. Study on the progress of conductive polyaniline. Journal of Southwest University for Nationalities (Natural Science Edition), 2003, 29(5):544-547.

[77] Cao Feng, Li Dongxu, Guan Zisheng. Progress in study on conducting polymer polyaniline. Materials Review, 2007, 21(8):48-50, 55.

[78] Huang Hui, Xu Jinquan, Guo Zhongcheng. Research progress and prospect of conductive polyaniline. Electroplating and Finishing, 2008, 30(11):9-12.

[79] Liu Zhanqing. Reaserch progress of polyaniline conductive properties. Adhesion in China, 2010(6): 24-25.

[80] Li Xi, Wu Shuhong, Zhang Chaocan. Research on promotion of PANI/polymer composites by emulsion polymerization. Materials Review, 2006, 20(7):315-319.

[81] Xu H, Chen Y, Shi X W, et al. Synthesis and characterization of fibular PANI/nano-CeO_2 composites. Advanced Materials Research, 2011,128:1021-1025.

[82] Dupare D B, Ghoshp P, Datta K, et al. Synthesis and characterization of a novel ammonia gas sensor based on PANI-PVA blend thin films. Sensors & Transducers, 2008, 93(6):103-113.

[83] Pan Chunyue, Hu Huiping, Ma Chengyin, et al. Emulsion polymerization of aniline. Chinese Journal of Applied Chemistry, 2000, 17(5):491-494.

[84] Kang Ruzhen, Yang Shanwu, He Xinlai. Concentration effect on structure and properties of polyaniline. Journal of Liaoning Technical University, 2004, 23(6):852-854.

[85] Pan Chunyue, Zeng Yan. Effect of emulsion polymerization conditions on the properties of polyaniline. Polymer Materials Science and Engineering, 2001, 17(3):163-165.

[86] Tess R W. Solvents theory and practice. American Chemical Society, 1973,7(2): 89-90.

[87] Han D, Chu Y, Yang L, et al. Reversed micelle polymerization: a new route for the synthesis of DBSA-polyaniline nanoparticles. Colloids and Surfaces A: Physicochemical and Engineering Aspects, 2005, 259(1):179-187.

[88] Shreepathi S, Holze R. Spectroelectrochemical investigations of soluble polyaniline synthesized via new inverse emulsion pathway. Chemical Material, 2005, 17(16):4078-4085.

[89] Chen Xiongluo, Kang Weimin. Electrospun nano non-woven materials and application. Chemical Fiber & Textile Technology, 2009, 1:25-29,37.

[90] Xu Ming. Electro-spinning Technique and Structure and Performance of Electro-spinning Products. Shuzhou: Soochow University, 2005.

[91] Chi Lei, Yao Yongyi, Li Ruixia, et al. Recent advance in manufacture of nano-fibers by electrospinning. Progress in Textile Science & Technology, 2004, 5: 1-5.

[92] Shih J H. A study of Composite Nanofiber Membrane Applied in Seawater Desalination by, Membrane Distillation. Taiwan University of Science and Technology, 2011.

[93] Sun Yat-Sen University. A Creep Resistant Nano Inorganic Particle/Polymer Composite Material and Preparation Method Thereof. CN 201310076127.7.

[94] Ma Lihua. Preparation and Properties of Water-soluble Conductive Polyaniline Complex Electrochromic Film. Chongqing:Southwest University, 2007.

Chapter 8 Development of Graphene Based Textiles

Jiali Yu, Binjie Xin* and Zhuoming Chen

8.1 Introduction

With the development of society and the improvement of life, functional and smart textiles are attracting increasing attention, and smart textiles become the development direction of textile and clothing industry in the future[1]. Smart materials, widely used in wearable electronic devices, flexible intelligent sensors, aerospace, military and other fields[2-3], become the hotspot in new material field. In 21st century, physicists found a two-dimensional carbon material graphene and tried further study, which made graphene conductive composite materials get fast development. How to apply graphene with perfect conductive performance into the textile field to improve the comprehensive performance of textile materials becomes another hotspot to expand the application category of textile materials.

8.1.1 Introduction of Graphene

Graphene is a novel two-dimensional material, and its appearance subverts the theory proposed by Landau and Peierls that absolute two-dimensional crystal is unstable in thermodynamics and cannot exist alone. The first study of graphene made public in 1970s[4]. Two professors from Manchester University, Geim and Novoselov, repeatedly exfoliated highly oriented pyrolytic graphite to prepare two-dimensional single stable graphene

---

Corresponding Author:
Binjie Xin
School of Fashion Technology, Shanghai University of Engineering Science, Shanghai 201620, China
E-mail: xinbj@sues.edu.cn

by mechanical exfoliation in 2004[5]. They were awarded the Nobel Prize for physics in 2010, which brings graphene to a research culmination in material science.

8.1.1.1 Structures and Properties of Graphene

Graphene is a single two-dimensional carbon material made of carbon atoms with the sp_2 hybridization, carbon atoms tightly being packed into honeycomb-like sheets[6]. Graphene is considered as the thinnest carbon material, and its thickness and C—C bond length are 0.335 and 0.142 nm respectively, as shown in Fig. 8-1. The honeycomb-like structure of graphene is considered as the basic unit of the other carbon allotropes. There are five-member ring lattice in the graphene lattice, which results in warp. The zero dimensional fullerenes can be formed when there are more than 12 five-member ring lattices. Graphene can also warp to form one-dimensional cylindrical carbon nanotubes, and a large number of graphene sheets can be stacked to form three-dimensional graphite[7].

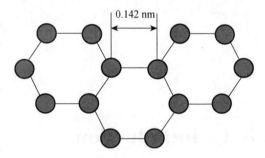

Fig. 8-1 Basic structure diagram of graphene

Each atom has one s orbital and two p orbitals in two planes. The π-electrons are in the rest p orbital form and the π-orbits are in the direction vertical to the plane. The lattice of graphene consists of two triangle interpenetrative sublattices, and the atoms of a lattice are in the triangle center of another. Each carbon atom connects with other three adjacent carbon atoms by σ bond. s, p_x and p_y hybrid orbitals at a 120° bond angle can form a strong covalent bond to make up sp_2 hybrid structure, which makes graphene sheets stable and contributes to the formation of conduction band and valence band, which makes graphene conductive in the two-dimensional[8]. The band energy mainly depends on the momentum of carriers in brillouin zone. Two bands with low-energy cross each other and form conical band structure near K and K', as shown in Fig. 8-2. With the limit of low energy, according to the theory of relativity Dirac equation, the distribution between energy and momentum is linear, and the carriers can be regarded as relativistic particles with zero quality, the effective speed of light is 10^6 m/s. Owing to the special local electronic structure, many phenomenon based on the typical characteristics of two-dimensional Dirac fermions can be observed[9]. Electron can be changed into hole, and the hole can also become electron near the Dirac point. With the limit of high energy, the linear relationship between energy and momentum is no longer valid, and the appearance of deformation results in anisotropy, namely triangle warp[10]. Six-member ring carbon atoms of graphene can form three-dimensional graphite with layered structure and anisotropy by the interaction of π-electrons[5].

Compared with other semiconductor materials, graphene has unique properties owing to its band structure with zero gap. Graphene has the ability of ballistic electron transport, which is embodied as follows: in theory, it has high electron mobility at room temperature and the quantum hall effect can be observed. The fractional quantum Hall Effect can even be observed at low temperature. Graphene has wonderful transmittance especially the monolayer graphene and large specific surface area, good thermal conductivity as well as excellent mechanical properties, and its mechanical property is better than steel and diamond[11].

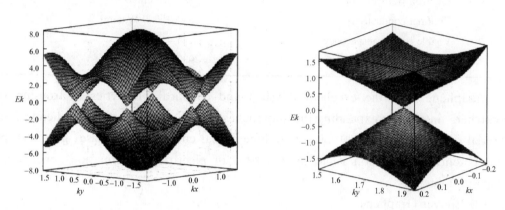

Fig. 8-2　Diagram of the band structure of graphene [12]. (*Ek*—the intrinsic value of the energy of graphene; *kx*, *ky*—the component of the inversion vector *k* on (*x*, *y*))

According to the difference of structure, graphene can be divided into monolayer graphene, bilayer graphene and few-layer graphene.

(1) Monolayer Graphene

For metals and semiconductors, monolayer graphene with different properties belongs to a typical semimetal[13-14]. Monolayer graphene has dual polarity at room temperature, and carriers can be transformed between the electrons and holes with the change of grid voltage. The study showed monolayer graphene has abnormal (half integer) quantum hall effect both at low temperature and room temperature[6,14]. Monolayer graphene with those special properties is suitable for the application in the electronics industry and has become the best material to the study of quantum physics phenomena. Another important property of monolayer graphene is that it has gas sensing ability. The resistance of graphene changes with the change of carrier concentration as graphene adsorbs gas molecules. By the gas sensing ability of monolayer graphene, the micron grade gas sensor used to detect the absorption and stripping of monatomic gases such as CO, H_2O, NH_3 and NO_2 was prepared by Schedin et al[15]. Graphene is an excellent low-noise material and one of high strength materials; it can be used to prepare two-dimensional nano-electronic mechanical system. Monolayer graphene can also be used in nano-imprinting technology and atomic simulation model, while its Young's modulus reaches 1.0

TPa. The highest Young's modulus (1100 GPa) and breaking strength (125 GPa) have been measured so far. The graphene paper with good biological compatibility prepared by Chen et al could be attributed to excellent mechanical properties of graphene[16].

Main properties of monolayer graphene are shown in Tab. 8-1.

Tab. 8-1 Main properties of monolayer graphene

Properties	Parameters
Young's modulus	1.0 TPa
Electron transfer rate	>10 000 $cm^2/(V \cdot s)$
Thermal conductivity	3000~5000 $W/(m \cdot K)$
Visible light transparency	>95%
Specific surface area	2630 m^2/g

Graphene with the excellent physical and chemical properties is arousing the researchers' interest in expanding the approaches to apply it into physical, chemical and material field. As a kind of typical two-dimensional enhancement phase, graphene has many potential application values in the field of electronic information, energy and composite materials.

(2) Bilayer Graphene

Compared with the conical band of monolayer graphene, the band of bilayer graphene is like a parabola without band gap in near K and K'. The carrier with a certain quality in bilayer graphene is regarded as Dirac fermions with quality. Compared with monolayer graphene, bilayer graphene also has abnormal quantum hall effect, but it remains metal properties at electric neutrality. However, the concentration of charge carrier changes by the changing grid voltage, which causes the asymmetry between the two layers, makes bilayer graphene has produced semiconductor band gap and the quantum hall effect return to be normal[17]. SiC epitaxial method was used to grow graphene by Zhou et al, and the band gap was measured to be 0.26 eV[18]. This limited band gap structure makes graphene suitable for application in the electronics industry. The study found that the band gap of graphene decreases with increase of layer numbers. When the layer numbers is more than 4, the band gap is close to zero. The reason why the band gap opens may be that the interaction of graphene and substrate destroys the symmetry of sublattice. The related research also found that the graphene grew by 4H-SiC epitaxial method has different stack orders. No matter how many layers (less than 10) of graphene with this structure, the electronic structure and properties of it are similar to monolayer graphene. Therefore, the preparation method of graphene plays an important role in determining its structure and properties.

(3) Few-layer Graphene

The band structure of few-layer graphene has no band gap. With the increase of the layer number, the graphene tends to be metal attribute. Few-layer graphene has large specific surface area, which can be comparable to monolayer graphene. It also has good gas adsorption properties. Fewer-layer graphene can absorb 3% hydrogen at 300 K and 10 MPa. Few-layer graphene has good functional ability; it can be dissolved in various solvents through different covalent and non-covalent modification. All the functionalization of covalent modification will affect the electronic structure and properties of the fewer-layer graphene. Therefore, it is necessary to achieve functionalization by non-covalent modification without the change of electronic structure. In addition, it makes sense that the modification of metal nano-particles on few-layer graphene to expand the application in electronics, optics, biology and other fields. The experimental results show that few-layer graphene can be modified with Pt, Au and Ag nanoparticles by simple chemical treatment. The application of few-layer graphene in optoelectronics is enlarged by these modifications. The change of chemical modification results in the changing magnetic properties by edge character of graphene. Edge states of graphene causing new magnetism features include ferromagnetic attracted much attention in recent years. The study found that the edge states of graphene state can be changed by changing the type of molecules adsorbed. Monolayer graphene, bilayer graphene and few-layer graphene all have potential advantages in the applications of electronic, storage, biology, sensors and energy storage devices.

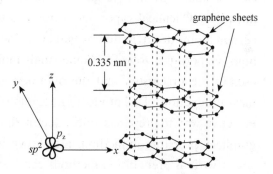

Fig. 8-3 Diagram of several layers structure for graphene [19]

8.1.1.2 Preparation Methods of Graphene

Preparation methods of graphene are usually can be classified into two parts: physical methods and chemical methods. Physical methods means that graphene is obtained from graphite or other similar materials with high lattice, including mechanical exfoliation method, SiC epitaxial growth method, etc; while chemical method means graphene is prepared through synthesis of small molecules or solution separation methods, including graphite intercalating exfoliation method, chemical vapor deposition method, oxidation-reduction method, electrochemical method, etc[20].

(1) Preparation Methods of Common Graphene

Mechanical exfoliation method is an approach to pull off graphene sheets from the surface of graphite lattices by mechanical force. This method is first used for physical preparation of graphene with perfect crystal structure and less defects. The disadvantage of this method is low efficiency, time-consuming, bad laborious repeatability and unsuitable for large-scale industrial production, so that this method is only applied on basic laboratory researches[20]. SiC epitaxial growth method is used to obtain graphene by ultra-high vacuum to heat single-crystal SiC to refactor Si and C atom to generate monolayer or multilayer graphene, but it is difficult to achieve large-scale preparation. The preparation condition is strict (ultrahigh vacuum and high temperature). It needs high cost, and it's hard to only corrodes SiC and not destroy the structure of graphene, which does not favor the subsequent transfer of graphene[21]. Oxidation-reduction method is that graphite disperses firstly in the solution by ultrasonic shaking and high speed centrifugal to get graphene precursor, then monolayer or multilayer graphene can be obtained by reduction process[4]. The short comings of this method are that the graphite treated by a strong oxidizer cannot restore completely, and the physical and chemical properties of graphene may be destroyed in oxidation-reduction process. In addition, the graphene prepared by extension method has uneven thickness and the graphene layers obtained by thermal expansion method may fold and deform[20]. Ultrasonic stripping method is difficult to expand graphite to get high quality graphene, but multilayer graphene can be obtained[20]. Chemical vapor deposition is a technology to generate graphene by growth of carbon atoms disintegrated by gaseous, liquid or solid carbon source on metal substrate at high temperature zone[22]. Some small molecular hydrocarbons or polymer materials is used as the precursor, and the obtained graphene has good quality and is suitable for mass production. This production technology is perfect and has wide application; it is an effective way to prepare semiconductor thin film materials on a grand scale[23]. Common graphene preparation methods are shown in Tab. 8-2. The advantages and disadvantages of various preparation methods can be visually observed.

Tab. 8-2 Comparison of preparation methods of common graphene

No.	Preparation methods	Growth conditions	Number of layers	Characteristics
1	Mechanical exfoliation method	Room temperature	1~10 uncontrollable	Graphene with high-quality, low productivity, limit to mass production
2	SiC epitaxial method	>1200 ℃	Monolayer, few-layer	Graphene with high-quality, high costs
3	Oxidation-reduction method	<50 ℃	1~10 uncontrollable	Simple method, high productivity, graphene with many defects
4	Chemical vapor deposition method	400~1000 ℃	1~4 controllable	Graphene with high-quality and large specific surface area, limited by the substrates
5	Graphite intercalating exfoliation method	Room temperature	1~10 few-layer	High preparation efficiency, low costs

(2) Graphite Intercalating Exfoliation Method

Graphite intercalating exfoliation method means that tens of nanometers thickness graphene nano-sheets are obtained by heating expansion or ultrasonic oscillation of expanded graphite prepared through inserting some molecules, ions or groups of atoms into natural graphite layers. The advantage of this technology is that the production process is relatively simple, it is suitable for mass production and the graphene is available on the market.

Thermal expansion reduction method refers to the approach that the oxygen-containing groups of graphite oxide decompose and release carbon dioxide while it is heated in a short period of time up to more than 1000 ℃, the pressure produced during the gas release can make graphene oxide effectively separated to obtain graphene[24].

There are some related reports about graphene preparation by intercalation of expanded graphite. Cui[25] adopted graphite as raw materials to prepare expanded graphite though chemical method including oxidation and acidification, washing, drying and heat expansion process, then a few-layer graphene by ultrasonic stripping was gained. Guo et al[26] also used natural flake graphite as raw materials, expandable graphite achieved by the oxidation of graphite layers, after microwave pyrolysis expansion, the secondary oxidation of expanded graphite and microwave expansion were conducted, and a few-layer graphene was prepared by ultrasonic stripping method. Experimental results show that the distance among graphite oxide layers increases, so that expandable graphite is easier to be introduced with oxygen-containing groups. With the action of microwave, the oxygen-containing groups' pyrolysis gas increases graphite layer spacing and exfoliation of graphite layer. Graphene with less than 5 layers could be obtained by ultrasonic flaking of the secondary expansion graphite.

8.1.1.3 Preparation of Graphene Composites

Graphene is a very valuable filler for polymers owing to its excellent electrical conductivity, high thermal stability and mechanical properties. The final properties of composites are not only determined by properties of both the filler and the polymer itself, those factors closely related to the process technology, such as filler dispersion, distribution status and the interface bonding strength between filler and matrix, they are the key to achieve excellent properties of the composites. The preparation methods of graphene/polymer composites have been reported in literatures as follows: solution blending method, which is known as a wet process; melt blending method; in-situ polymerization method.

(1) Solution Blending Method

Graphene uniformly dispersed graphene/polymer composites can be prepared by solution blending method though blending polymer solution with low viscosity and graphene. Solution blending method is one of the most common methods to prepare graphene/polymer composites because of its simple process without special equipment. The preparation of graphene/polyimide composites[27] and graphene oxide/WPU composite membranes[28] by this method have been reported. The key of this process is to prepare graphene uniformly dispersed solution. Graphene has great specific surface area and it is easy to occur agglomeration phenomena. Actually, the special effect of graphene in nanoscale large specific surface area cannot be achieved if graphene cannot disperse uniformly in a solvent or polymer matrix.

(2) Melt Blending Method

Melt blending method is another main technology to industrially produce composites by blending melted polymers through high shearing force. Compared with solution blending method, melt-blending method would not involve the use of a large amount of solvents, it is more suitable for the large-scale production with low cost. Kim et al[29] prepared graphene/polylactic-acid composites by blending polylactic-acidand expanded graphite at 175~200 ℃, then studied the thermal properties, electrical properties and mechanical properties of the composites. The problem is that dispersion degree of graphene in matrix by melt blending method cannot achieve same effect compared with that of in-situ polymerization method. The thermal expansion of graphite oxide (TEGO) or graphene derivatives with small volume density will appear the feed problem in the process. In order to solve this problem, Steurer et al[30] found a valid approach that TEGO was firstly blended with polymer solution to prepare composites with graphene uniform dispersion, and then fed into the extruder with melt blending. In addition to the non-uniform dispersion of fillers, high shearing force in the process may lead to damage of graphene structure with high length to diameter ratio and polymer molecule chains, the final properties of the composites would be affected.

(3) In-situ Polymerization Method

In-situ polymerization is a method to prepare graphene nano composites while the polymer monomer is added into the solvent to make graphene disperse uniformly and the polymerization reaction is induced with the introduction of initiator at a certain condition. Lee et al[31] firstly prepared graphene materials by thermal reduction of expanded graphene, poly (caprolactone alcohol) and diisocyanate were added, and then the graphene/waterborne polyurethane nano-composites can be prepared by in-situ polymerization. Research results showed that the conductive properties and modulus of waterborne polyurethane could be improved, and the reaction between graphene and waterborne polyurethane was enhanced.

8.1.1.4 Current Researches on Graphene/Polymer Conductive Composites

Preparation of graphene/polymer conductive composites has attracted increasing attention because of the excellent conductive properties of graphene. The common graphene/

polymer conductive composites include graphene/silicone rubber (PDMS) conductive composites, graphene/polyurethane (PU) conductive composites, graphene/polyethylene terephthalate (PET) conductive composites, graphene/polyethylene (PVA) conductive composites and expanded graphene/polymethylmethacrylate (PMMA) conductive composites, etc.

(1) Graphene/PDMS Conductive Composites

Zhao et al[32] prepared a room temperature vulcanized silicone rubber composite with graphene and graphene oxide as a filler by solution blending. The result showed that, as the mass fraction of graphene was 3%, the tensile strength of the composites increased by 200%, the conductivity increased by 9 orders of magnitude and the percolation threshold reached to 1%.

(2) Graphene/PU Conductive Composites

Wang et al[28] prepared graphene oxide/PU membranes by solution blending method. Natural graphite was oxidized and ultrasonic shaked with improved Hummers method to get graphene oxide. Orthogonal experiment was conducted to study and analyze the effect of graphene on mechanical properties, electrical conductivity, thermal stability and water-resistance of the composites. Results showed that the addition of graphene oxide improved the electrical conductivity, tensile strength and thermal stability of the composites.

(3) Graphene/PET Conductive Composites

Zhang et al[33] prepared graphene/PET composites by melt blending method. Graphene can be evenly dispersed in the PET matrix at monolayer or few-layer state. The crimp and drape of graphene and the interaction between the layers contribute to the formation of conductive network, greatly improve the conductive properties of composites. The electrical conductivity of the composites can reach 2.11 S/m with the graphene volume fraction of 3%.

(4) Graphene/PVA Conductive Composites

Zhou et al[34] obtained graphene oxide/PVA composite films by solution casting, then the film was impregnated and restored in a mixture of sodium per sulfate and sodium hydroxide. This method inhibited the agglomeration of graphene in polymer matrix during the reduction process. The conductivity of the composites could reach 8.9×10^{-3} S/m when the mass fraction of graphene is 3%. When the mass fraction of graphene increased from 0.3% to 0.7%, the electrical conductivity of the material increased from 1.3×10^{-8} to 2.5×10^{-5} S/m.

(5) Graphene/PMMA Conductive Composites

Zheng et al[35] compared the electrical conductivity of natural graphite with that of graphene, and expanded graphite in PMMA matrix. The results showed that the conductivity of PMMA/expanded graphite composites was better than that of PMMA/natural graphite composites. McAllister et al[36] found that the pressure between the layers was

300 MPa at 300 ℃ when the temperature upto 1000 ℃ and the interlayer pressure reaches 130 MPa. The Hamaker constant is pointed out that the pressure of 2.5 MPa could make two pieces of neighboring graphene oxide separated. The specific surface area of graphene prepared by thermal reduction could reach 600~900 m^2/g.

8.1.2 Graphene Dip Coating Method

Domestic and foreign researchers have made a lot of work on finishing technology of textile materials, including sizing processing, coating processing, etc, and these studies have been relatively mature.

Wang et al[37] conducted fabric coating on nanotubes, such as polyaniline, with oxidative polymerization between the opening carbon nanotubes as a template and polyaniline. Using single factor experiment for the dosage of finishing agent, the surface resistivity and microwave shielding properties of the coated fabrics were analyzed. The results showed that the minimum surface specific resistance of the coated fabric reached 16 Ω, which means the conductivity of the treated fabrics greatly improved. Waveguide method is used to test the microwave electromagnetic shielding effectiveness of the treated fabrics and the SE (shield electromagnetic) value was 48 dB.

Wang Chaoxia from Jiangnan University invented a graphene oxide derivative, which is mainly used for textiles with multi-functional finishing. The fabric was impregnated in the aqueous solution of graphene oxide derivatives, then dried and restored, so that a layer of graphene oxide derivatives was covered on the surface of fabric to obtain multi-functional functions with excellent conductivity, UV resistance and water-repellency. However, strong oxidant and reducing agent introduced in the oxidation-reduction process could destroy the structure of polysulfone amide (PSA), and affect the original excellent properties of PSA[38].

This research put forward a novel method based on the present study, namely graphene dip coating. Graphene is firstly evenly dispersed in adhesive with the action of mechanical force and ultrasonic shaking, and then a graphene size can be obtained. PSA yarn or PSA fabric is impregnated in this graphene size where graphene and adhesion agent mixed in a certain proportion. A layer of graphene dip coating was fixed on the surface of PSA yarn or polysulfone amide fabric to solve the problem of polysulfone amide on antistatic aspect and UV resistance. However, graphene ranges along the surface of them vertically by the padding-drying-ruring process. The diagram of preparation process of functional textile materials by graphene dip coating method is shown in Fig. 8-4.

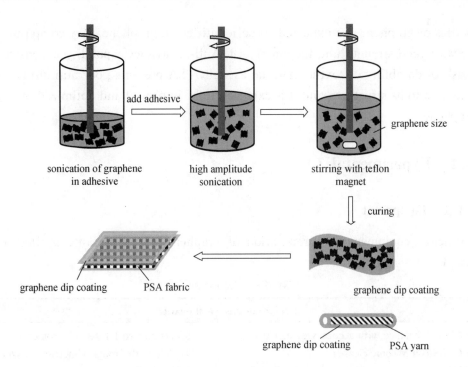

Fig. 8-4　Diagram of preparation process of graphene/PSA composites by graphene dip coating method

8.2　Preparation and Characterization of Graphene/PSA Composite Membrane

The application of graphene in preparing PSA composites can improve and optimize the properties and widen the scope of application of PSA textile material because of the unique and excellent properties of graphene. Usually, the thermal stability of the fiber and fabric refers to the stability of structural morphology and composition under a continuously heating temperature, and it is considered as one of the significant properties of heat-resisting material. PSA fiber is considered as an organic heat-resisting material with excellent heat resistance and thermal stability, the graphene/PSA composite membranes with different mass fractions of graphene are fabricated via the blending of graphene solution. It is supposed that it has certain effects on the following properties of PSA: (1) It has certain effects on the thermal stability of PSA, if the membrane is fabricated by adding graphene into PSA spinning solution due to the excellent thermal properties of graphene; (2) Ordinary PSA material has high specific resistance and exhibits static electricity, while graphene exhibits excellent electrical properties, so it is expected that the conductivity of PSA can be improved and optimized by blending of graphene with PSA; (3) Certain mechanical strength is required for textile material and mechanical properties are the basis of durability of the products, therefore, the superior mechanical properties of graphene are expected to be identified in graphene/PSA composite membrane with good strength and toughness; (4) Anti-ultraviolet capability is considered as the basis of durability of textile material as well, PSA presents poor anti-ultraviolet capability due to its structure, and it is expected to be improved and optimized by addition of graphene.

8.2.1 Experimental

8.2.1.1 Equipment

Experimental equipment for preparation of graphene/PSA composite is illustrated in Tab. 8-3.

Tab. 8-3 Equipment list

Equipment name and model	
FA1104A balance with accuracy of 0.000 1	Spectrum-two FT-IR spectrometer
KQ-700B ultrasonic cleaner	TGA4000 thermogravimetric analyzer
85-2 constant-temperature magnetic stirrer	YG065C electronic strength tester
WD-5000 electric blast drying oven	ZC36 high-insulation resistance meter
S-3400N scanning electron microscope	UT70A general digitalmultimeter
JNOEC XS-213 optical microscope	UV-1000F textile anti-ultraviolet tester
D2 PHASER X-ray diffraction	

8.2.1.2 Preparation of Graphene/PSA Composite Membrane

To improve the dispersion of graphene in DMAC solvent, a certain amount of graphene powders was added into a conical flask filled with DMAC solvent for 60 min ultrasonic treatment. Subsequently, the graphene/PSA composite spinning solutions with different mass fraction of graphene were prepared by adding corresponding amount of PSA spinning solution into the conical flask for mechanical stirring and 120 min ultrasonic stirring. Similarly, the required PSA spinning solution could be achieved by pouring the PSA spinning solution into a certain amount of DMAC solvent for dilution and uniform ultrasonic stirring. Proper amount of PSA spinning solution (after the standing defoaming treatment) and the graphene/PSA composite spinning solution were dripped on the substrate of desktop digital-display spin coater for rotation at a low speed of 1000 r/min for 10 s. Subsequently, the liquid-state membrane with uniform thickness was formed on the substrate by rotating at a high speed of 3000 r/min for 30 s. The fabricated membrane was soaked into the water to extract the solvent and then placed in drying oven at 100 ℃ for 2 h. PSA membrane and graphene/PSA composite membranes (mass fraction of graphene is 0.1%, 0.4%, 0.7% and 1.0%, respectively) were fabricated, as illustrated in Fig. 8-5.

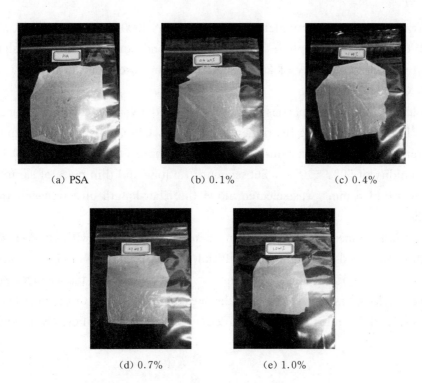

Fig. 8-5　PSA membrane and graphene/PSA composite membranes with different mass fraction of graphene

8.2.1.3　Characterization

Metal spraying by E-1010 ion sputter was performed on the surface of PSA sample with a layer of metal membrane (thickness is 10 nm). The micro topography of graphene and graphene/PSA composite samples were characterized by SEM (S-3400N, Japan) with a resolution ratio of 4 nm, magnification of 300 000 and acceleration voltage of 5~15 kV.

The dispersion and surface morphological structure of graphene sheet in the composite membrane was observed by an optical microscope equipped with CCD digital color camera (JNOEC XS-213, Wuhan Aisipei Scientific Instrument Co., Ltd., China) under the circumstances: 220~240 V, 50/60 Hz, 57 mA, magnification of 40~1000.

Crystallization of the composite membranes was tested by a X-ray diffraction (D2 PHASER, Bruker, Germany) with a radioactive source of CuKα, wavelength of 0.154 056 nm, working voltage of 30 kV and electric current of 10 mA.

Tests on chemical composition and molecular structure of the graphene/PSA composite membrane were performed by spectrum-two FT-IR (Perkin Elmer, US) under the circumstances: attenuated total reflection, scanning times for 16, data interval of 2 cm^{-1}, resolution ratio of 16 cm^{-1}, sample dimension of 1 cm×1 cm.

Non-isothermal decompositions of PSA and the graphene/PSA composite membrane were analyzed by a thermogravimetric analyzer (TGA 4000, Perkin Elmer, US) under a protection of nitrogen atmosphere and the testing conditions are: sample weight of 5~10 mg, nitrogen flow-velocity of 20 mL/min, temperature from 30 to 700 ℃, heating rate of 20 ℃/min.

Tensile mechanical properties were investigated by an electronic strength tester (YG065C, Laizhou Electronic Instrument Co., Ltd., China) after 24 h treatment in the constant-temperature and constant-humidity atmosphere (temperature of 20 ℃ ± 2 ℃, relative humidity of 50% ± 10%). Subsequently, a long and thin piece of sample (length of 8 cm, width of 5 mm) was selected along the fiber-length direction and the sample was weighed.

The arithmetic mean value of the parameters was analyzed with clamping length of 50 mm, drawing speed of 10 mm/min, initial tension of 0.1 cN and force measurement accuracy of 0.01 mm. Each sample is testing for 10 times to obtain the average. A long and thin sample is selected as a fasciculus and the breaking strength of the composite membrane could be calculated according to the surface density of the sample by Eqs. (8-1) and (8-2).

$$\sigma = \frac{m}{S} = \frac{m}{L \times W} \tag{8-1}$$

Where, σ refers to surface density (g/m^2), m refers to sample weight (g), S refers to sample area (m^2), L refers to sample length (m), and W refers to sample width (mm).

$$P = \frac{F}{10 \times \sigma \times W} \tag{8-2}$$

Where, P refers to breaking strength (cN/dtex), F refers to breaking force (cN), σ refers to surface density (g/m^2), and W refers to sample width (m).

Conductivity of the samples was studied by a high-insulation resistance meter (ZC36, Shanghai Anbiao Electronics Co., Ltd., China). The composite membrane should be placed in the constant-temperature and constant-humidity atmosphere for 24 h in advanced to satisfy the requirements, including test voltage of 10~100 V, maximum value of 10^{17} Ω, temperature of test environment of 20 ℃ ± 2 ℃, relative humidity of 35% ± 10%, when the specific resistance was within the range of 10^6~10^{17} Ω. The test was performed by a general digital multimeter (UT70A, Shanghai Yizhu Electric Co., Ltd., measurement range of 2×10^2~2×10^9 Ω, 10 times test for each sample) when the specific resistance of composite membrane is lower than 10^6 Ω. The conductivities of composite materials with different mass fraction were compared on the basis of arithmetic mean value.

Anti-ultraviolet capability of the samples (thickness of 0.001 m, dimension of 5 cm × 5 cm) was investigated by a textile anti-ultraviolet tester (UV-1000F, Labsphere, US) and the testing conditions are: wavelength of 250~400 nm, measurement transmissivity of 0%~100%, absorbency of 0~2.5 A, scanning time ≤ 5 s, data interval of 1 nm, diameter of beam sample of 10 mm.

8.2.2 Results and Discussion

8.2.1.1 Microstructure

SEM image of graphene is shown in Fig. 8-6 and the graphene with multi-layer structure is observed. Graphenes with less-layer structure (2-layer, 4-layer, 5-layer, 6-layer, 8-layer, 10-layer) are shown in Fig. 8-7 and they are selected in this experiment.

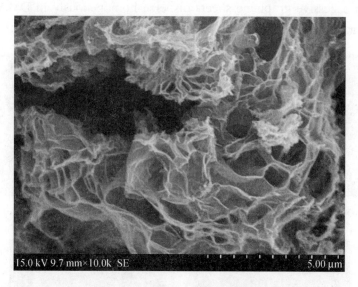

Fig. 8-6　SEM image of graphene

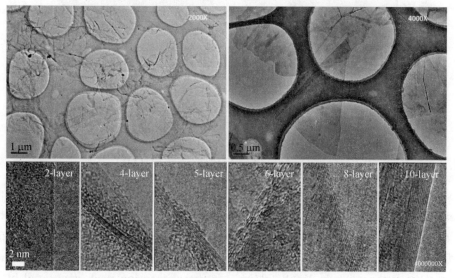

Fig. 8-7　TEM images of graphene. (TEM images are provided by Deyang Olefinic Carbon Science & Technology Co., Ltd.)

Effects of graphene with different mass fraction on microstructure of graphene/PSA composite system are illustrated in Fig. 8-8. Graphene particles are dispersed in PSA matrix uniformly when the mass fraction of graphene is 0.1%. A slight agglomeration of graphene particles in PSA is observed when the mass fraction of graphene is increased to 0.4%. Increasingly serious agglomeration of graphene particles in PSA is obversed as the mass fraction of graphene raised from 0.4% to 1.0% because of the sufficiently strong surface polarity of graphene. The results indicate that a conductive network of graphene can be formed when the mass fraction of graphene is high, and it also reveals that it's difficult for an excessive graphene sheets disperse homogenously in DMAC, which obstructs the formation of smooth membrane.

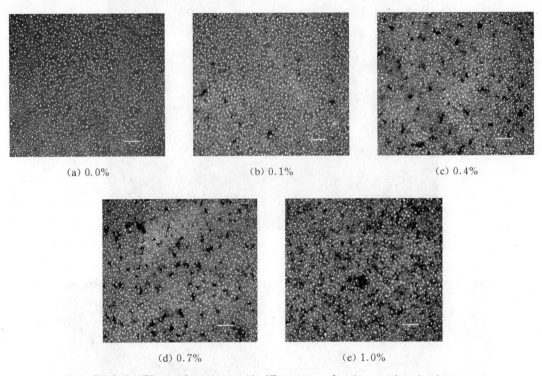

Fig. 8-8　Effects of graphene with different mass fraction on microstructure of graphene/PSA composite system

8.2.1.2　Fourier Transform Infrared Spectrometry (FTIR)

FTIR spectra of graphene/PSA composite membranes with different mass fraction of graphene are illustrated in Fig. 8-9. As can be seen in the figure, the addition of graphene has very little effect on the position and shape of PSA characteristic absorption peak. The absorption peak at 3320 cm^{-1} corresponds to the N—H stretching vibration in PSA; the absorption peak at 1 662.79 cm^{-1} is the absorption peak generated by C=C stretching vibration in PSA; the absorption peak at 1 590.85 cm^{-1} corresponds to C—N stretching vibration; the absorption peaks at 1500~1300 cm^{-1} are mainly caused by

C—H in-plane bending vibration; the absorption peaks at 1300~1000 cm^{-1} and 1000~650 cm^{-1} are correspond to C—C skeletal vibration and C—H out-of-plane bending vibration, respectively, and the absorption peak at 1 148.88 cm^{-1} is the characteristic absorption peak of —SO$_2$— stretching vibration. With the increasing mass fraction of graphene, the absorption peak of pure PSA membrane at 3 321.29 cm^{-1} moves towards the short-wave direction gradually and then the blue shift occurs. That is because the quantum size effect of graphene makes the electronic energy level around the Fermi level change from continuous energy level to discrete energy level, and then the band gap increases. Coulomb interaction takes effect on the electron and hole generated by illumination, and the absorption peak of electron and hole moves towards the short-wave direction since it is tightly bounded in certain space[39]. It can be seen in the figure that the reflectivity of PSA the infrared light in the wavelength range of 4500~1720 cm^{-1} decreases due to the addition of nano-fibers. Furthermore, the reflectivity of composite membrane to the infrared light gradually decreases with the increasing mass fraction of graphene, which indicates that the absorption of composite membrane to the infrared light enhances by adding the graphene into PSA matrix.

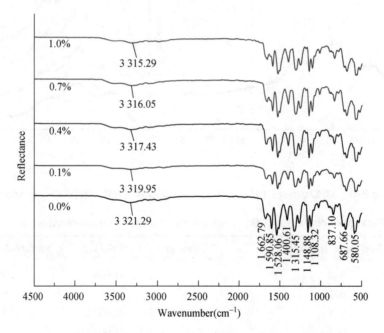

Fig. 8-9 FTIR spectra of graphene/PSA composite membranes with different mass fraction of graphene

8.2.1.3 Crystal Structure and Mechanical Properties

Crystallinity is considered as a significant parameter for characterizing the polymer properties, certain physical and mechanical properties of the polymer are closely linked to its crystallinity. The crystallization has the functions of closely connecting the polymer molecular chain and enhancing the intermolecular force, thus improving its strength and hardness. Generally, the higher the crystallinity is, the larger the crystalline region will be and then leading to a strong strength and hardness of polymer as well as a superior dimensional stability.

XRD curves of graphene and graphene/PSA composite membranes are shown in Fig. 8-10. Tab. 8-2 shows the basic parameters of graphene/PSA composite membranes at different mass fraction of graphene. As illustrated in Fig. 8-10 (a), XRD curve at 26° is the characteristic diffraction peak of graphene. As illustrated in Fig. 8-10 (b), the diffraction peak with high intensity occurs at the right side of 26° and becomes sharper with the increasing mass fraction of graphene. In addition, the graphene sheets are dispersed in the PSA matrix uniformly which may be arising from the heterogeneous nucleation of graphene particles, as a result, the density of crystal nucleus of polymer matrix increases resulting in an improving crystallinity of PSA.

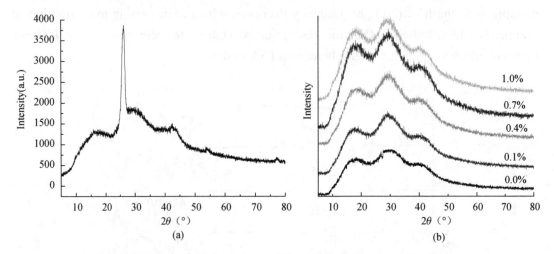

Fig. 8-10 XRD curves: (a) graphene; (b) graphene/PSA composite membranes

Tab. 8-2 Parameters of graphene/PSA composite membranes with different mass fraction of graphene

Mass fraction of graphene(%)	Area(m^2)	Weight(g)	Surface density(g/m^2)
0.0		0.003 2	8.00
0.1		0.003 3	8.25
0.4	0.08×0.005	0.003 2	8.00
0.7		0.002 9	7.25
1.0		0.003 1	7.75

Nucleation barrier of the polymer can be decreased and the crystallinity of the polymer can be improved after the addition of graphene into the polymer because graphene is a low-dimension heterogeneous nuclei, therefore, the mechanical properties of the polymer can be improved correspondingly[40]. The mechanical properties of graphene/PSA composite membranes with different mass fraction of graphene are demonstrated in

Fig. 8-11. It can be observed that the breaking strength of the composite membrane can be enhanced with increasing mass fraction of graphene. The improved breaking strength of composite membrane can be attributed to enhanced crystallinity of the composite, which is in accordance with the conclusion drawn by Xu et al[40] and Rafiee et al[41] that the mechanical properties of the polymer can be increased effectively by adding very few of graphene into the polymer. However, the severe graphene agglomeration can hinder the crystallization of polymer and the improvement rate of breaking strength of the composite membrane decreases (however, its strength is still better than that of pure PSA) when quite a lot of graphene powders are added.

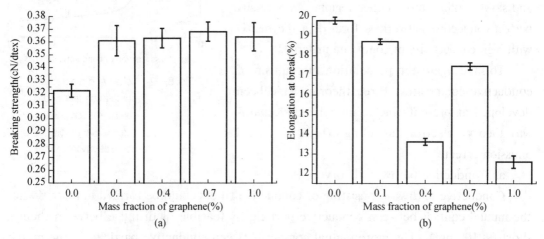

Fig. 8-11　Physical properties of PSA composite membranes with different mass fraction of graphene:
(a) Breaking strength; (b) Elongation at break

8.2.1.4　Conductive Mechanism and Conductive Property

(1) Formation Mechanism of Conductive Network

The dispersion of conductive filler in the matrix has significant effects on formation of conductive network. Usually, only a little additive amount is required for the conductive filler, actual contact between the particles to form the conductive pathway seems to be difficult as there is a long distance between them after being dispersed in the composite material. Therefore, most of research focuses on the dispersion of conductive filler to investigative the theory formation of conductive network. There are quite a lot of factors can affect the dispersion of conductive filler in the matrix, such as compatibility of conductive filler and polymer, and processing conditions. Considerable researches have been done by scholars for the development of dispersion of conductive particle based on the dynamics theories and thermodynamics, and the percolation theory has become the most widely used one. In this case, the formation of conductive pathway in the polymer is investigated and the conductive percolation phenomenon of composite material is firstly studied. Conductive filler has small effect on the improvement of conductivity of polymer-based composites when the mass fraction of conductive filler is low. However, the conductivity of the composite rapidly increases when the mass fraction of conductive filler increases to a certain degree. This phenomenon is called the conductive percolation phenomenon, and the current mass fraction of conductive filler is called the conductive percolation threshold, as illustrated in Fig. 8-12. After passing through the percolation region, the conductivity of composite material changes a little with increasing mass fraction of conductive filler and slowly enters into another region. As a result, perfect conductive network is formed in the matrix with high conductivity of composite material.

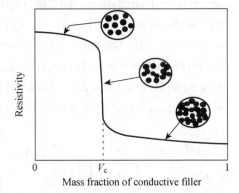

Fig. 8-12 Schematic diagram of percolation theory of conductive composites [42]. (V_c—critical point of conductivity)

To investigate the percolation mechanism of conductive composites, three theories have been developed at present, including conductive pathway theory, electron tunneling effect, and field emission effect.

a. Conductive pathway theory

Conductive pathway consisting of conductive fillers could be formed in matrix when the mutual contact between conductive particles is feasible or distance between them is short (<100 nm). The more mutual contact between conductive particles in the matrix is, the shorter distance between the conductive particles will be, and resulting in an easier formation of conductive pathway (consisting of conductive fillers) in the matrix[43]. It is certainly an oversimplification that the conductivity of composite material is explained on the basis of conductive pathway theory, as there are only very few mutual-contact conductive particles in most of the conductive composites, which makes the formation of conductive pathway seems to be difficult and results in poor conductivity of composite material, thus failing to satisfy specified requirements.

b. Electron tunneling effect

The formation of conductive pathway could be achieved through the mutual contact of a certain amount of conductive particles in the polymer-based conductive composites. However, most of the conductive particles fail to contact directly and the distance between them is short, some of them just be isolated by few polymers whose electrons will stride across the potential barrier to achieve the transition from a particle to adjacent one. Eventually, the tunnel current is formed and the conducting behavior occurs which is called the electron tunneling effect[44]. The tunnel current is generated by the adjacent particles whose distance between each other is approximate 10 nm and it seems impossible for them to form the conductive pathway once the distance becomes longer. In this case, the electron transition is unavailable.

c. Field emission effect

Conducting behavior of conductive fillers is available even if it presents relatively low content under a sufficiently strong applied electric field and high temperature because of the field emission effect. This effect can be explained that electron emission and electron transition to long-distance particles are feasible in the electric field, and the powerful electric field between these particles can induce the field emission (resulting in the current) when the distance is less than 10 nm. Similar to the conductive mechanism of capacitor, the conductive pathway is formed by field emission current.

Usually, the actual conduction behavior of conductive composites is under the combined action of three mechanisms instead of the individual action of a single mechanism.

(2) Formation of Conductive Network of Graphene/PSA Composite Membrane

Conductivity, mass fraction in the matrix and dispersion of graphene are considered as significant parameters affecting the conductivity of conductive composites. By formula derivation and theoretical analysis, Simmoms[45], Sheng et al[46], Ezquerra et al[47] and Van Beek et al[48] proposed that there was certain conductive relationship between two adjacent conductive particles, namely, the distance between the particles shortened gradually with the increasing concentration of conductive filler. Graphene exhibits excellent quantum tunneling effect[5, 11], because the free electrons exist in the six-member ring-like structure of graphene, and then the density of free charge can be increased and the electron tunneling effect can be enhanced. These result in the improvement of conductivity of composite membrane with the increasing mass fraction of graphene in the matrix. The formation diagram of conductive network in graphene/PSA composite system is illustrated in Fig. 8-13.

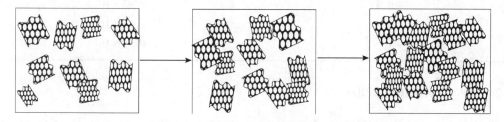

Fig. 8-13 Formation diagram of conductive network in graphene/PSA composite system

(3) Testing of Conductivity of Graphene/PSA Composite Membrane

Surface specific resistance and the magnitude order on the surface resistance of graphene/PSA composite membranes with different mass fraction of graphene are demonstrated in Fig. 8-14. The relevant data are shown in Tab. 8-3. Local conduction resulting from the local contact of particles and tunnel conduction effect is achieved when the mass fraction of graphene is low (0.1%). However, it seems difficult for the particles to achieve the uniform distribution in the matrix, which results in the unavailable complete conductive pathway. The distance between particles decreases as the mass fraction of graphene increased to 0.4%. In this case, tunnel conduction effect becomes more obvious and the local conductive network in composite system expands although the particles in the matrix fail to contact with each other completely. Therefore, the specific resistivity of composite membrane decreases and the conductivity of the composite can be remarkably improved. The conductive network becomes increasingly close and the network structure becomes increasingly prosperous with the increased mass fraction of graphene to 0.7%. In this case, the composite membrane exhibits superior conductivity. However, the improvement rate of conductivity of composite membrane decreases as the further increasing mass fraction of graphene. That is because of the stacking and aggregation of graphene particle with its continuous increasing mass fraction.

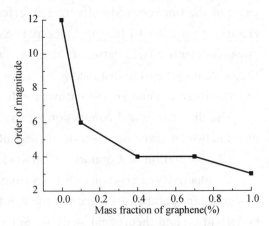

Fig. 8-14 Order of magnitude on surface resistance of graphene/PSA composite membranes with different mass fraction of graphene

Tab. 8-3 Surface specific resistances of graphene/PSA composite membranes with different mass fraction of graphene

Mass fraction of graphene(%)	Average surface specific resistance(Ω)
0.0	3.10×10^{12}
0.1	1.40×10^{6}
0.4	8.95×10^{4}
0.7	2.71×10^{4}
1.0	9.87×10^{3}

8.2.1.5 Thermal Stability

DSC curves of graphene/PSA composite membranes with different mass fraction of graphene are shown in Fig. 8-15. As observed in the figure, the thermal decomposition of graphene/PSA composite membrane could be divided into three intervals: the first stage is the slight weight-loss stage (30~400 ℃); the second stage is the thermal decomposition stage (400~600 ℃); the third stage is the carbon stabilization stage (600~700 ℃).

At the stage of slight weight-loss, the weight loss is mainly caused by the volatilizations of bound moisture among polymer molecules and diverse auxiliaries, such as the DMAC solvent, when the temperature rises from 30 to 100 ℃. The TG curve decline sharply when the temperature is increased to 170 ℃, and then become smooth and enters to a certain platform as the temperature further raise to 400 ℃. The reason for the weight loss of the composite can be attributed to the decomposition of small-molecular-weight oligomer. During the preparation of graphene by expansion and stripping of intercalation, potassium permanganate is used as the oxidant and the concentrated sulfuric acid is used as the intercalation agent. However, the thermal stability of these oxygen-containing groups, such as —COOH, —C=O and —OH, is poor at high temperature, and the gases, such as CO_2 and CO, would be generated when they are heated. These will lead to a weight loss of graphene under the heating atmosphere.

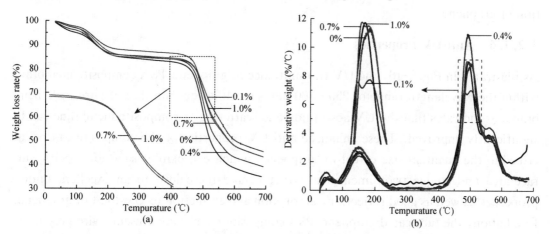

Fig. 8-15 DSC curves of graphene/PSA composite membranes with different mass fraction of graphene: (a) TG; (b) DTG

Tab. 8-4 Parameters of graphene/PSA composite membranes with different mass fraction of graphene during thermal decomposition

Mass fraction of graphene(%)	T_0(℃)	$d\alpha/dt$	T_{max}(℃)	α(%)
0.0	415.32	9.014	493.13	35.18
0.1	417.58	7.015	493.76	37.83
0.4	418.82	10.945	494.87	39.30
0.7	419.47	11.034	494.23	43.45
1.0	421.04	9.817	494.18	43.51

T_0—The initial decomposition temperature;
$d\alpha/dt$—The maximum decomposition rate;
T_{max}—The temperature corresponding to the maximum decomposition rate;
α—The residual weight rate at 700 ℃.

At the stage of thermal decomposition, the weight loss is mainly caused by the gradually increased movement velocity, which can break the macromolecular chain of PSA with the small molecule being released in the form of gas, and that results in the weight loss of the composite. The decomposition of PSA at 500~600 ℃ with nitrogen atmospheremainly occurs at the C—N part of acylamino. Based on bond energy analysis and PSA structural formula, it can be concluded that the weight loss at this stage can be attributed to the increasing generation of gases, such as SO_2, NH_3 or CO_2. As shown in Tab.8-4, the initial decomposition temperature of PSA increases as the raising mass fraction of graphene, which indicates that the graphene can postpone the decomposition time of the composite membrane.

At the stage of carbon stabilization, the composite membrane exhibits superior residual rate than that of pure PSA membrane when the temperature reaches about 700 ℃, and it indicates that the heat resistance of PSA can be remarkably increased by the addition of graphene.

8.2.1.6 Anti-UV Properties

As illustrated in Fig.8-16, the UV transmittance of graphene/PSA composite membrane within the wavelength range of 250~400 nm is much lower than that of pure PSA membrane. It indicates that the UV absorption and scattering of composite membrane can be significantly improved. These enhanced anti-UV properties of the composite can be ascribed to the quantum size effect of graphene, which can exhibit superior specific surface area and unique two-dimensional structure and these lead to an excellent optical properties as well as good reflection, absorption and extinction to UV-light of graphene. In addition, the fabricated graphene/PSA composite membrane presents with grey-black color, and the color becomes darker with the increasing mass fraction of graphene. The results indicate the graphene plays an important role in enhancing the anti-UV properties of graphene/PSA composite membrane and relieving the aging properties resulting from the UV-light irradiation.

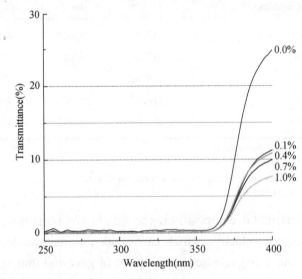

Fig.8-16 UV transmittance of graphene/PSA composite membranes with different mass fraction of graphene

8.2.3 Summary

In this section, graphene/PSA composite membranes with different mass fraction of graphene from 0.0% to 1.0% were prepared by using a desktop digital-display spin coater. The experimental results indicate that the graphene sheets with a low mass fraction (0.0%~0.4%) could be uniformly dispersed in the PSA matrix, however, agglomeration of graphene was observed in the composite membrane with the increasing mass fraction of the graphene from 0.4% to 0.7%. Crystallization properties and breaking strength of the pure PSA can be improved after the addition of graphene, and the thermal stability of graphene/PSA composite membrane enhances with the increasing mass fraction of the graphene. It is also discovered that there is no significant influence of the graphene on the molecular structure and chemical composition of PSA. The conductivity of graphene/PSA composite membrane can be remarkably increased after the addition of graphene. However, the improvement degree of conductivity of the composite decreases as the mass fraction of graphene increased from 0.1% to 0.7% because of the severer agglomeration of graphene. The graphene/PSA composite membrane exhibits significantly superior anti-UV properties than that of pure PSA membrane and it presents excellent absorption and scattering to the UV-light.

8.3 Effects of Plasma Treatment on Properties of Graphene/PSA Composite Membrane

The moisture absorption properties and the conductivity of graphene/PSA composite membrane were improved by plasma treatment. The effects of plasma treatment parameters, such as the discharge power, treatment time and argon flow, on surface morphology, wettability (tests of wicking and contact angle), and conductivity of the graphene/PSA composite membrane are investigated. In addition, the optimum plasma treatment parameters are used to treat the graphene/PSA composite membranes with different mass fraction of graphene including 0.1%, 0.4%, 0.7% and 1.0%, and the properties of the graphene/PSA composite membranes with different mass fraction of graphene were studied under the action of plasma.

8.3.1 Experimental

8.3.1.1 Experimental Materials

The experimental materials include pure PSA membrane, graphene/PSA composite membranes with different mass fraction of graphene (0.1%, 0.4%, 0.7% and 1.0%), argon (purity ⩾ 99.999%, provided by Shanghai Jiulong Industrial Gases Co., Ltd., China), dyeing agent and deionized water.

8.3.1.2 Characterization

The micro-morphology of sample was investigated by using a scanning electron microscope(SEM) (S-3400N, Hitachi, Japan, resolution ratio = 4 nm, magnification = 300 000, acceleration voltage = 5 kV~15 kV) in the vacuum environment.

Wettability testing was carried out by using an optical contact angle measurement Instrument (OCA15EC, Dataphysics, Germany, measurement range of contact angle = 0~180°, measurement accuracy = ±0.1°). Before the testing, the samples (dimension = 2 cm × 2 cm) should be placed in a constant-temperature and constant-humidity environment for 24 h (temperature = 20 ℃ ± 2 ℃, relative humidity = 50% ± 10%).

8.3.1.3 Plasma Treatment

The samples with a dimension of 5 cm × 5 cm were treated by plasma under different parameters by using full-automatic plasma chemical vapor deposition equipment (PECVD-601, Beijing Chuangshiweina Science & Technology Co., Ltd., China). The diagram of plasma treatment device is illustrated in Fig. 8-17. Prior to the treatment, the samples were fixed on the lower electrode by a dedicated metal bar in the vacuum chamber. The base pressure of the vacuum chamber is 5×10^{-4} Pa and then the argon was connected with the chamber to keep the constant pressure in the deposition chamber at 15 Pa.

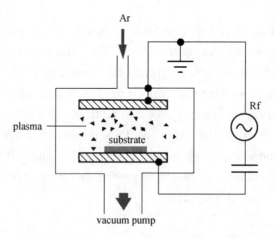

Fig. 8-17 Diagram of plasma treatment device. (Rf—power supply)

Different plasma treatment parameters for graphene/PSA composite membrane with mass fraction of graphene of 0.7% and plasma treatment parameters for graphene/PSA composite membranes with different mass fraction of graphene are listed in

Tab. 8-5 and Tab. 8-6 respectively.

Tab. 8-5 Plasma treatment parameters for graphene/PSA composite membrane with mass fraction of graphene of 0.7%

Argon flow rate (mL/min)	50	50	10, 20, 30, 40, 50
Discharge power (W)	300	50, 100, 200, 300, 400	300
Treatment time (min)	1, 5, 10, 15, 20	5	5

The experiment was conducted at following conditions: background vacuum is 5×10^{-4} Pa, substrate temperature is room temperature, constant pressure is 15 Pa.

Tab. 8-6 Plasma treatment on pure PSA membrane and graphene/PSA composite membranes with different mass fraction of graphene

No.	Sample	Plasma treatment parameters	
1	Pure PSA membrane	Technology parameter	Value
2	graphene/PSA composite membrane with mass fraction of graphene of 0.1%	Background vacuum (Pa)	5×10^{-4}
3	graphene/PSA composite membrane with mass fraction of graphene of 0.4%	Substrate Temperature (℃)	Room temperature
		Argon flow rate(mL/min)	100
4	graphene/PSA composite membrane with mass fraction of graphene of 0.7%	Constant pressure (Pa)	15
		Discharge power (W)	300
5	graphene/PSA composite membrane with mass fraction of graphene of 1.0%	Treatment time (min)	5

8.3.2 Results and Discussion

8.3.2.1 Morphological Structure

SEM images of graphene/PSA composite membrane (mass fraction of graphene being 0.7%) before and after plasma treatment are shown in Fig. 8-18, while the plasma treatment parameters are shown in Tab. 8-6. As observed in Fig. 8-18 (a), the surface of the composite membrane is smooth before plasma treatment. Obviously, lots of dense holes are discovered in the membrane after plasma treatment and the surface becomes rough as illustrated in Fig. 8-18 (b).

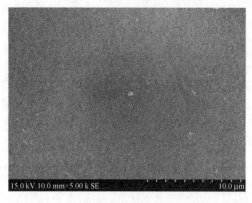 

(a) Before plasma treatment　　　　　　　　(b) After plasma treatment

Fig. 8-18　SEM images of composite membrane before and after plasma treatment

8.3.2.2　Moisture Absorption and Wicking Properties

(1) Moisture Absorption Properties

The wicking height measurements which used to evaluate the moisture absorption properties of the samples (sample dimension of 0.5 cm × 5 cm, testing temperature at 20 ℃, testing time for 1 h) before and after plasma treatment were conducted by using a self-developed device.

(2) Wicking Properties

As can be observed in Fig. 8-19, the wicking properties of graphene/PSA composite membrane obviously change after plasma treatment. Compared with pure PSA membrane, the moisture absorption properties of the composite membrane are superior because of the increased water channels in the composite membrane. No significant difference is observed in the moisture absorption properties of the composite membrane with the increasing mass fraction of graphene as shown in Fig. 8-19 (a). It can be observed in Fig. 8-19 (b) that the moisture absorption properties of composite membrane enhance

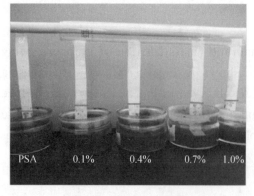 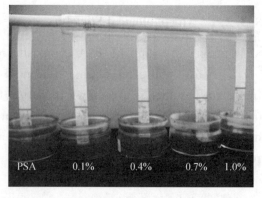

(a) Before plasma treatment　　　　　　　　(b) After plasma treatment

Fig. 8-19　Moisture absorptions of composite membranes before and after plasma treatment with the increasing mass fraction of graphene because of the increased dense holes in the composite membrane as well as the increased micro-space around the graphene in the PSA membrane. Although the structural integrity of PSA membrane is destroyed because of the graphene agglomeration resulting from high mass fraction of graphene, it indicates that micro-gap exists among the graphene agglomeration, which indicates enhanced moisture absorption of the composite membrane. In this case, the anti-static properties of the composite membrane can be also improved.

8.3.2.3 Wettability and Contact Angle

(1) Theoretical Model of Contact Angle (CA)

The wettability, which is generated during the process of the solid interface changing from the solid-gas interface into the solid-liquid interface, depends on the chemical composition and microstructure of solid and it is considered as one of the significant chemical properties of the solid material. The contact angle (θ) is defined as the angle between the solid surface and a tangent drawn on the drop-surface, passing through the triple-point atmosphere-liquid-solid. The contact angle can be used to exhibit the wettability of the solid material. The greater surface energy of the solid is, the smaller contact angle will be. The small contact angle indicates a good wettability of the solid material. Generally, the solid material is classified into the hydrophilic material when the contact angle is smaller than 90°, while it is classified into the hydrophobic material when the contact angle is greater than 90°[49-50]. Models of contact angle on material surface mainly consist of theoretical models, such as Young, Wenzel, Cassie and Cassie-Baxter.

(2) Young's Equation[51]

Very little liquid may be changed into droplets after being dripped on the solid surface. Furthermore, the liquid on the solid surface maintains a certain liquid shape when it is in a balanced state, as demonstrated in Fig. 8-20.

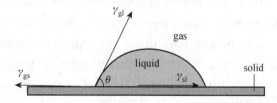

Fig. 8-20 Young's contact angle and mechanical relations of liquid on solid surface

The droplet maintaining certain shape on the solid surface is the balanced effect of three interface tensions, which is called Young's equation as shown in Eq. (8-3).

$$\gamma_{gs} = \gamma_{gl} \cos \theta + \gamma_{ls} \tag{8-3}$$

The contact angle in the Young's equation is a contact angle under the circumstances of smooth solid surface, chemical homogeneity, rigidity, isotropy, no chemical reaction, and negligible line tension of three-phase contact line[52]. This formula is suitable only for droplet contacting with smooth surface rather than rough surface.

(3) Wenzel Model[53]

By the law of thermodynamics, it is deduced that the relationship between the apparent contact angle (θ_r) and the Eigen contact angle (θ) is shown in Eq. (8-4).

$$\cos \theta_r = r \cos \theta \qquad (8-4)$$

Where, θ refers to the Eigen contact angle arising from the droplet contacting with smooth surface, r refers to the surface roughness (i.e., ratio of contact area and value of vertical projection, generally, it is higher than 1), θ_r refers to the apparent contact angle resulting from the droplet contacting with rough surface.

The Wenzel model is shown in Fig. 8-21. It can be concluded from Eq. (8-2) and Fig. 8-21 that the hydrophilicity or the hydrophobicity of the solid material can be enhanced by increasing the surface roughness of the solid material.

(4) Cassie-Baxter Model[54-55]

It is possible for the droplet to stay on the pores between the micro-rough chain and the micro-structure rather than penetrate into the interior of rough structure because of the surface tension when the droplet contacts with the rough surface. In this case, a compounded contact model is

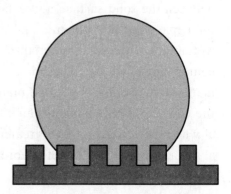

Fig. 8-21　Schematic diagram of contact angle of Wenzel model

proposed by Cassie and Baxter on the basis of Wenzel model as shown in Fig. 8-22. Supposing that the ratio of area arising from droplet contacting with solid surface and total area is f_1 and the ratio of area arising from droplet contacting with pore and total area is f_2, and $f_1 + f_2 = 1$. In addition, the contact angle of droplet and smooth hydrophobic solid surface is named as θ and the contact angle of droplet and gas is named as θ_g. The contact angle arising from the droplet contacting with the surface is linked to proportion of two elements (solid and gas), therefore, the Cassie-Baxter is expressed as follows.

$$\cos \theta_c = f_1 \cos \theta + f_2 \cos \theta_g \qquad (8-5)$$

The contact angle θ_g of droplet is equal to 180° since it is ball-like in the air. The above formula can be simplified as follows since $\cos \theta_g = -1$:

$$\cos\theta_c = f_1(\cos\theta+1)-1 \qquad (8\text{-}6)$$

The value of $\cos\theta_c$ becomes more close to -1 when the contact area of solid surface and droplet becomes smaller (i.e., f_1 comes to be lower) under the condition of unchanged contact area of droplet and solid surface. In this case, the contact angle on the solid surface is more close to 180° and the droplet is still ball-like, namely, the hydrophobicity of hydrophobic solid surface can strengthen effectively by deducting the direct contact area of solid surface and droplet as well as increasing the pores.

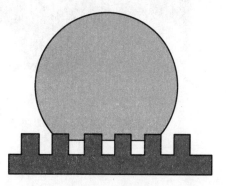

Fig. 8-22 Schematic diagram of contact angle of Cassie-Baxter model

(5) Test Results of Wettability

The effects of plasma treatment with different time on the contact angle of composite membrane are investigated and the results are shown in Fig. 8-23. The relationship between the contact angle and the treatment time is illustrated in Fig. 8-24. During the treatment, the discharge power and argon flow were 300 W and 50 mL/min, respectively. As demonstrated in Fig. 8-24, the contact angle decreased with the treatment time increased from 1 to 5 min, and then increased with the raising time from 5 to 15 min and decreased with the further increasing time from 15 to 20 min. The minimum contact angle is observed when the sample was treated for 5 min, which indicates that the sample can be easily wetted by deionized water and the surface energy of the sample is large. That is because the continuous coverage of PSA on the membrane surface was destroyed by plasma treatment, and the active group on the membrane surface increased when the treatment time is increased to 5 min. Fluctuant change of contact angle is observed with the treatment time increased from 5 to 20 min because of the saturation on the surface of composite membrane which is generated by a long plasma treatment time[56].

1 min

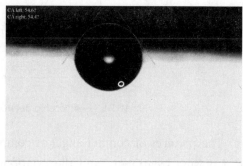

5 min

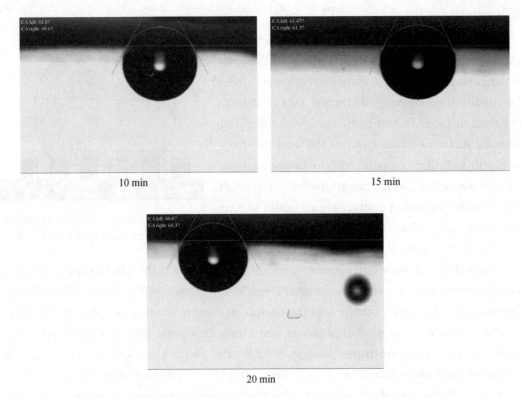

Fig. 8-23 Contact angle of samples treated by plasma with different treatment time

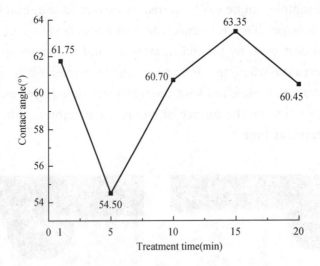

Fig. 8-24 Relationship between contact angle and treatment time

The pictures of contact angel of composite membrane influencing by discharge power of plasma treatment, and the relationship between the contact angle and discharge power are demonstrated in Fig. 8-25 and Fig. 8-26, respectively. During the plasma treatment, the treatment time was set to be 5 min and argon flow rate was set to be

50 mL/min. It can be observed that the contact angle of the composite membrane increased as the discharge power raised from 50 to 100 W, and then decreased with the further increased power from 100 to 300 W. The minimum contact angle is discovered when the discharge power is 300 W. The significantly decreasing contact angle of the composite membrane from 61.80° to 54.50° indicates an improved wettability[57] and which can be attributed to the increasing destroying continuous layer of PSA on the membrane surface, which is because the density of high-energy particle of plasma in the chamber increased with the raising power.

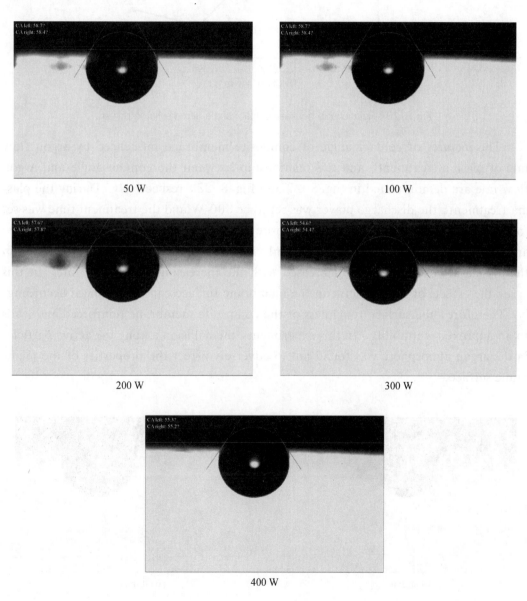

Fig. 8-25 Contact angle of samples treated by plasma with different discharge power

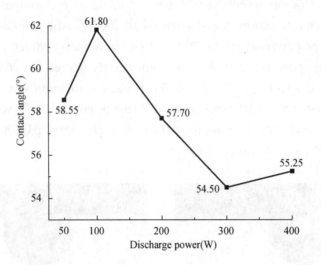

Fig. 8-26 Relationship between contact angle and discharge power

The pictures of contact angle of composite membrane influenced by argon flow rate of plasma treatment, and the relationship between the contact angle and argon flow rate are demonstrated in Fig. 8-27 and Fig. 8-28, respectively. During the plasma treatment, the discharge power was set to be 300 W and the treatment time was set to be 5 min. The contact angle of composite membrane gradually decreased when the argon flow rate increased from 10 to 50 mL/min. The concentration of gas particles in the plasma atmosphere can be increased with the increasing argon flow rate. In this case, the effects of the particles on the membrane surface can be enhanced accordingly. Therefore, the surface roughness of the composite membrane reinforced and leads to an improved wettability. In this experiment, the collision among the active particles in the argon atmosphere was found not to adversely affect the properties of the membrane surface.

10 mL/min

20 mL/min

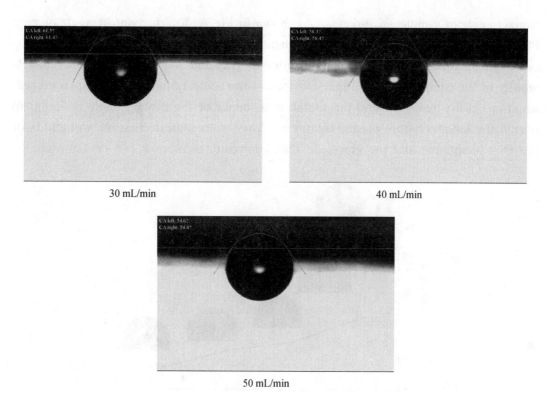

Fig. 8-27 Contact angle of samples treated by plasma with different argon flow rate

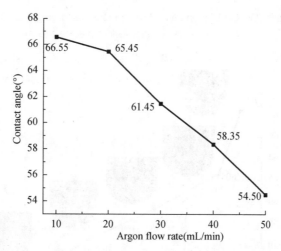

Fig. 8-28 Relationship between contact angle and argon flow rate

The pictures of contact angle of the samples, including PSA membrane and graphene/PSA composite membranes with different mass fraction of graphene (0.1%, 0.4%, 0.7%, 1.0%), before and after plasma treatment are demonstrated in Fig. 8-29 and Fig. 8-30, respectively. During the treatment, the discharge power was set to be 300 W, treatment time was set to be 5 min, and the argon flow rate was set to be 50 mL/min. As illustrated in Fig. 8-29, forward treatment was carried out since the contact angle of sample was greater. As shown in Fig. 8-30, the contact angle of pure PSA membrane is larger than that of graphene/PSA composite membrane, and the contact angle decreases with raising mass fraction of graphene suggesting an improved wettability of the composite membrane. Obviously, the contact angle of all samples decreases after plasma treatment and the variation tendency of the contact angle is similar to that of the samples before plasma treatment. The results indicate that the wettability of the PSA membrane and the graphene/PSA composite membrane can be improved by

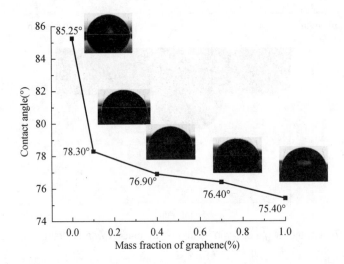

Fig. 8-29 Contact angle of PSA membrane and graphene/PSA composite membranes before plasma treatment

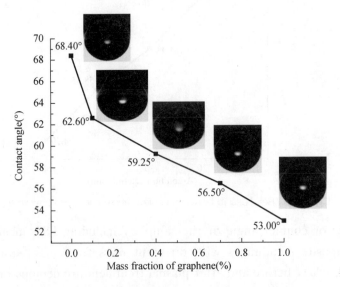

Fig. 8-30 Contact angle of PSA membrane and graphene/PSA composite membranes after plasma treatment

plasma treatment. The gap between graphene sheet increases with more graphene stacking, and hence the capability of water conservation improves. It can also be concluded that the etching action of the plasma help moisture penetration and binding on the surface of membrane, which contributed to the wettability on the membrane surface. The increased capacity of the membrane surface to hold moisture will lead to a more moisture-rich surface and will also make the conductivity of the membrane enhanced and make surface resistivity decreased[58], so the goal of improving and optimizing the static electricity of PSA can be realized.

8.3.3 Summary

In this section, the PSA membrane and graphene/PSA composite membranes with different mass fractions of graphene were treated by plasma chemical vapor deposition with different conditions, including treatment time, discharge power and argon flow rate.

The experimental results indicate that the wicking properties of graphene/PSA composite membrane before plasma treatment can be improved by the addition of graphene.

The smooth surface of graphene/PSA composite membrane becomes rough with quite a lot of micro-pore structures after argon plasma treatment. The moisture absorption properties of the composite membrane can be improved after argon plasma treatment and the moisture absorption properties enhanced with the increasing mass fraction of graphene from 0.1% to 1.0%.

The contact angle of the sample decreases after plasma treatment indicating an enhanced wettability. The effects of different plasma parameters on the contact angle of composite membrane can be summarized as follows:

(1) The contact angle of composite membrane decreased firstly as the treatment time increased from 1 to 5 min, and then increased as the time raised from 5 to 15 min, and decreased again as the further increased time from 15 to 20 min. The change of the contact angle of the composite membrane is obviously fluctuant.

(2) The contact angle of composite membrane increased firstly with the discharge power raised from 50 to 100 W and then decreased as the power increased from 100 to 300 W, and the contact angle slightly increased with the further increased power from 300 to 400 W.

(3) The contact angle of the composite membrane decreased significantly with the increased argon flow rate from 10 to 50 mL/min. The decreasing contact angle of the composite membrane indicates an improved wettability because rough surface of the membrane can be generated after plasma treatment, and the rough surface can lead to an increased surface energy of the composite membranes.

8.4 Preparation and Characterization of Graphene/PSA Yarn

8.4.1 Experimental

8.4.1.1 Experimental Materials

The experimental materials include PSA double-ply yarn (yarn fineness of 115 dex, Shanghai Tanlon Fiber Co., Ltd., China), graphene powders (Sichuan Deyang Ene Carbon Technology Co., Ltd., China) and acrylic adhesion agent (Shanghai Tuebingen Chemical Factory Co., Ltd., China).

8.4.1.2 Preparation of Graphene/PSA Yarn

Graphene/PSA yarn was prepared by impregnating PSA double-ply yarn into graphene sizing, and the preparation process is illustrated in Fig. 8-31. The graphene sizing was firstly fabricated by adding a certain amount of graphene powders into adhesion agent. The dispersion uniformity of graphene powders in the adhesion agent was improved by magnetic stirring for 2 h and ultrasonic blending for 2 h. Subsequently, the graphene sizing was poured into a self-developed sizing device, and then the padding treatment by yarn guide roller was carried out. The graphene/PSA yarn was obtained after drying and shaping treatment. The effects of the concentration of graphene sizing on the morphological structure, molecular structure and chemical composition, thermal stability, conductivity and mechanical properties of the graphene/PSA fabric were systematically investigated.

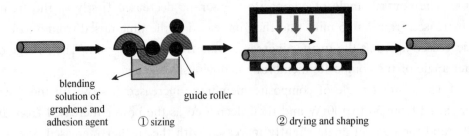

Fig. 8-31 Preparation process of graphene/PSA yarn

8.4.2 Results and Discussion

8.4.2.1 Morphological Structure

The adhesion effect of graphene on the surface of PSA yarn is demonstrated in

Fig. 8-32. The graphene sheets are uniformly dispersed in the adhesive uniformly and attached to the PSA surface when the mass fraction of graphene is set to be 1.5%. Graphene sheets on the PSA surface gradually connect with each other and the effective conductive pathway between graphene sheets increases with the increasing graphene concentration. As a result, the resistance on the surface of PSA yarn deducts and the conductivity enhances. However, agglomeration of graphene is observed when its mass fraction is increased to 2.0%.

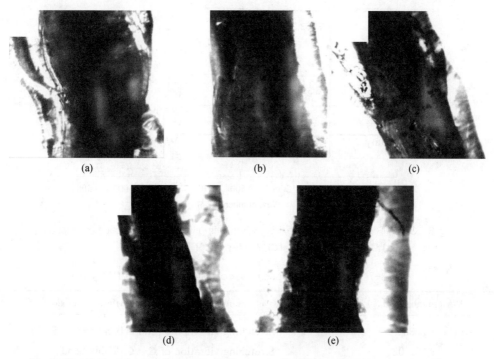

Fig. 8-32 Optical image of graphene/PSA yarns with different mass fraction of graphene:
(a) 0.0 %; (b) 1.5%; (c) 2.0%; (d) 2.5%; (e) 3.0%

8.4.2.2 Fourier Transform Infrared Spectroscopy (FTIR)

FTIR curves of graphene/PSA yarns with different mass fraction of graphene are illustrated in Fig. 8-33. It can be clearly observed that the addition of graphene has obvious effects on the change of position and shape of characteristic absorption peak of PSA yarn. The details of the absorption peaks of PSA and graphene/PSA yarn are shown in Tab. 8-7. The characteristic absorption peak of PSA at 3 339.12 cm^{-1} gradually moves towards the short-wave direction called the blue shift with the increasing content of graphene. That is because of the quantum size of graphene, and the electronic energy level around the Fermi level of the particle changes from the continuous energy level into the discrete energy level. In this case, the band gap widens, and the absorption peak of electron-hole pair moves towards the short-wave direction because of the strong space bound. As can be seen in Fig. 8-33, the infrared reflectivity at 4500~1720 cm^{-1} of PSA decreases after the addition of graphene and the value reduces with the increasing mass fraction of graphene. This indicates the infrared absorption of PSA yarn enhanced by attached graphene.

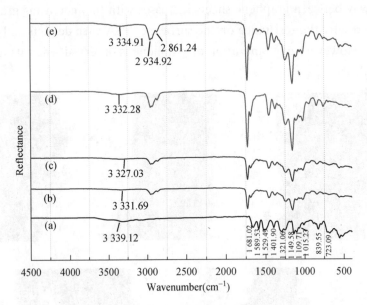

Fig. 8-33 FTIR curve of graphene/PSA yarns with different mass fraction of graphene:
(a) 0.0%; (b) 1.5%; (c) 2.0%; (d) 2.5%; (e) 3.0%

Tab. 8-7 Details of absorption peak of PSA and graphene/PSA yarn

Wavenumber(cm^{-1})	Functional group and the vibration mode
3330	Stretching vibration of amide bond (—N—H—)
1 661.02	Stretching vibration of C=C double bond
1 589.53	Stretching vibration of C—N
1500~1300	In-plane bending vibration of C—H
1300~1000	Skeletal vibration of C—C
1000~650	Out-of-plane bending vibration of C—H
1 149.58	Stretching vibration of —SO$_2$—

8.4.2.3 Thermal Stability

DSC curves of graphene/PSA yarn with different mass fractions of graphene are shown in Fig. 8-34. As observed in the figure, the thermal decomposition of graphene/PSA yarn could be divided into three stages, including the slight weight-loss stage (30~400 ℃), the thermal decomposition stage (400~600 ℃) and the carbon stabilization stage (600~700 ℃). The thermal decomposition behavior of graphene/PSA yarn from room temperature to 700 ℃ is similar to that of graphene/PSA composite membranes discussed in Section 8.3.1.5. The thermal decomposition data of graphene/PSA yarn are shown in Tab. 8-8. As shown in this table, the initial decomposition temperature and the temperature corresponding to the maximum decomposition rate of PSA yarn are postponed after the impregnation of graphene with a relatively high mass fractions. In addition, the residual rate of the sample at 700 ℃ increases significantly after the impregnation of graphene with mass fraction of 1.5%~3.0%.

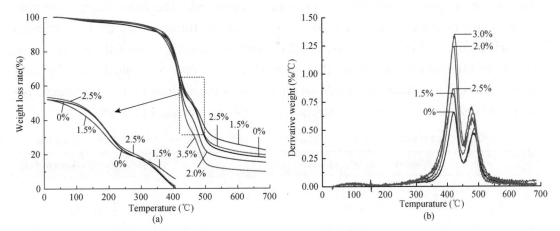

Fig. 8-34 DSC curves of graphene/PSA yarn with different mass fraction of graphene: (a) TG; (b) DTG

Tab. 8-8 Basic parameters of graphene/PSA yarn during thermal decomposition

Mass fraction of graphene(%)	T_0(℃)	$d\alpha/dt$	T_{max}(℃)	α(%)
0.0	402.72	0.702	435.08	17.98
1.5	399.53	8.237	432.65	22.24
2.0	405.69	1.269	431.02	23.09
2.5	406.03	0.894	436.80	23.95
3.0	406.50	1.347	442.47	25.46

T_0—The initial decomposition temperature;
$d\alpha/dt$—The maximum decomposition rate;
T_{max}—The temperature corresponding to the maximum decomposition rate;
α—The residual weight rate at 700 ℃.

8.4.2.4 Conductivity

Graphene as a new type of conductive material has an ideal quantum tunneling effect and excellent conductivity, and the conductivity of single-layer graphene electron can reach 200 000 cm^2/(V·s)[59]. Therefore, the conductivity of PSA yarn can be reinforced by the addition of graphene. The surface specific resistance of graphene/PSA yarns with different mass fraction of graphene is demonstrated in Tab. 8-9. It is can be observed that it seems to be impossible to form complete conductive pathway due to the local conduction arising from the local contact of particles and tunnel conduction effect when the mass fraction of graphene is extremely low (1.5%). The distance among the particles in the matrix becomes to decrease when the mass fraction of graphene is increased to 2.0%, and then the tunnel conduction effect enhances and the local conductive network in the composite system expands although the particles fail to contact with each other completely. In this case, the resistance of graphene/PSA yarn can be remarkably decreased and the conductivity of the yarn can be enhanced. The increasingly close conductive network and prosperous network structure can be formed when the mass fraction of graphene is raised to 2.5% because of the increasing directly contacted particles in the matrix. In this case, the graphene/PSA yarn can exhibit superior conductivity. The conductive network is further improved and optimized as the agglomeration of graphene becomes to be severer with the mass fraction of graphene increased to 3.0%.

Tab. 8-9 Surface specific resistance of graphene/PSA yarn with different mass fraction of graphene

Mass fraction of graphene(%)	Surface specific resistance(Ω)
0.0	3.28×10^{12}
1.5	1.09×10^{8}
2.0	9.17×10^{7}
2.5	2.58×10^{7}
3.0	8.87×10^{6}

8.4.2.5 Mechanical Properties

The tensile properties of the graphene/PSA yarns were investigated by an electronic tensile strength tester (YG065C, Laizhou Electronic Instrument Co., Ltd., China). Prior to the testing, the samples were placed at an environment with constant-temperature of (20 ± 2) ℃ and constant-humidity of $(50 \pm 10)\%$ for 24 h. All of the samples were measured for 10 times and the average would be used for analysis. The other testing conditions are shown in Tab. 8-10.

Tab. 8-10 Testing condition of tensile properties of graphene/PSA yarn

Item	Testing condition
Clamping length	50 mm
Drawing speed	10 mm/min
Initial tension	0.1 cN
Force measurement accuracy	0.01 cN
Elongation measurement accuracy	0.01 mm

The average breaking strengths and elongations at break of graphene/PSA yarns with different mass fraction of graphene are illustrated in Fig. 8-35. The relevant data are shown in Tab. 8-11. As depicted in Fig. 8-35, the breaking strength of graphene/

PSA yarn gradually increases with the increasing mass fraction of graphene from 1.5% to 2.0% and then decreases as the further increased mass fraction of graphene form 2.0% to 3.0%. The decreasing breaking strength of graphene/PSA yarn can be attributed to the agglomeration of graphene.

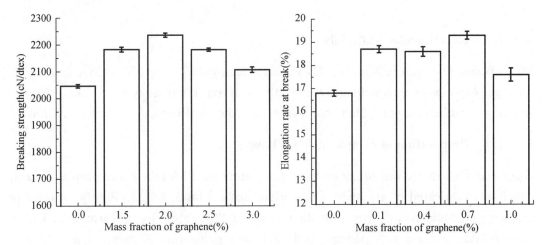

Fig. 8-35 Breaking strength and elongation rate at break of graphene/PSA yarns with different mass fraction of graphene

Tab. 8-11 Mechanical properties of graphene/PSA yarns with different mass fraction of graphene

Mass fraction of graphene(%)	Average breaking strength (cN)	Average elongation at break (mm)	Average elongation rate at break (%)
0.0	2 045.9	37.8	16.8
1.5	2 182.4	40.9	18.7
2.0	2 236.0	40.4	18.6
2.5	2 181.9	42.6	19.3
3.0	2 107.3	39.5	17.6

8.4.3 Summary

Graphene/PSA yarns with different mass fraction of graphene are fabricated by impregnating the PSA double-ply yarn into the graphene sizing. The results show that the graphene sheets with a low mass fraction of 1.5% can be uniformly dispersed in the graphene sizing, while agglomeration of graphene is observed when the mass fraction of graphene is high. The impregnation of graphene has no significant effects on the molecular structure and chemical composition of PSA. The thermal stability and mechanical properties of graphene/PSA yarns can be improved when compared with pure PSA yarns.

8.5 Preparation and Characterization of Graphene/PSA Fabric

8.5.1 Experimental

8.5.1.1 Experimental Materials

The experimental materials include PSA fabric (Shanghai Tanlon Fiber Co., Ltd., China), graphene powders (Sichuan Deyang Ene Carbon Technology Co., Ltd., China) and acrylic adhesion agent (Shanghai Tuebingen Chemical Factory Co., Ltd., China).

8.5.1.2 Preparation of Graphene/PSA Fabric

Graphene/PSA fabric was prepared by impregnating the PSA fabric into graphene sizing which was fabricated by the method described in Section 8.4.1.2. Firstly, a layer of graphene membrane was formed on the surface of PSA fabric by performing a double-impregnation and a double-padding finishing with sizing rate of 80%. Then the graphene/PSA fabric was obtained after drying in oven at 80 ℃ for 3 minutes followed by baking at 120 ℃ for 1 minute under atmosphere. The effects of the concentration of graphene sizing on the morphological structure, conductivity, thermal stability, mechanical properties and anti-UV properties of the graphene/PSA fabric were systematically investigated.

8.5.1.3 Characterization

The morphological structure of PSA fabric coated with graphene membrane was characterized by a scanning electron microscope (S-3400N, Hitachi, Japan) and an optical microscope. Conductivity of PSA fabric coated with graphene membrane was studied by a high-insulation resistance meter (ZC36, Shanghai Anbiao Electronics Co., Ltd., China). The non-isothermal decomposition of the graphene/PSA fabric was investigated by a thermogravimetric analyzer (TGA4000, US) under the protection of nitrogen atmosphere. Mechanical properties of the graphene/PSA fabric were studied by an electronic single-fiber strength tester (YG006). The anti-UV properties of the graphene/PSA fabric were characterized by a textile anti-ultraviolet tester (UV-1000F, Labsphere, US).

8.5.2 Results and Discussion

8.5.2.1 Dispersion of Graphene in Graphene Sizing

Graphene sizing is a two-phase structure system and it consists graphene powder which is presented in the form of dispersed-phase, as well as acrylic adhesion agent which is presented in the form of continuous-phase. It is easy for graphene and acrylic adhesion agent to generate a new interface layer in composite system because of the high specific surface area and high surface free energy of graphene. The compatibility between the graphene and acrylic adhesion agent can be influenced by the mass fraction of graphene and it is poor when the mass fraction of graphene is high due to the agglomeration of graphene. In addition, the compatibility of these two phases has direct effects on the properties of graphene/PSA fabric. Interfacial compatibility models of graphene sizing system are shown in Fig. 8-36, including incompatible system, partially compatible system and completely compatible system. Thermodynamically speaking, the two-phase-structure composite system made via adding nano-particle into the continuous-phase matrix belongs to partially compatible system which is between the incompatible system [Fig. 8-36 (a)] and the complete separation system [Fig. 8-36 (c)], and the macroscopic homogeneous phase and microcosmic heterogeneous phase are available. As demonstrated in Fig. 8-36 (b), a transition layer can be produced between the two phases when the graphene powder (I) begin to mix with adhesion agent (II). The generation of transition layer can prove that there is compatibility between the two phases which are compatible in the transition layer; however, they are separate in the whole blending system[60-61].

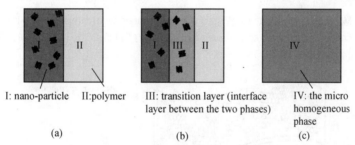

Fig. 8-36　Interfacial compatibility model of graphene sizing system: (a) Incompatible system; (b) Partially compatible system; (c) Completely compatible system

8.5.2.2　Morphological Structure

Morphological structure of graphene membranes with different mass fraction of graphene powder on surface of PSA fabric is illustrated in Fig. 8-37. As shown in Fig. 8-37 (a), pure membrane of acrylic adhesion agent with a good transparency was observed on the PSA surface. The graphene sheets in the membrane coated on the PSA surface are uniform when the mass fraction of graphene is 1.0%. The graphene sheets in the membrane coated on the PSA surface gradually connect with each other and effective conductive pathways among the graphene sheets increase when the mass fraction of graphene is increased to 1.5%. It can be deduced that the resistance of graphene/PSA fabric decreases and its conductivity increases. However, it is difficult for graphene to disperse uniformly in the adhesion agent and agglomeration of graphene was observed in the membrane.

The schematic diagram of graphene membrane is demonstrated in Fig. 8-38.

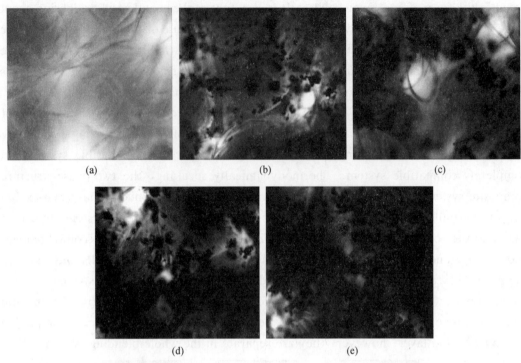

Fig. 8-37　Morphological structure of graphene membranes with different mass fraction of graphene impregnated on surface of PSA fabric: (a) 0.0%; (b) 1.0%; (c) 1.5%; (d) 2.0%; (e) 2.5%

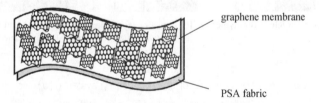

Fig. 8-38　Schematic diagram of graphene membrane on the surface of PSA fabric

8.5.2.3　Conductivity

The surface specific resistance of graphene/PSA fabric with different mass fraction of graphene powder is illustrated in Fig. 8-39. The surface specific resistance of graphene/PSA fabric is obviously low compared with that of pure PSA fabric, indicating that the conductivity of graphene/PSA fabric is significantly improved after being coated with graphene membrane. In addition, the resistance slightly reduced as the increased mass fraction of graphene powder from 1.0% to 2.5%. The gradually decreased resistance suggests an increasingly improved conductivity of graphene/PSA fabric and it is because of the gradually established conductive network in the impregnated fabric which is discussed in **Section 8.2.1.4.**

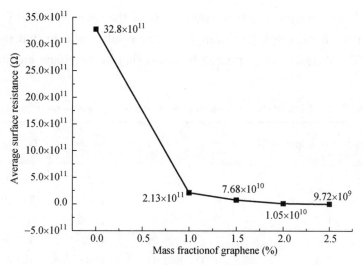

Fig. 8-39 Surface specific resistance of graphene/PSA fabric with different mass fraction of graphene powder

8.5.2.4 Thermal Stability

DSC curves of graphene/PSA fabric with different mass fraction of graphene are shown in Fig. 8-40. As observed in the figure, the thermal decomposition of graphene/PSA fabric includes slight weight-loss stage (30～400 ℃), thermal decomposition stage (400～600 ℃) and carbon stabilization stage (600～700 ℃). The thermal decomposition behavior of graphene/PSA fabric at room temperature to 700 ℃ is similar to the samples discussed in Section 8.3.1.5 and Section 8.4.2.3. The thermal decomposition data of graphene/PSA fabric are shown in Tab. 8-12. As shown in this table, the initial decomposition temperature and the temperature corresponding to the maximum decomposition rate of PSA fabric are postponed after coating the graphene membrane on the PSA fabric and the temperature value is generally increased with the mass fraction of graphene increased

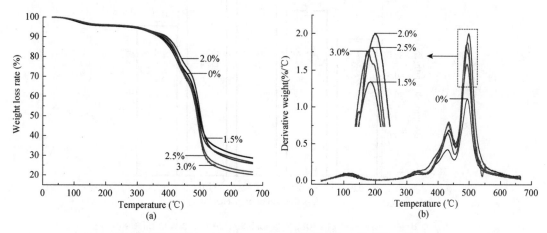

Fig. 8-40 DSC curves of graphene/PSA fabric with different mass fraction of graphene: (a) TG; (b) DTG

from 1.0% to 2.5%. Moreover, the residual rate of the samples at 700 ℃ increases significantly after the impregnation of graphene. The results indicate that the thermal stability of the PSA fabric can be improved by impregnating graphene membrane onto the fabric surface.

Tab. 8-12 Parameters of graphene/PSA fabrics during thermal decomposition

Mass fraction of graphene(%)	T_0(℃)	$d\alpha/dt$	T_{max}(℃)	α(%)
0.0	382.34	1.204	503.02	28.42
1.0	384.58	1.678	506.89	32.67
1.5	395.74	2.023	505.94	34.93
2.0	398.09	1.936	507.65	35.55
2.5	403.41	1.905	513.47	36.02

T_0—The initial decomposition temperature;
$d\alpha/dt$—The maximum decomposition rate;
T_{max}—The temperature corresponding to the maximum decomposition rate;
α—The residual weight rate at 700 ℃.

8.5.2.5 Mechanical Properties

The breaking strength of graphene/PSA fabric with different mass fraction of graphene is illustrated in Fig. 8-41. As can be observed in the figure, the breaking strength of graphene/PSA fabric was improved after coating with graphene membrane because of the excellent mechanical properties of graphene. In addition, the adhesion agent in graphene membrane can also enhance the breaking strength of graphene/PSA fabric. Furthermore, the breaking strength of graphene/PSA fabric gradually increases with the in-

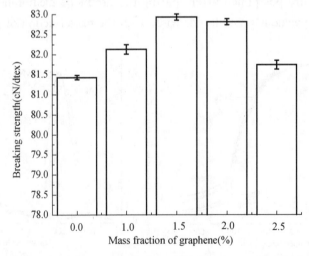

Fig. 8-41 Breaking strength of graphene/PSA fabric with different mass fraction of graphene

creasing mass fraction of graphene from 1.0% to 1.5% and then decreases as the further increased mass fraction from 1.5% to 2.5%. The decreasing breaking strength of graphene/PSA fabric can be ascribed to the agglomeration of graphene when blended with a relatively high mass fraction.

8.5.2.6 Anti-UV Properties

As illustrated in Fig. 8-42, the UV transmittance of graphene/PSA fabric decrease after coating with graphene membrane and the value gradually reduced as the increasing mass fraction of graphene from 1.0% to 2.5%. The results indicate that the UV light absorption and scattering properties of graphene/PSA fabric enhance and the anti-ultraviolet capability reinforce. Moreover, the shielding coverage of UV light widens remarkably after coating with graphene membrane. The improved UV resistance of graphene/PSA fabric can be attributed to the quantum size of graphene. In addition, the fabricated graphene/PSA fabric presents grey-black color and the color becomes darker with the increasing mass fraction of graphene, thus greatly enhance the UV shielding effect of graphene/PSA fabric. Therefore, the graphene plays an important role in enhancing the anti-UV properties of graphene/PSA fabric and relieving its aging properties because of the UV-light irradiation.

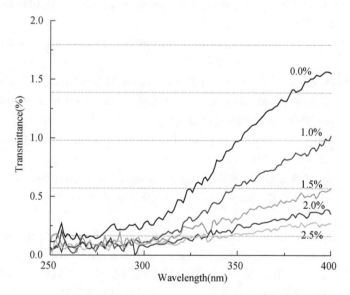

Fig. 8-42 UV transmittance of graphene/PSA fabric with different mass fraction of graphene

8.5.3 Summary

Graphene/PSA fabric was prepared by impregnating PSA fabric into graphene sizing. The effects of the mass fraction of graphene sizing on the morphological structure, conductivity, thermal stability, mechanical properties and anti-UV properties of graphene/PSA fabric were systematically investigated. The results show that uniform distribution of graphene sheets in sizing can be obtained when the mass fraction of graphene is relatively low at 1.0%. However, agglomeration of graphene was observed when the mass fraction of graphene is high at 2.5%. The conductivity, thermal stability and anti-UV properties of graphene/PSA fabric can be improved after coating with graphene membrane and the improvement increased as the mass fraction of graphene rising from 1.0% to 2.5%. The mechanical properties of graphene/PSA fabric are also enhanced by impregnating the PSA fabric into graphene sizing and the breaking strength firstly increased with the mass fraction of graphene raising from 1.0% to 1.5% and then decreased with the mass fraction further increased from 1.5% to 2.5%.

References

[1] Shi M W, Xiao H. The present state and perspectives of the smart textiles. Hi-Tech Fiber & Application, 2010, 35(4): 5-9.

[2] Eswaraiah V, Balasubramaniam K, Ramaprabhu S. Functionalized graphene reinforced thermoplastic nanocomposites as strain sensors in structural health monitoring. Journal of Materials Chemistry, 2011, 21(34): 12626-12628.

[3] Wang Y, Yang R, Shi Z, et al. Super elastic graphene ripples for flexible strain sensors. ACS Nano, 2011, 5(5): 3645-3650.

[4] Xu X J, Qin J G, Li Z. Research advances of Graphene. Process in Chemistry, 2009, 21(12): 2559-2567.

[5] Geim A K, Novoselov K S. The rise of graphene. Nature Materials, 2007, 6: 183-191.

[6] Novoselov K S, Geim A K, Morozov S, et al. Two-dimensional gas of massless Dirac fermions in graphene. Nature, 2005, 438(7065): 197-200.

[7] Neto A H C, Guinea F, Peres N M, et al. The electronic properties of grapheme. Rev. Mod. Phys., 2009, 81(1-3): 109-162.

[8] Kelly B T. Physics of Graphite. London: Applied Science Publishers, 1981.

[9] Geim A H, MacDonald A K. Graphene: exploring carbon flatland. Physics Today, 2007, 60(8): 35-41.

[10] Zhang X M. Fabrication and Photoelectric Properties of Graphene by Chemical Vapor Deposition Method. Jilin: Changchun University of Science and Technology, 2012.

[11] Novoselov K S, Geim A K, Morozov S V, et al. Electric field effect in atomically thin carbon films. Science, 2004, 306(5696): 666-669.

[12] Liang X Q. Tightly bound approximate calculation of graphene band structure. Guangxi Physics, 2011, 32(1): 7-10.

[13] Soldano C, Mahmood A, Dujardin E. Production, properties and potential of grapheme. Carbon, 2010, 48(8): 2127-2150.

[14] Novoselov K S, Jiang Z, Zhang Y, et al. Room-temperature quantum hall effect in graphene. Science, 2007, 315(5817): 1379.

[15] Schedin F, Geim A K, Morozov S V, et al. Detection of individual gas molecules adsorbed on gra-

phene. Nature Materials, 2007, 6(9): 652.

[16] Chen H, Müller M B, Gilmore K J, et al. Mechanically strong electrically conductive and biocompatible graphene paper. Advanced Materials, 2008, 20(18): 3557.

[17] Mccann E. Asymmetry gap in the electronic band structure of bilayer graphene. Physical Review B, 2006, 74(16).

[18] Zhou S Y, Gweon G H, Fedorov A V, et al. Substrate-induced band gap opening in epitaxial graphene. Nature Materials, 2007, 6(10): 770.

[19] Sengupta R, Bhattacharya M, Bandyopadhyay S, et al. A review on the mechanical and electrical properties of graphite and modified graphite reinforced polymer composites. Progress on Polymer Science, 2011, 36(5): 638-670.

[20] Shi Y S, Li X H, Ning Q J. Preparation and research status of graphene. Electronic Composites and Materials, 2010, 29(8): 70-73.

[21] Wang L, Tian L H, Wei G D, et al. Epitaxial growth of graphene and their applications in devices. Journal of Inorganic Materials, 2011, 26(10): 1009-1019.

[22] Sun Z Z, Yan Z, Yao J, et al. Growth of graphene from solid carbon sources. Nature, 2010, 468(7323): 549-552.

[23] Kim K S, Zhao Y, Jang H, et al. Large-scale pattern growth of graphene films for stretchable transparent electrodes. Nature, 2009, 457(7230): 706-710.

[24] Hou Y. Study on The Fabrication and Piezoresistive Property of Electrically Conductive Graphene Based Polymer Composites. Beijing: Beijing University of Chemical Technology, 2012.

[25] Cui W J. Feasible study on graphite preparation by expanded graphite. Frontier Science, 2011, 3(5): 40-47.

[26] Guo X Q, Huang J, Wang Y K, et al. Preparation of graphene nanosheets by exfoliating secondary expanded graphite. Journal of Functional Materials, 2013, 44(12): 1800-1803.

[27] Huang W J, Zhao Y, Wang X L. Mechanical and tribological properties of graphene/polyimide composites. Journal of Functional Materials, 2012, 43(24): 3484-3488.

[28] Wang Q Q, Wang Q J. Research on the performance of the blended membrane with graphene oxide and water-based polyurethane. Leather Science and Engineering, 2012, 22(5): 27-31.

[29] Kim I H, Jeong Y G. Polylactide/exfoliated graphite nanocomposites with enhanced thermal stability, mechanical modulus, and electrical conductivity. Journal of Polymer Science Part B: Polymer Physics, 2010, 48(8): 850-858.

[30] Steurer P, Wissert R, Thomann R, et al. Functionalized graphenes and thermoplastic nanocomposites based upon expanded graphite oxide. Macromolecular Rapid Communications, 2009, 30(4-5): 316-327.

[31] Lee Y R, Raghu A V, Jeong H M, et al. Properties of waterborne polyurethane/functionalized graphene sheet nanocomposites prepared by an in situ method. Macromolecular Chemistry & Physics, 2009, 210(15): 1247-1254.

[32] Zhao J, Liu J, Sun G W, et al. Preparation and properties of RTV silicon rubber/graphite composites. Special Purpose Rubber Products, 2013, 34(6): 5-10.

[33] Zhang H B, Zheng W G, Yan Q, et al. Electrically conductive polyethylene terephthalate/graphene nanocomposites prepared by melt compounding. Polymer, 2010, 51(5): 1191-1196.

[34] Zhou T N, Chen F, Tang C Y, et al. The preparation of high performance and conductive poly(vinyl alcohol)/graphene nanocomposite via reducing graphite oxide with sodium hydrosulfite. Compos-

ites Science and Technology, 2011, 71(9): 1266-1270.

[35] Zheng W, Wong S C. Electrical conductivity and dielectric properties of PMMA/expanded graphite composites. Composites Science and Technology, 2003, 63(2): 225-235.

[36] McAllister M J, Li J L, Adamson D H, et al. Single sheet functionalized graphene by oxidation and thermal expansion of graphite. Chemistry of materials, 2007, 19: 4396-4404.

[37] Wang J M, Zhu Z C, Li Y, et al. Coating for fabrics with nanotube-PANIs and the properties of conducting and shielding microwave. Journal of Textile Research, 2005, 26(4): 10-13.

[38] Wang C X, Cao J L, Wang K Z, et al. A New Method of Oxidized Graphene Derivatives for Multi-Functional Finishing of Textiles. CN 104328653A, 2015.

[39] Zhang L D. Nano Materials and Nano Structures. Beijing: Science Press, 2001.

[40] Xu J Z, Tang H, Li Z M. Low-dimensional carbon nanofillers induced the crystallization of polymers. Polymer Bulletin, 2011(1): 16-23.

[41] Rafiee M A, Rafiee J, Wang Z, et al. Enhanced mechanical properties of nanocomposites at low graphene content. ACS Nano, 2009,3(12): 3884-3890.

[42] Ruschau G R, Yoshikawa S, Newnham R E. Resistivities of conductive composites. Journal of Applied Physics, 1992, 72(3): 953-959.

[43] Rajagopal C, Satyam M. Studies on electrical conductivity of insulator-conductor composites. Journal of Applied Physics, 1978, 49(11): 5536-5542.

[44] Carmona F. Conducting filled polymers. Physica A: Statistical Mechanics and Its Applications, 1989, 157(1): 461-469.

[45] Simmoms J G. Genearlized formula for the electric tunnel effect between similar electrodes separated by a thin insulating film. Journal of Applied Physics, 1963, 34(6): 1793-1803.

[46] Sheng P, Klafter J. Hopping conductivity in granular disordered systems. Physical Review B Condensed Matter, 1983, 27(4): 2583-2586.

[47] Ezquerra T A, Kulescza M, Cruz C S. Charge transport in polythylene-graphite composite materials. Advanced Materials, 1990, 2: 597-600.

[48] Beek L K H V, Pul B I C F V. Internal field emission in carbon black-loaded natural rubbervulcanizates. Journal of Applied Polymer Science, 1962, 6(24): 651-655.

[49] Premkumar, Rajan J, Khoo S B. Electrochemically generated super-hydrophilic surfaces. Chem. Comm., 2005(5): 640-642.

[50] Pan Q M, Cheng Y X. Superhydrophobic surfaces based on dandelion-like ZnO microspheres. Applied Surface Science, 2009, 255(6): 3904-3907.

[51] Yong T. An essay on the cohesion of fluids. Philosophical Transactions of the Royal Society of London, 1805, 95: 65-87.

[52] Shalel-Levanon S, Marmur A. Validity and accuracy in evaluating surface tension of solids by additive approaches. Journal of Colloid and Interface Science, 2003, 262(2): 489-499.

[53] Wenzel R N. Resistance of solid surfaces to wetting by water. Industrial & Engineering Chemistry, 1936, 28(8): 988-994.

[54] Baxter S, Cassie A B D. The water repellency of fabrics and a new water repellency test. Journal of the Textile Institute Transactions, 1945, 36(4): T67-T90.

[55] Cassie A B D, Baxter S. Wettability of porous surfaces. Transactions of the Faraday Society, 1944, 40: 546-551.

[56] Temmerman E, Leys C. Surface modification of cotton yarn with a DC glow discharge in ambient

air. Surface & Coating Teachnology, 2005,200(1): 686-689.
[57] Gao Z Q, Peng S, Sun J, et al. Influence of processing parameters on atmospheric pressure plasma etching of polyamide 6 films. Applied Surface Science, 2009, 255(17): 7683-7688.
[58] Bhat N V, Benjamin Y N. Surface resistivity behavior of plasma treated and plasma grafted cotton and polyester fabrics. Textile Research Journal, 1999, 69(1): 38-42.
[59] Zhu L X, Li Y Z, Zhao X, et al. Preparation of graphene by electrochemical method and its electrical conductivity. Chemical Journal of Chinese University, 2012, 33(8): 1804-1808.
[60] Dong M, Wang R M, Yao M. Advances in research on improving interfacial compatibility of polymer blends. Coating Painting & Electroplating, 2006, 4(5): 15-19.
[61] Zheng Y T, Chen J J, Cao D R. Advances in interfacial compatibility of plant fiber/thermoplastic composites. Journal of Cellulose Science and Technology, 2005, 13(1): 45-55.

Chapter 9 Sputtering Technology for Textiles

Xiaoxia Liu*, Zhuoming Chen, Shan He and Mingyu Zhuang

9.1 Introduction

There are strong demands of heat-insulated textiles in the fields of human and industrial protection. People wear protective clothing made of fabrics with heat-radiation-reflection performance to protect their skin. In high-temperature environments, tailor-made heat-protection clothing can be used for decreasing the heating rate of human body surface, thus providing the wearer with more reaction and escape time, as well as avoiding or reducing the damage of heat source to human body[1]. For the same reason, robot working in the poor industrial production environment also requires proper protection so as to effectively avoid the machine damage and prolong its service life, thus increasing the production efficiency of the enterprise.

With the rapid development of science and technology, application of the robot in both life and production fields has been increasing continuously. In 2011, CEO of FOXCONN (the largest OEM enterprise in the world)-Guo Taiming announced "A Million Robots Plan" and assembled 1 000 000 robots within 3 years. According to some statistics, the sales of industrial robots in 2012 was CNY 2.4 billion; in 2015, it was expected to reach up to CNY 5.4 billion[2]. All the information shows that more and more industrial robots have been applied in the industrial production. In addition, higher requirements on the quality and performance of productions as well as the processing conditions in the industrial production increase continuously. However, it is difficult for the human

---

Corresponding Author:
Xiaoxia Liu
School of Fashion Technology, Shanghai University of Engineering Science, Shanghai 201620, China
E-mail: liuxiaoxialucky@126.com

to execute the short-distance operation due to the terrible conditions. For the above-mentioned reasons, robots instead of human are required in the industrial production. In order to prolong the service lives of robots in the long-term and extremely poor production environment, protective measures should be executed[3]. At present, different types of robots have been applied in the industrial production, such as spray-painting robots, high-temperature-resistance robots, welding robots, cleaning robots and transfer robots[2]. Fig. 9-1 shows some robots for the preparation of protective clothing.

The traditional sense of human protection refers that people face with the physical, chemical and biological threats in people's work and life, the functional protective clothing is developed to counter these threats. The robot works in the poorer industrial environment, although it is not with the abilities of perceiving the environment and making a physiological reaction which is different from human being, the service life and working efficiency of the robot is limited by the poor working environment as well[3]. Nowadays, robots have been applied in such industries as automobile industry, metal products, food industry, medical industry and microelectronics, etc. In these industries, working environments of the robots are extremely poor, for example, the welding robot faces with not only local high temperature and heat radiation, but also the spattered sparks during the welding process. If no proper protection measures are taken, the wires and components of currently used robot will be damaged due to high-temperature and burning. Take the cleaning robot used for cleaning work piece for another example, since humidity and temperature of the cleaning environment is high and the cleaning solution is with certain chemical corrosion, water will enter into the circuit and machine components will be corroded due to long-term usage of the cleaning robot. As a result, the service life of the cleaning robot greatly decreases and the safety risk exists. Therefore, considering not only the safety but also the economic benefit, special protection for the robot is essential.

(a) Coating robot   (b) Cleaning robot   (c) Welding robot

Fig. 9-1 Different types of robots for preparation of protective clothing

At present, several overseas companies have engaged in developing protective clothing for robots, including Altra-Tech (US), ASP (France), HERMAMN BOESE (Germany) and RJ HANLON (US), etc. They are able to customize functional material and clothing structure for every robot according to its functions.

In recent years, several domestic companies have been also engaged in the production of protective clothing of the robot, including Shanghai Yinling Biological Technology Co., Ltd., Shanghai Fujing Industry & Trade Co., Ltd., and Chengdu Leierte Science & Technology Co., Ltd., etc. However, the research ability is still at the initial stage. At present, a large amount of robots has been applied in the automobile manufacturing industry, only the robot that wears the protective clothing is allowed to execute the production task in the major companies (mainly referring to the sino-foreign joint ventures). Foreign products have been widely used due to the short supply of protective clothing of the robot in domestic market.

Different from the ordinary protective clothing for human being, protective clothing for robot should be specially designed and tailor-made according to the dimension, movement range and working environment of the robot. For example, the following factors should be considered for the production of protective clothing for welding robot, including the temperature, the highest temperature around the welding gun, spattering intensity of the spark, and the position of the wear-resistance point. The protective clothing for robot has extremely high benefit and great potential application in the industrial development in future.

9.1.1 Protective Fabric for Robot

There are different requirements on protective fabrics for human and robot. The main differences can be summarized as follows.

(1) Comfort. Comfort requires the fabric with effective air and moisture permeability, softness as well as warmth retention property. As for the protective fabric to human, comfort is more important than protection, while for the protective fabric to robot, it mainly considers the protection effect[3] instead of comfort.

(2) Wear-resisting property. Because the internal surface of protective fabric directly contacts with human skin, it has high requirement on softness, skin-friendly property and no stimulation but low requirement on wear-resisting property. Robot has high requirements on physical properties of fabric, such as strength, wear-resisting property and high-temperature-resistance. Friction is easily generated since the internal surface of protective fabric frequently contacts with metal components, and a larger amount of robots has to work for 24 hours continuously. Therefore, a good wear-resisting property is an essential and important to protective fabric for robot.

(3) High-temperature-resistance. Protective fabric for human is required to have high-temperature protection in a short time, such as firefighter uniform, while the protective fabric for robot must keep proper high-temperature-resistance and heat-insulating property during a long-term continuous usage.

Generally, obviously different requirements can be discovered between protective fabrics for human and robot. Most of the protective fabrics for robots are made of high-performance fibers due to the poor working environment. The development of protective clothing for industrial robot has great economic benefit. Protective fabric for robot is required in several working conditions, such as high-temperature, high-humidity and clean room.

9.1.2 Heat Insulating Material

Heat transfer modes mainly include heat conduction, heat convection and heat radiation. Usually, three modes of heat transfer simultaneously exist, but the modes of heat transfer are different in various environments[4]. Heat conduction can be defined as the heat transfer through a solid when a temperature difference exists across the solid. Heat convection is defined as the heat transfer occurring when water (or air) flows above a streambed of dissimilar temperature. Heat radiation is the radiation resulting from oscillation of electric charges of a body that has a temperature higher than absolute zero. Certain medium is required when heat conduction and heat convection occur, while no medium is required in the heat radiation[5]. Radiating capacity of the object is related to the temperature of the object. Heat insulating material can be classified into three types based on the heat insulating modes, including heat barrier, heat-reflection and heat-radiation[6].

9.1.2.1 Heat-barrier Type

Heat insulating material of heat-barrier type as the traditional thermal insulation material has been widely applied in external thermal insulation system. The insulating function can be executed depending on the properties and special structure of the material of heat-barrier type. The lower the thermal conductivity is, the better the heat-insulating property will be. The heat insulating material can block high-temperature conduction when it is made by the material with extremely-low thermal conductivity, such as organic polymer foam, inorganic mineral and the material with special macro vesicle structure which can make the material confine more still air, thus minimizing the thermal conductivity of the material and increasing the heat-insulating property[6]. The innovative aerogel can be filled in the internal gap of the fabric or coated on the fabric surface to increase the heat-insulating property of the fabric.

9.1.2.2 Heat-reflection Type

Heat insulating material of heat-reflection type refers that the specular reflection ability which can reflect most of the heat radiation existing on the material surface, thus realizing the heat insulation. This kind of material mainly includes metallic film with high reflectivity and coating material mixed with heat-reflection material, which can be coated onto the fabric surface.

9.1.2.3 Heat-radiation Type

Heat insulating mode of heat-radiation type is active and its nature is to transmit the absorbed heat of the far infrared ray with a wavelength of 8~13.5 μm to reduce temperature. Heat insulating mode of heat-radiation type has higher efficiency than that of the heat barrier and heat-reflection types. Ceramics technology is mainly adopted for preparing infrared radiation materials with high emissivity. At present, the widely used infrared radiation filler is infrared radiation ceramic powder.

9.1.3 High-temperature Protective Fabric

Excellent heat-insulating protective fabric is designed with several heat insulating modes instead of a single mode to improve the heat-insulating property of the fabric. Since the development of the heat-insulating coating material referring to the heat-radiation type is late and the heat dissipation technology is still in the initiate development, current heat-insulating protective fabric is mainly prepared via the compound of heat insulating materials of heat-barrier and heat-reflection types. Heat insulating properties can be endowed to fabric by post-finishing process, and the fabric with high insulating properties is usually used as substrate, such as aramid fiber, PSA fiber, basalt fiber and modified flame-retardant fiber. Heat-barrier performance of the fabric substrate is mainly affected by three parameters, including its air permeability, thermal conductivity and thermal stability.

9.1.3.1 Air Permeability

Air permeability has an important influence on the heat convection of fabric. Heat convection in the fabric mainly refers to the heat convection of air, and decreasing flow rate and efficiency of air in the fabric can limit the heat convection. Air permeability of fabric is mainly affected by the fabric structure, and it will decline when the fabric has a dense structure. In this case, the heat-insulating property of the fabric can be greatly enhanced[7]. In addition, air permeability of the fabric will also decrease after coating treatment.

9.1.3.2 Thermal Conductivity

Thermal conductivity has an important influence on heat-insulating property of fabric. Thermal conductivity can affect the transfer ability of heat in fabric, and a low thermal conductivity can increase the heat-insulating property of fabric. Generally, there are two methods can be used to decrease the thermal conductivity of fabric. The first one is to use inorganic fiber with low thermal conductivity to prepare fabric, such as basalt fiber and ceramics fiber. The second one is to fill the fabric with substance, which has low thermal conductivity, such as SiO_2 aerogel[8].

9.1.3.3 Thermal Stability

For a high-temperature working environment, fabric is not only required to impede various modes of energy transfer, but also have excellent thermal stability (i.e., maintaining stable mechanical properties and morphology in the high-temperature environment). In the practical application, it has high requirement on the mechanical properties of fabric used in the high-temperature environment. For example, a protective clothing of robot, which continuously works in a long-term and high-temperature environment, must keep good wear-resisting property and mechanical properties. Organic polymers of high-performance fiber with high strength and high-temperature resistance mainly include aramid, PSA, PBO and PTFE fibers, and the inorganic fibers of this kind mainly include glass fiber, basalt fiber, carbon fiber and metal fiber, etc[9].

9.1.4 Finishing of High-temperature Protective Fabric

In order to improve the performance of the protective fabric, appropriate finishing mode should be selected based on the fabric substrate and desired performance. There are several finishing technologies can be used to increase the heat-radiation-reflection ability of protective fabric, including coating, laminate the fabric substrate with aluminum foil and magnetron sputtering.

9.1.4.1 Coating

Coating is a popular post-treatment method to improve heat-radiation-reflection ability of fabric. Generally, materials with good heat-reflection properties can be added into the adhesive and then coated onto the fabric surface to endow the fabric with heat-insulating property. As a high-temperature protective material, it is demanded that the adhesive must have good adhesion property when used in the high-temperature environment.

Aluminum paste is a kind of coating adhesiveand it can be prepared by mixing with PU syrup in a proper proportion. Because aluminum powder has smooth flake shape, it can effectively prevent the penetration of IR and UV. In addition, the coating surface of the aluminum powders with very fine diameter is similar as a mirror face and can reflect heat radiation[10].

Rutile-type TiO_2 is also an excellent heat-insulating material and it has a series of advantages, such as high covering ability, strong coloring power, high lightness value and good weather resistance, and it can be widely used as white pigment. It has the maximum refraction coefficient among the currently used pigments, so it has high reflectivity and it is the promising candidate as filler with heat-insulating function. After testing by spectrograph, it is found that its reflectivity in the visible region is almost 100%, its reflectivity in the near-infrared visible region is higher than 85%, and its absorptivity in the near-ultraviolet region (200~400 nm) is higher than 85%[6].

According to different surface status of fabrics, suitable surface treatments on fabrics are required. For example, surface modification is required on the aramid before coating due to its poor hydrophilic performance and poor associativity with adhesive. Surface treatment on aramid can be classified into physical and chemical modification. Physical treatment with plasma can introduce polar functional groups on aramid surface to increase its specific surface area, surface energy and wettability, in this case, the adhesion strength of the aramid with other material can be improved[11-12]. Tao et al[13] grafted alkoxysilane on the aramid molecular chain to greatly increase the adhesion between the fiber and matrix resin, and the LISS value increased about 57%. However, chemical modification on the aramid requires long reaction time, and it is not suitable for the continuous preparation of the high-temperature protective fabric and only suitable for the theoretical research of the composite material. What's more, it is difficult to realize its continuous industrial finishing.

Phenomenon such as poor metal effect, air bubble, leakage, threshing and non-uniform distribution usually occurs when coating the metal power syrup. In addition, the operation technology is complicated and obviously affected by the environment. Moreover, the coating is limited by the high-temperature performance of the adhesive, and it is difficult to meet corresponding requirement if the temperature is too high.

9.1.4.2 Aluminum Foil Lamination

As a kind of routine metal, aluminum foil has ultra-strong heat-radiation-reflection ability in visible region and it can be used in a high-temperature environment. The aluminum foil can be glued onto fabric substrate, and then laminate processing can be conducted to prepare the laminated fabric. Although the aluminum foil is compounded on the fabric, its extensibility is low and certain pre-stretching force should be applied to the aluminum foil during the laminating to prevent its breaking under a tensile strength. In addition, the pre-stretch can also ensure the smoothness of the fabric.

Compared with the coated fabric, laminated fabric with aluminum foil has better heat reflection effect and surface cooling effect. During the laminating processing of aluminum foil, the used adhesive should maintain good performance at a high temperature. In this case, laminating technology of aluminum foil is depending on the end-use temperature of the adhesive.

9.1.4.3 Magnetron Sputtering

Magnetron sputtering is a type of physical vapor deposition (PVD) and has been widely applied in electronics, construction and automobile industry, and it is especially suitable for the large-area coating. Magnetron sputtering has some advantages, such as high-speed and low temperature. Moreover, it can sputter many kinds of materials and has become one of the most widely used methods to prepare thin films. The basic principle of magnetron sputtering is shown in Fig. 9-2. In the vacuum chamber, a target (or cathode) plate is bombarded by energetic ions generated in glow discharge plasma, situated in front of the target. The bombardment process causes the removal of target atoms, which may then condense on a substrate to form a thin film[14-15].

In the field of textile, magnetron sputtering is generally used to endow the fabric with excellent properties, such as conductivity, electromagnetic shielding performance, anti-bacterial property and anti-ultraviolet property, by coating metal thin films on the fabric. Li et al[16] coated copper thin films onto carbon fiber by using magnetron sputtering technology, the experimental results revealed that the deposition rate increases with an increasing of power.

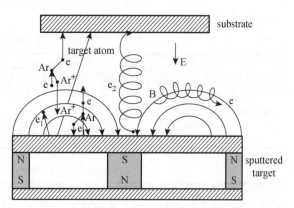

Fig. 9-2　Schematic of magnetron sputtering. (E—electric field)

In addition, the adhesive fastness between the fiber and the film increases after plasma treatment is executed on the fiber surface. Chu et al[17] fabricated anti-bacterial PBT/PET fabric by magnetron sputtering technology through depositing nano TiO_2 thin films onto the fabric. The experimental results showed that the deposited films have no significant influence on the air permeability and comfort of the fabric, and it can still maintain anti-bacterial properties after 30 times of washing and 6 minutes of friction. Ehiasarian, et al[18] prepared an anti-bacterial fabric by coating copper thin films onto textile fabric via magnetron sputtering, and the result showed that the anti-bacterial property of the coated fabric is 3 times better than that of the un-coated fabric. However, there are few studies on the preparation of functional fabric with heat insulation through coating metallic film onto fabric by using magnetron sputtering.

9.2　Preparation and Characterization of High-temperature Protective Fabric by Coating with Silicone

In this section, high-temperature protective fabrics were prepared by magnetron sputtering technology through coating silicone thin films onto aramid and PSA fabric substrates. Physical and thermal properties of the coated fabrics are studied. The influences of the coating parameters on the heat-insulating properties of the coated fabrics are also investigated.

9.2.1 Experimental

Specifications of aramid and PSA fabrics used in the experiment are illustrated in Tab. 9-1. 800-mesh aluminum paste as a finishing agent with diameter of 15 μm is coated on the fabrics prior to the sputtering. Specifications of liquid silicone (LSR) are illustrated in Tab. 9-2.

Tab. 9-1 Fabric specification

Fabric	Warp density [piece/(10 cm)]	Weft density [piece/(10 cm)]	Surface density (g/m^2)	Thickness (mm)
Aramid 1313	240	200	266.22	1.31
PSA	300	300	213.24	0.73

Tab. 9-2 Specification of liquid silicone

Item	Value	Item	Value
Viscosity (25 ℃, MPa·s)	8000~1000	Tensile strength (MPa)	2.0
Specific gravity	1.2~1.35	Elongation rate (%)	≥200
Hardness (Shore A)	40±5	Curing ratio	A∶B=100∶(2~5)

9.2.2 Characterization

Testing of vertical flammability is conducted and the limiting oxygen index (*LOI*) of the pure and coated fabrics is detected to analyze the flame resistance.

Thermal stabilities of the pure and coated fabrics are investigated by TC3000 Hotwire Thermal Conductivity Measuring Instrument according to the standard—GB 10297—1998 *Test Method for Thermal Conductivity of Nonmetal Solid Materials by Hotwires Method*. Thermal conductivity, thermal diffusion coefficient, and specific heat of the samples can be automatically detected by Hotwire 3.0 Thermal Conductivity Measuring Software of the instrument, which is characterized by high accuracy, fast speed and easy operation. Anti-thermal radiation properties of the pure and coated fabrics are also analyzed by using the self-designed device, which mainly consists of a thermal radiation source, two inductive thermometers, an iron bracket, a timer and several clips, as illustrated in Fig. 9-3.

Mechanical properties of the pure and coated fabrics, including tensile strength, wear-resistance and bursting properties, are also studied.

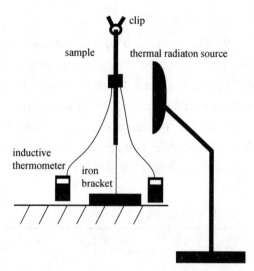

Fig. 9-3 Schematic of self-designed device for anti-thermal radiation properties analysis

9.2.3 Results and Discussion

9.2.3.1 Flame Resistance

(1) Vertical Flammability

Tab. 9-3 shows the testing results of vertical flammability of the pure and coated fabrics. As depicted in the table, although pure PSA fabric presents a longer damaged length than that of pure aramid fabric, the afterflame and afterglow time of PSA fabric is shorter than that of aramid fabric, and the results indicate that PSA fabric has a superior flame resistance. However, the afterflame and afterglow time as well as the damaged length of the coated aramid fabric are significantly lower than that of the coated PSA fabric indicating that the flame resistance of aramid fabric can be obviously improved by coating with silicone thin films. The flame resistance of the coated aramid fabric is higher than the highest protection level required for the electric-welding protective fabric. However, the afterflame time of the coated PSA fabric exceeds the national standard (2 s) for electric-welding protective fabric.

Tab. 9-3 Testing results of vertical flammability

Item	Sample			
	Aramid fabric	PSA fabric	Coated aramid fabric	Coated PSA fabric
Ignition time (s)	12	12	12	12
Afterflame time (s)	1.4	1.3	1.9	16
Afterglow time (s)	0.8	0.5	1.2	1.4
Damaged length (mm)	5	8	11	57

(2) Limited Oxygen Index (*LOI*)

Tab. 9-4 shows the *LOI* of the pure and coated fabrics. As can be observed in the table, the pure PSA fabric and aramid fabric have the similar *LOI*, and that suggests they have the analogous flame resistance. Theoretically, self-extinguishing may occur when the *LOI* of the textile material exceeds 21%. In addition, inflaming retarding can be achieved when the *LOI* of the textile material exceeds 27%[19]. As revealed in Tab. 9-4, the aramid fabric, PSA fabric and coated aramid fabric can satisfy inflaming-retarding requirements. However, the *LOI* of the coated PSA fabric decreases to 25.89%.

Tab. 9-4 *LOI* of the pure and coated fabrics

Samples	LOI(%)	Standard deviation
Aramid fabric	31.51	0.664
PSA fabric	30.87	0.923
Coated aramid fabric	31.10	0.329
Coated PSA fabric	25.89	0.943

9.2.3.2 Thermal Stabilities

Tab. 9-5 shows the thermal conductivities of the pure and coated fabrics. As can be seen in the table, the thermal conductivities of the four fabrics increase with increasing temperature. High thermal conductivity proves superior thermal conductivity and poorer heat-insulating properties. Therefore, the heat-insulating properties of the four fabrics decrease with the increasing ambient temperature. The thermal conductivities of PSA fabric in four temperature gradients are lower than that of aramid fabric. The result indicates that PSA fabric can exhibit better heat-insulating properties than aramid fabric. However, the heat-insulating properties of the PSA fabric and aramid fabric decrease after coating treatment. Compared with the coated aramid fabric, the coated PSA fabric exhibits poorer heat-insulating properties.

Tab. 9-5 Thermal conductivities of the pure and coated fabrics

Samples	Thermal conductivity[W/(m · K)]			
	50 ℃	100 ℃	150 ℃	200 ℃
Aramid fabric	0.085 1	0.093 9	0.095 8	0.096 6
PSA fabric	0.083 8	0.084 8	0.087 2	0.089 3
Coated aramid fabric	0.103 7	0.115 1	0.116 5	0.138 3
Coated PSA fabric	0.154 1	0.170 1	0.185 1	0.193 1

Fig. 9-4 shows the anti-thermal radiation properties of the pure and coated fabrics. As can be observed in the figure, the four samples have similar changing trend that the temperatures on obverse and reverse side of the fabric remarkably increase in the first three minutes and then slightly increase with the further raising testing time. The detected high temperature reveals poor anti-thermal radiation properties. Temperature difference between the two sides of the fabrics is discovered, and a higher temperature of the obverse side of the fabric can be detected than the reverse side. Obviously, temperatures of the two sides of the aramid fabric and PSA fabric are higher than that of the coated fabrics. The results indicate that the anti-thermal radiation properties of the fabrics can be remarkable improved after coating treatment and two reasons are given below.

(1) Aluminum paste with extremely fine aluminum powder particles is selected as the filler of coating agent in this study, and that can provide a mirror effect to present excellent reflection to thermal radiation. In this case, the heat on the fabric surface can decrease by thermal radiation.

(2) The fabric surface becomes smooth and the porosity remarkably decreases aftersputtering. In this case, it is difficult for the heat to transmit by air and penetrate into the fabric, and the heat conduction effect of heat convection decreases significantly, thus decreasing the temperature on the fabric.

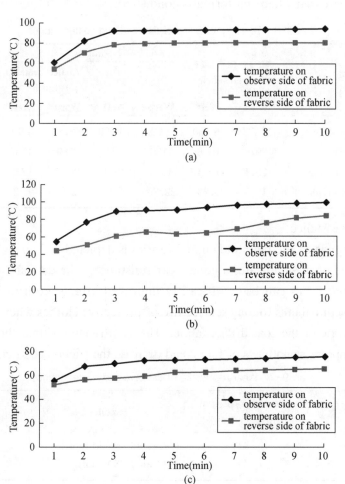

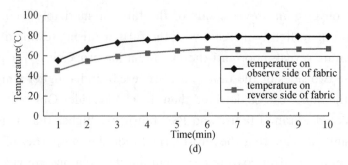

Fig. 9-4 Anti-thermal radiation properties of: (a) Aramid fabric; (b) PSA fabric; (c) Coated aramid fabric; (d) Coated PSA fabric

9.2.3.3 Mechanical Properties

(1) Tensile Properties

Parameters of tensile properties of the pure and coated fabrics are demonstrated in Tab. 9-6. It is can be observed that the breaking strength and breaking force of the fabrics decreased after sputtering. On the contrary, elongation and elongation rate at break of the fabrics increased when sputtering is conducted.

Tab. 9-6 Tensile properties of the pure and coated fabrics

Item	Samples							
	Aramid fabric		PSA fabric		Coated aramid fabric		Coated PSA fabric	
	Warp	Weft	Warp	Weft	Warp	Weft	Warp	Weft
Breaking strength (N/m)	59 530	36 160	23 720	15 680	53 780	31 740	18 920	11 820
Breaking force (N)	2976	1808	1186	784	2689	1587	946	591
Elongation at break (mm)	25.4	10.4	45.5	45.1	26.7	12.4	33.6	30.5
Elongation rate at break (%)	12.7	5.23	22.75	22.55	13.35	6.2	16.8	15.25

(2) Wear-resistance

Wear-resistance of the pure and coated fabrics is depicted in Tab. 9-7. A high value of rotation speed of disk indicates a good wear-resistance. The data show that the wear-resistance of the fabrics can be significantly improved after sputtering. The increased wear-resistance contributes to long service life of protective clothes when prepared by the coated aramid fabric or the coated PSA fabric. This is attributed to that the coating silicone thin films can improve smoothness of fabric and decrease the friction coefficient.

Tab. 9-7 Wear-resistance of the pure and coated fabrics

Samples	Rotation speed of disk (r/min)	Samples	Rotation speed of disk (r/min)
Aramid fabric	320	Coated aramid fabric	398
PSA fabric	78	Coated PSA fabric	193

(3) Bursting Properties

Bursting strength of the pure and coated fabrics is shown in Tab. 9-8. The results indicates that the bursting properties of the aramid and PSA fabric decrease after sputtering.

Tab. 9-8 Bursting strength of the pure and coated fabrics

Samples	Bursting strength (N)	Samples	Bursting strength (N)
Aramid fabric	1950	Coated aramid fabric	1850
PSA fabric	1160	Coated PSA fabric	950

9.2.4 Summary

In this section, high-temperature fabrics were prepared via magnetron sputtering technology by coating silicone thin films onto aramid and PSA fabric substrates. The experimental results showed that wear-resistance and anti-thermal radiation properties of aramid fabric and PSA fabric improve after sputtering, however, the flame resistance, heat-insulating properties, tensile properties and bursting properties of the two fabrics decrease after sputtering. Compared with the coated PSA fabric, the coated aramid fabric exhibits superior flame resistance, anti-thermal radiation properties and mechanical properties. The coated aramid fabric can satisfy the protection requirements of electric-welding arm manipulator and can be selected as the top-grade protective fabric; while the coated PSA fabric can meet the basic requirements of protective fabric and can be selected as the mid-grade protective fabric.

9.3 Study on Al-film Coated PSA via Magnetron Sputtering

In this section, aluminium thin film was coated on PSA fabric by using magnetron sputtering. Flame resistance, thermal reflection properties and mechanical properties of the coated fabric are studied, and the optimal experimental processing for excellent properties is investigated based on the single-factor and orthogonal experiments.

9.3.1 Experimental

9.3.1.1 Experimental Materials

Materials used for sputtering aluminium thin film include acetone (purity≥99.5%), absolute alcohol (purity≥99.7%), aluminium target with diameter of $\phi 80$ mm × 6 mm (purity: 99.99%), and argon gas (concentration: 99.99%). Specifications of PSA fabric are shown in Tab. 9-9.

Tab. 9-9 Specifications of PSA fabric

Item	Warp density [piece/(10 cm)]	Weft density [piece/(10 cm)]	Surface density(g/m^2)
PSA fabric	312	237	173.73

9.3.1.2 Experimental Solution

A new heat-insulation protective PSA fabric was prepared by magnetron sputtering through coating aluminium thin film onto the PSA fabric surface. Thermal reflection properties of the coated PSA fabric were investigated by single-factor experiment and orthogonal experiment. Appropriate coating parameters for good thermal reflection properties of the coated PSA fabric were initially determined based on the single-factor experiment, and then coating parameters for improved thermal reflection properties of the coated PSA fabric were investigated in the orthogonal experiment.

(1) Single-factor Experimental Solution

Single-variable experimental solution is illustrated in Tab. 9-10. In this section, influences of several important factors on the thermal reflection properties of the coated fabric were investigated by Fourier Transform Infrared spectrometer (FT-IR spectrometer), including argon flow, constant gas pressure, sputtering power, sputtering time and substrate temperature. This single-factor experiment could be performed according to Tab. 9-10.

Tab. 9-10 Single-variable experimental program

Experimental No.	Argon flow rate (mL/min)	Constant gas pressure (Pa)	Sputtering power (W)	Sputtering time (min)	Substrate temperature (℃)
1	10	0.2	600	20	Room temperature
2	10	0.2	600	20	100
3	10	0.2	600	20	200
4	10	0.2	600	20	300
5	10	0.2	600	10	Room temperature
6	10	0.2	600	30	Room temperature
7	10	0.2	600	40	Room temperature
8	10	0.2	200	20	Room temperature
9	10	0.2	400	20	Room temperature
10	10	0.2	800	20	Room temperature
11	10	0.4	600	20	Room temperature
12	10	0.6	600	20	Room temperature
13	10	0.8	600	20	Room temperature
14	5	0.2	600	20	Room temperature
15	15	0.2	600	20	Room temperature
16	20	0.2	600	20	Room temperature

(2) Multi-factor Orthogonal Experimental Solution

Orthogonal experimental solution is designed based on the results of the single-factor experiment as shown in Tab. 9-11. The constant gas pressure in the orthogonal experiment is fixed at 0.2 Pa. In this section, thermal reflectivity of the coated fabric was analyzed and sputtering parameters for the optimal thermal reflection properties were studied.

Tab. 9-11 Orthogonal experimental program

Experimental No.	Sputtering power (W)	Substrate temperature (℃)	Sputtering time (min)	Argon flow rate (mL/min)
1	100	room temperature	10	5
2	100	150	20	10
3	100	300	30	15
4	300	room temperature	20	15
5	300	150	30	5
6	300	300	10	10
7	500	room temperature	30	10
8	500	150	10	15
9	500	300	20	5

9.3.1.3 Preparation of Al Film

Prior to sputtering, PSA fabric (15 cm×15 cm) was ultrasonic cleaned in acetone solution for 30 min and then washed with deionized water for 3~5 times followed by drying in oven at 80 ℃ to remove moisture. The pre-treated sample was fixed on the substrate holder with a rotation speed of 8 r/min in vacuum chamber of the sputtering system. Subsequently, the chamber was evacuated at 5.0×10^{-4} Pa and the substrate holder was heated to the desired temperature. The sputtering was conducted for 3 min at 200 W (DC power supply).

9.3.1.4 Characterization

(1) Thermal Reflection Properties

Thermal reflectivity, which refers to the ratio of heat reflected by object surface and total energy transmitted to the object (ie, heat rays transmitted to the object), can exhibit the thermal reflection properties of the coated fabrics. Thermal reflection properties of the coated fabrics were studied by fourier transform infrared spectrometer (FT-IR spectrometer) with the infrared reflection spectrum of 2.5~25 μm, and then the reflectivity of the coated fabric can be calculated by Eq. (9-1).

$$R_n = \sum_{2.5 \mu m}^{25 \mu m} G_\lambda \rho_\lambda \qquad (9-1)$$

In where, ρ_λ is unit wavelength and G_λ is reflectivity of unit wavelength.

(2) Flame Resistance

Flame resistance of the coated fabrics was investigated according to the standard GB/T 17951—2006 *Flame Retardant Fabric Burning Test* that the sample (6 cm×15 cm) was placed in a clip and the flame height was kept within the range of (4±0.2) cm after igniting the sample. Then move the clip to make the sample contact with the flame, and record the damaged length of the sample after stable flame was obtained and igniting for 12 s.

(3) Wear Resistance

Wear resistance of the coated fabrics was detected based on the standard GB/T 21196—2007 *Textiles-Determination of Abrasion Resistance of Martindale Fabrics*. Stop rubbing the sample and observe the wear on fabric surface if the sample (15 cm×15 cm) is worn out after being rubbed for 10 000 times. No. 600 waterproof abrasive paper was used in the friction experiments on the coated fabrics.

(4) Tensile Properties

Tensile properties of the coated fabrics were analyzed according to the standard GB/T 3923 *Fabric Tensile Properties*. The samples (5 cm×15 cm) were drawn in the mode of constant-speed elongation by an electronic fabric strength tester.

(5) Tearing Strength

Tearing strength of the coated fabrics (5 cm×15 cm) was investigated based on the standard GB/T 3917.2—2009 *Fabric Tearing Performance* by an electronic fabric strength tester (YG026B, clamping length is set to be 5 cm, drawing speed is set to be 100 mm/min).

9.3.2 Results and Discussion

9.3.2.1 Thermal Properties

Influences of sputtering parameters on the thermal reflection properties revealed by thermal reflectivity of the coated fabrics are studied by using single-factor experiment, including gas pressure, argon flow rate, sputtering power and time, substrate temperature.

(1) Effects of Gas Pressure in Chamber on Thermal Reflectivity

Relationship between gas pressure and thermal reflectivity is illustrated in Tab.9-12. It can be clearly observed that the reflectivity of the coated fabrics increases with the intensity of gas pressure increased from 0.2 to 0.4 Pa, and then decreases as the gas pressure further increased from 0.4 to 0.8 Pa. However, the thermal reflectivity corresponding to 0.8 Pa is still higher than that at 0.2 Pa indicating improved thermal reflection properties can be obtained when sputtered at a suitable high gas pressure. During sputtering, an increasing amount of Ar atoms can change into Ar^+ with a raising working gas pressure, in this case, the electric field can accelerate and strike the target to form thin film with small particles and smooth surface. Resultantly, the mirror effect is enhanced and the reflectivity is strengthened. Unfortunately, secondary sputtering on the target may occur when the gas pressure further increases to 0.6~0.8 Pa and then causes a decreasing thermal reflectivity.

Tab. 9-12 Relationship between gas pressure and thermal reflectivity

Gas pressure (Pa)	0.2	0.4	0.6	0.8
Thermal reflectivity (%)	76.43	78.32	78.19	77.72

(2) Effects of argon flow rate on Thermal Reflectivity

Relationship between argon flow rate and thermal reflectivity is depicted in Tab. 9-13. Although a slightly decreasing thermal reflectivity is observed when the argon flow rate increases from 5 to 10 mL/min, it can be seen that the reflectivity of the coated fabric increases with a raising argon flow rate from 5 to 15 mL/min, and the changing trend becomes stable as the argon flow rate further raises from 10 to 20 mL/min. The reason for this phenomenon is similar to the section "Effects of Gas Pressure in Chamber on Thermal Reflectivity".

Tab. 9-13 Relationship between argon flow rate and thermal reflectivity

Argon flow rate (mL/min)	5	10	15	20
Thermal reflectivity (%)	77.26	76.43	78.84	78.85

(3) Effects of Sputtering Power on Thermal Reflectivity

Sputtering power, which has significant effects on the structure and properties of membrane, is considered as a key parameter during magnetron sputtering. Relationship between sputtering power and thermal reflectivity is shown in Tab. 9-14. It is can be seen that the thermal reflectivity of the coated fabric slightly decreases as the sputtering power raised from 400 to 600 W and then obviously increases as the power further increases from 600 to 800 W. An increased sputtering power can endow Ar ions with high energy to strike the target and form an uniform thin film onto the fabric. With the sputtering power further increased, the number of activated sites (acting as nucleation center of new phase of membrane) on the substrate surface increases. In this case, a diffusion layer of the integration between sputtered atom and substrate atom is formed, which contributes to a better contact between the membrane and substrate as well as the increasing of smoothness and reflectivity of membrane.

Tab. 9-14 Relationship between sputtering power and thermal reflectivity

Sputtering power (W)	400	600	800
Thermal reflectivity (%)	77.18	76.43	78.86

(4) Effects of Sputtering Time on Thermal Reflectivity

Relationship between sputtering time and thermal reflectivity is demonstrated in

Tab. 9-15. The result shows that the deposition time has no significant influence on the thermal reflectivity of the coated fabric. It can be possibly because a complete and compact metallic reflective film forms on the fabric surface after depositing for 10 min, and the further increased time would not affect the thermal reflectivity.

Tab. 9-15　Relationship between sputtering time and thermal reflectivity

Sputtering time (min)	10	20	30
Thermal reflectivity (%)	76.56	76.43	76.90

(5) Effects of Substrate Temperature on Thermal Reflectivity

Relationship between substrate temperature and thermal reflectivity is illustrated in Tab. 9-16. It can be clearly observed that the thermal reflectivity of the coated fabric improved with an increased substrate temperature from 50 to 200 ℃ and then decreases with the temperature further increased from 200 to 300 ℃. It can be attributed to that an improved morphology structure of the thin film can be obtained when the thin film is prepared at a suitable high temperature, which contributes to an increased thermal reflectivity. However, surface diffusion of the deposited particles would occur and cause a rough thin film surface when the thin film is formed at a relatively high temperature. Consequently, the thermal reflectivity of the coated fabric decreases.

Tab. 9-16　Relationship between substrate temperature and thermal reflectivity

Substrate temperature (℃)	50	200	300
Thermal reflectivity (%)	76.43	79.99	75.56

Based on the analysis above, the coating parameters for good thermal reflection properties of the coated PSA fabric can be initially determined in the single-factor experiment (gas pressure = 0.4 Pa, argon flow rate = 15 mL/min, sputtering power = 400 W, sputtering time = 10 min, substrate temperature = 200 ℃). According to the single-factor experiment, the optimal coating parameters for further improved thermal reflection properties of the coated PSA fabric were investigated by orthogonal experiment, as shown in Tab. 9-17. The results indicate that the sputtering parameters for the optimal thermal reflection properties of the coated fabric are: gas pressure = 0.4 Pa, argon flow rate = 10 mL/min, sputtering power = 500 W, sputtering time = 30 min and substrate temperature = 150 ℃.

Tab. 9-17　Orthogonal experimental results

Experimental No.	Sputtering power (W)	Substrate temperature (℃)	Sputtering time (min)	argon flow rate (mL/min)	Thermal reflectivity (%)
1	100	room temperature	10	5	72.48
2	100	150	20	10	78.24
3	100	300	30	15	75.97

(continued)

Experimental No.	Sputtering power (W)	Substrate temperature (℃)	Sputtering time (min)	argon flow rate (mL/min)	Thermal reflectivity (%)
4	300	room temperature	20	15	75.70
5	300	150	30	5	81.34
6	300	300	10	10	76.98
7	500	room temperature	30	10	79.95
8	500	150	10	15	78.05
9	500	300	20	5	77.40
K_1	226.70	228.13	227.51	231.22	
K_2	234.02	237.63	236.34	235.17	
K_3	235.41	230.35	237.26	229.72	
R	8.71	9.5	9.72	5.45	
Influencing sequence	Sputtering power → heating temperature → sputtering power → Argon flow rate				
Optimal parameters	500	150	30	10	

Horizontal burning testing results of the PSA fabric and the magnetron-sputtered PSA fabric are shown in Tab. 9-18. Based on the data in Tab. 9-18, both the pure and coated fabric can reach the highest requirement of the standard GB 17591—2006 *Flame Retardant Fabric*. As can be seen in the table, the damaged length of the PSA fabric decreases after sputtering. The results indicate that the flame resistance of the PSA fabric can be improved by coating with aluminium thin film. The improved flame resistance of the coated PSA fabric can be attributed to that aluminum thin film with non-flammable properties can isolate the fabric substrate from air and then enhance the flame resistance of the coated PSA fabric.

Tab. 9-18 Horizontal burning testing results of PSA fabric and coated PSA fabric

Sample	PSA fabric	Coated PSA fabric
Ignition time (s)	12	12
Afterflame time (s)	0	0
Damaged length (mm)	8	5

9.3.2.2 Mechanical Properties

(1) Wear-resistance

The friction times were recorded when the breaking of the yarn occurred in the fabric, and the abrasion observed on the fabric surface could be used to investigate the wear-resistance of the fabric. The appearance of PSA fabric before and after sputtering is illustrated in Fig. 9-5. Based on the testing record, the yarns of the PSA fabric were broken when the friction times reached 6500; while the warp yarns of the coated PSA fabric were broken and the weft yarns were slightly broken with the same friction times for 6500. The results indicate that the wear-resistance of the PSA fabric can be improved after sputtering with aluminium thin films, which can increase the smoothness of the surface of the PSA fabric and decrease the friction coefficient.

 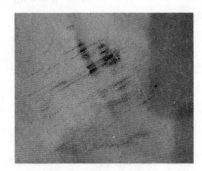

Before sputtering　　　　　　　　　　After sputtering

Fig. 9-5　Picture of PSA fabric before and after sputtering

(2) Tensile Properties

Parameters of the tensile properties of PSA fabric and coated PSA fabric are shown in Tab. 9-19. As can be observed in the table, although the CV of breaking force of the coated PSA fabric in the warp direction increases indicating a slightly decreased uniformity of the testing result, the breaking strength, breaking force, elongation at break and elongation rate at break of the fabric increase when deposited with aluminium thin film. The result reveals that the tensile properties of the PSA fabric can be improved after coating with aluminium thin film.

Tab. 9-19　Tensile properties of PSA fabric and coated PSA fabric

Index	Warp-wise of PSA fabric	Weft-wise of PSA fabric	Warp-wise of coated PSA fabric	Weft-wise of coated PSA fabric
Breaking strength (N/m)	17 925	15 205	18 575	15 350
Breaking force (N)	896.25	760.25	928.75	767.5
Elongation at break (mm)	31.45	23.3	32	26
Elongation rate at break (%)	31.45	23.3	32	26
Coefficient of variation (CV) of breaking force (%)	3.25	4.76	4.99	3.69

(3) Tearing Strength

Tearing strength of PSA fabric and coated PSA fabric is illustrated in Tab. 9-20. The results show that the average tearing strength of PSA fabric decreases after sputtering with aluminium thin film and the increasing *CV* of tearing strength of coated PSA fabric indicates a poor uniformity of the testing results. The reducing tearing strength of coated PSA fabric can be ascribed to the increasing slippage resistance among the fabric yarns after coating with aluminium thin film.

Tab.9-20　Tearing strength of the pure and coated PSA fabric

Index	Warp-wise of PSA fabric	Weft-wise of PSA fabric	Warp-wise of coated PSA fabric	Weft-wise of coated PSA fabric
Maximum (N)	71.59	86.06	32.23	53.63
Minimum (N)	66.69	81.22	29.93	49.41
Average (N)	69.14	83.64	31.08	51.52
CV (%)	5.02	4.09	5.22	5.79

9.3.3　Summary

In this section, aluminium thin film was coated on PSA fabric by using magnetron sputtering. Thermal reflection properties of the coated PSA fabric are investigated through single-factor and multi-factor orthogonal experiments, and the flame resistance and mechanical properties of the coated PSA fabric are also studied. The experimental results can be summarized as below.

Sputtering parameters for good thermal reflection properties of the coated PSA fabric can be initially determined in the single-factor experiment, and the value of gas pressure, argon flow rate, sputtering power, sputtering time and substrate temperature are 0.4 Pa, 15 mL/min, 400 W, 10 min and 200 ℃, respectively. The optimal sputtering parameters of further improved thermal reflection properties of the coated PSA fabric can be obtained by multi-factor orthogonal experiment, and the value of gas pressure, argon flow rate, sputtering power, sputtering time and substrate temperature are 0.4 Pa, 10 mL/min, 500 W, 30 min and 150 ℃, respectively. In addition, flame resistance, wear-resistance and tensile properties of the PSA fabric can be improved after sputtering with aluminium thin film, while the tearing strength of the coated PSA fabric decreases.

9.4　Study on Al-film Coated Aramid via Magnetron Sputtering

In this section, aluminium thin film was coated on aramid fabric by using magnetron sputtering. The relationship between thermal reflectivity and deposition parameters, including sputtering time and substrate temperature, is investigated by single-factor experiment. In addition, anti-thermal radiation properties, flame resistance, wear resistance, film flexibility, film hardness and adhesive fastness of the coated aramid fabric are also studied.

9.4.1 Experimental

9.4.1.1 Experimental Materials

Materials used for sputtering aluminium thin film include acetone (purity≥99.5%), absolute alcohol (purity≥99.7%), aluminium target with diameter of $\Phi 80 \times 6$ mm (purity: 99.99%), and argon gas (volume fraction: 99.99%). Aramid fabric is selected as substrate in this study and its specifications are shown in Tab.9-21.

Tab.9-21 Specification of aramid fabric

Material	Structure	Warp density [piece/(10 cm)]	Weft density [piece/(10 cm)]	Surface density (g/m²)
Aramid fabric	Plain	88	88	200

9.4.1.2 Experimental Solution

Sputtering parameters are selected based on the previous studies as illustrated in Tab.9-22.

Tab.9-22 Sputtering parameters of single-factor experiment

Experimental No.	Sputtering power (W)	Sputtering time (min)	Substrate temperature (℃)	argon flow rate (mL/min)	Gas pressure (Pa)
1	600	20	Room temperature	10	0.2
2	600	10	Room temperature	10	0.2
3	600	30	Room temperature	10	0.2
4	600	40	Room temperature	10	0.2
5	600	20	100	10	0.2
6	600	20	200	10	0.2
7	600	20	300	10	0.2

9.4.1.3 Preparation of Al Film

Prior to sputtering, aramid fabric (15 cm×15 cm) was ultrasonic cleaned in acetone solution for 30 min and then washed with deionized water for 3~5 times followed by drying in oven at 80 ℃ to remove moisture. Subsequently, plasma treatment for 6~7 min was conducted on the aramid fabric. The sputtering processing of Al film on aramid fabric can refer to section 9.3.1.3.

9.4.1.4 Characterization

(1) Anti-thermal Radiation and Thermal Reflection Properties

Anti-thermal radiation and thermal reflection properties of the coated fabric are studied by the method as referred to section 9.2.2 and section 9.3.1.4, respectively.

(2) Flame Resistance

Flame resistance of the coated fabric was investigated according to the standard FZ/T 01028—1993 *Textile Fabrics Burning Behavior Determination Horizontal Burning*. Horizontal burning was conducted on the coated fabric for 15 s to record the flame spreading distance, afterglow time and damaged length.

(3) Wear Resistance

Wear resistance of the coated fabric was studied by a disc-type surface friction tester. In this testing, a carbonized grinding wheel of A-100 and a pressurized counterweight of 1000 g are conducted on the samples. The number of revolution is then recorded when yarns in fabric break during the friction. 5~10 piece of samples with diameter of 125 mm were tested and the distance between the testing point and fabric edge should be greater than 100 mm.

(4) Flexibility

Flexibility of fabric is the capability of deformation without damage (cracking or spalling) under external force. The flexibility of the coated fabric was studied according to the standard GB/T 1731—1993 *Determination of Flexibility of Films*.

(5) Hardness

Hardness, which is considered as a significant index of mechanical properties, can be used to judge the damage resistance of thin film. The hardness of the aluminium thin film coated onto the aramid fabric was analyzed according to the standard GB/T 6739—2006 *Determination of Film Hardness by Pencil Test*.

(6) Adhesive Fastness

The influence of plasma treatment on the coated fabrics was studied by two methods based on the previous studies that the adhesive fastness of thin film can be identified according to the damage degree of thin film caused by friction. The two methods are described as below.

Method 1. ISO 12947-2: 1998 *Textiles-Determination of Abrasion Resistance of fbrics by the Martindale Method—Part 2: Determination of Specimen Breakdown*. This method is suitable for all textile fabrics except the fabrics with short wear life. Pilling tester (Martindale method) and No. 600 waterproof abrasive paper were selected in the friction damage experiment performed to the coated fabric before and after plasma treatment, with friction revolution of 500.

Method 2. GB/T 3920—2008 *Textiles—Tests for Color Fastness Color Fastness to Rubbing*.

a. Sample preparation: two pieces of samples with a dimension of 50 mm×200 mm (warp-wise × weft-wise) were selected and then fixed on the base plate of rubbing fastness tester. Prior to the testing, keep the length direction of the sample parallel to the traverse direction of the equipment.

b. Sample test: the rubbing cloth was fixed on the rubbing head of the tester, and kept the warp direction of the rubbing cloth parallel to the moving direction of rubbing head. Subsequently, 10 times of rubbing were performed along the length direction of the dry sample for 10 s.

c. Sample evaluation: the level was determined by comparing the color of rubbed cotton cloth with the standard color card.

9.4.2　Results and Discussion

9.4.2.1　Anti-thermal Radiation Properties

Anti-thermal radiation properties of aramid fabric and coated aramid fabric revealing by the temperature detected on the fabric surface are shown in Figs. 9-6 and 9-7, respectively. It can be clearly observed that the temperature detected on both obverse and reverse sides of the aramid fabric and coated aramid fabric remarkably increases with the increasing testing time. This is because the scattered heat of fabric is much lower than the absorbed heat arising from solar radiation with high intensity, in addition, much more heat can be reserved in fabric with the raising testing time, in this case, the detected temperature on the obverse and reverse sides of fabric increases significantly. Moreover, the both temperatures inspected on the observe side of the aramid fabric and coated aramid fabric are higher than that on the reverse side of the two fabrics. The temperature differences between both sides of the two fabrics can be attributed to continuously increasing absorbed heat which is greater than the scattered heat of fabric. What's more, the detected temperatures on both sides of the coated aramid fabric are obviously lower than that of the aramid fabric. The decreasing temperature of the coated aramid fabric indicates improving anti-thermal radiation properties of the aramid fabric after sputtering with aluminium thin film, which can increase the thermal reflectivity of the coated aramid fabric.

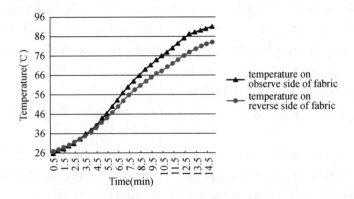

Fig. 9-6　Temperature on both sides of the aramid fabric

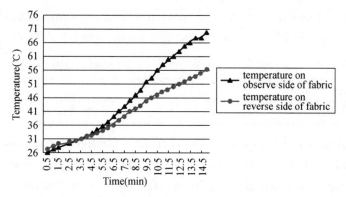

Fig. 9-7 Temperature on both sides of the coated aramid fabric

Comparison of temperature difference between both sides of the aramid fabric and coated aramid fabrics is shown in Fig. 9-8. As demonstrated in Fig. 9-8, the greatest temperature difference between the obverse and reverse sides of the aramid fabric reaches 9 ℃; whereas the greatest temperature difference between both sides of the coated aramid fabric reaches 14 ℃. The reason is that the coated aluminium thin film with high thermal conductivity almost covers the overall surface of the coated aramid fabric, therefore, the heat on the observe side of the coated aramid fabric absorbed from the thermal radiation can be greatly scattered into ambient environment and transmits only a little to the reverse side of the fabric. Consequently, the temperature difference between the obverse and reverse sides of the coated aramid fabric is greater than that of the aramid fabric.

By the analysis mentioned above, the anti-thermal radiation properties of the aramid fabric enhance remarkably after sputtering with aluminium thin film.

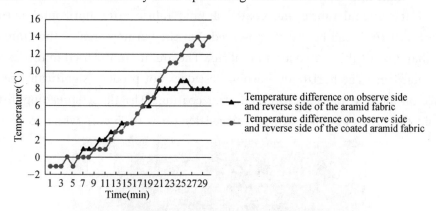

Fig. 9-8 Comparison of temperature difference between both sides of the aramid fabric and coated aramid fabric

9.4.2.2 Thermal Reflectivity

Influence of sputtering time on thermal reflectivity of the coated aramid fabric is shown in Tab. 9-23. It can be clearly observed that the thermal reflectivity of fabric increases with increasing sputtering time, that is because the film thickness increases due to the increasing sputtering time, and the more impact and complete film structure is formed in the film on fabric surface, which facilitates to increase the thermal reflectivity.

Tab. 9-23 Relationship between sputtering time and thermal reflectivity

Sputtering time (min)	20	30	40
Thermal reflectivity (%)	66.30	69.24	76.56

Influence of substrate temperature on thermal reflectivity of the coated aramid fabric is illustrated in Tab. 9-24. It can be seen that the thermal reflectivity increases with the substrate temperature raised from 50 to 200 ℃ and then slightly decreases with the sbustrate temperature further increased from 200 to 300 ℃. Diffusion of deposited Ar atoms or ions becomes vigorous when the substrate temperature increases, and then more activated sites acted as nucleation center of new phase of membrane can be generated on the substrate surface. In this case, more dense and smooth thin film can be formed and the thermal reflectivity of the coated fabric increases. However, further increased substrate temperature can lead to non-uniform structure of deposited film, and then the thermal reflectivity of the coated fabric decreases.

Tab. 9-24 Relationship between substrate temperature and thermal reflectivity

Substrate temperature (℃)	50	200	300
Thermal reflectivity (%)	66.30	67.09	66.90

9.4.2.3 Flame Resistance

Pictures of the aramid fabric and coated aramid fabric after horizontal burning are shown in Fig. 9-9. It can be clearly observed that the burning area of the aramid fabric is larger than that of the coated aramid fabric. The aramid fabric used in this experiment is flame retardant. The horizontal burning test does not produce significant combustion difference on the base fabric before and after coating. The flame spreading distance of the aramid fabric is 0.8 cm and its damaged length is 0.25 cm; while for the coated

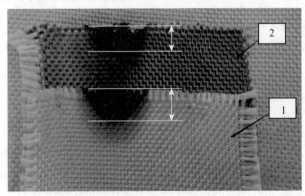

Fig. 9-9 Picture of fabrics after horizontal burning. (1—aramid fabric; 2—coated aramid fabric)

aramid fabric, the flame spreading distance is 0.6 cm and the damaged length is 0.15 cm. It is also reported that the damaged length of the aramid fabric reaches 0.8 cm when prepared by traditional coating. According to the analysis mentioned above, it can be concluded that the flame resistance of the aramid fabric can be improved after sputtering with aluminium thin film, and the magnetron sputtering can enhance the flame resistance of the aramid fabric when compared with the traditional coating technology.

9.4.2.4 Wear Resistance

Appearance of the aramid fabric and coated aramid fabric after friction experiment for 530 times is shown in Fig. 9-10. Severe abrasion can be observed on the aramid fabric surface after friction, while the surface of the coated aramid fabric maintains integrity with the same condition of friction. The results indicate that the anti-friction properties of the aramid fabric increases greatly after sputtering with aluminium thin film.

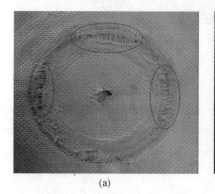
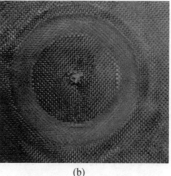

Fig. 9-10 Appearance after friction experiment for 530 times: (a) aramid fabric; (b) coated aramid fabric

Appearance of the coated aramid fabric after friction experiment for 1100 times is demonstrated in Fig. 9-11. It can be seen that the surface of the coated aramid fabric rubbed for 1100 times is similar to the surface of the aramid fabric rubbed for 530 times. The results suggest that the anti-friction properties of the coated aramid fabric are 2 times higher than that of the aramid fabric.

Fig. 9-11 Appearance of the coated aramid fabric after friction experiment for 1100 times

9.4.2.5 Flexibility

The experimental results of flexibility of the coated aramid fabric are supported by GB/T 1731—1993 *Determination of Flexibility of Films*. Flexibility effect of the coated aramid fabric is shown in Fig. 9-12. It can be observed that the aluminum-coated fabric presents a continuous and complete structure and exhibits perfect flexibility at the folding mode. The good flexibility of the coated aramid fabric is probably because that the deposited aluminium thin film on fabric is at nano-scale and it has no significant influence on the flexibility of fabric substrate.

Fig. 9-12　Flexibility effect of the coated aramid fabric

9.4.2.6　Hardness

The experimental results of hardness of the coated aramid fabric are supported by GB/T 6739—2006 *Determination of Film Hardness by Pencil Test*. Hardness of the coated aramid fabric is illustrated in Fig. 9-13. As seen in the figure, the scratched mark on fabric surface is caused by the pencil with 3H model. The results indicate that the magnetron-sputtered fabric is of moderate hardness and can properly maintain the original surface properties after coating treatment. Therefore, the magnetron-sputtered fabric can be applied in preparation of mechanical protective clothing for withstanding general damage.

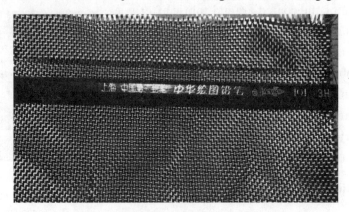

Fig. 9-13　Hardness of the coated aramid fabric

9.4.2.7　Adhesive Fastness

(1) Method 1. ISO 12947-2：1998 *Textiles Determination of Abrasion Resistance of Fab-*

rics by The Martindale Method—Part 2: Determination of Specimen Breakdown.

In this section, pilling tester and No. 600 waterproof abrasive paper as two processing methods were conducted for testing of adhesive fastness of the coated aramid fabric treated by plasma. The testing results of adhesive fastness of the coated aramid fabric are illustrated in Fig. 9-14. It can be clearly observed in Fig. 9-14 (a), severer pilling is observed on the fabric surface and the sputtered film is slightly fall off, whereas only slight pilling can be seen on the fabric surface. The aluminum film on the fabric surface goes grey with no fiber leakage, in general, the membrane structure on the fabric surface is complete.

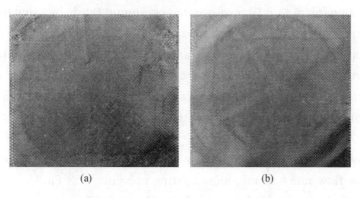

Fig. 9-14 Testing results of adhesive fastness of the coated aramid fabric by:
(a) Pilling tester; (b) No. 600 waterproof abrasive paper

(2) Method 2. GB/T 3920—2008 *Textiles—Tests for Color Fastness—Color Fastness to Rubbing*.

The testing results of adhesive fastness of the coated aramid fabric are shown in Fig. 9-15, and the comparison between the testing results and the standard grey card is shown in Fig. 9-16. Based on the comparison with the standards of GB/T 250—2008 *Grey Scale—for Assessing Change in Color*, it can be concluded that the friction-resistance fastness of the coated aramid fabric by Technology 1 is 4.5 while that by Technology 2 is 3.5; therefore, Technology 2 obviously surpasses Technology 1.

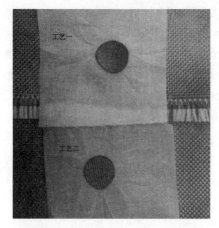

Fig. 9-15 Comparison of testing results

Fig. 9-16 Comparison with the standard grey card

From the experimental results of two above-mentioned methods, it can be concluded that the adhesive fastness of the film fabricated by Technology 2 increases remarkably during the pre-treatment of fabric substrate, therefore, the plasma treatment can strengthen the effects of magnetron-sputtered aramid fabric substrate significantly. Furthermore, the magnetron-sputtered film after plasma treatment exhibits excellent adhesive fastness.

9.4.3 Summary

In this section aluminium thin film was coated on aramid fabric by using magnetron sputtering. The experimental results are summarized as below.

(1) The optimized processing and parameters for preparation of coated aramid fabric by magnetron sputtering of substrate are given as below:

a. The pre-treatment process for fabric substrate: treating solution consisting of acetone and absolute alcohol→ultrasonic treatment for 30 min→drying for 30 min→treated by plasma for 7 min;

b. Coating parameters: coating power = 600 W, coating time = 20 min, room temperature, argon flow rate = 10 mL/min, coating pressure = 0.2 Pa.

(2) The anti-thermal radiation properties, thermal reflection properties, flame resistance, wear resistance and anti-friction properties of the aramid fabric can be remarkably improved after sputtering with aluminium thin film.

(3) The coated aluminium thin film on the aramid fabric has no significant influences on the flexibility and hardness of fabric substrate.

9.5 Preparation and Characterization of Al/SiO$_2$ Multi-layer Film Coated via Magnetron Sputtering

In this section, single-layer Al film, double-layer SiO$_2$/Al film, and three-layer SiO$_2$/Al/SiO$_2$ film with different membrane structures are fabricated on aramid and basalt fabrics by using magnetron sputtering. For different structures, Al is used as the metal reflective layer and SiO$_2$ is used as the dielectric and protective layer. In addition, morphological structure, crystal structure, ultraviolet reflectance, infrared reflection and anti-thermal radiation properties of the coated fabrics are systematically studied.

9.5.1 Experimental

9.5.1.1 Experimental Materials

Materials used in this study include acetone (purity≥99.5%), absolute alcohol (purity≥

99.7%), aluminium target with diameter of ϕ80 mm × 6 mm (purity: 99.99%), SiO_2 target with diameter of ϕ80 mm × 6 mm (purity: 99.99%), and argon gas (concentration: 99.99%). Specification of aramid fabric is illustrated in Tab. 9-25.

Tab. 9-25 Specification of aramid fabric

Fiber	Fabric weave	Warp density [piece/(10 cm)]	Weft density [piece/(10 cm)]	Surface density (g/m^2)
Aramid	Plain	312	237	173.73

9.5.1.2 Process Design for Single-layer Al Film

Al exhibits excellent reflectivity to both ultraviolet light (0.01~0.40 μm) and infrared light (0.7~500 μm). In addition, Al film can be protected by Al_2O_3 film (located on surface of Al film) with transparent, firm and stable properties. Therefore, Al film can be used for preparation of boards, foils and evaporated films with different shapes and also has been widely applied in preparation of reflective coating in various reflectors. The deposition parameters for fabrication of Al film are shown in Tab. 9-26. The sputtering processing of Al film on fabrics can refer to Chapter 9.3.1.3.

Tab. 9-26 Deposition parameters for fabrication of Al film

Item	Gas pressure (Pa)	Argon flow rate (mL/min)	Sputtering power (W)	Pre-sputtering time (min)	Sputtering time (min)
Parameter	0.8	40	200	3	15

9.5.1.3 Process Design for Double-layer SiO_2/Al Film

It should be noticed that Al is easy to oxidize when being coated on fabric and then the thermal reflection properties of Al film would be weaken. Dielectric materials with excellent light transmittance, such as SiO_2, ZnO and TiO_2, can be introduced to prevent Al from being oxidized. Aramid belonging to aromatic polyamide fiber exhibits poor binding force with inorganic metal materials, such as Al. However, SiO_2 as a semiconductor material presents excellent binding force with both metal materials and non-metal materials. In addition, SiO_2 film has been widely used in the fields of semiconductor, microwave, photoelectron, optical device and film sensor, due to its excellent properties, such as low refractivity ($n = 1.458$), superior light transmittance, as well as excellent insulativity, stability and mechanical properties[20], which prove that SiO_2 is particularly suitable used as the protective layer of metal reflective coating. Therefore, SiO_2 is considered as the middle layer, which can remarkably strengthen the binding force between Al and fabric substrate. The deposition parameters for fabrication of SiO_2 film are shown in Tab. 9-27. After the fabrication of SiO_2 film, Al film is coated on the SiO_2 film with deposition parameters shown in Tab. 9-26.

Tab. 9-27 Deposition parameters for fabrication of SiO$_2$ film

Item	Gas pressure (Pa)	Argon flow rate (mL/min)	Sputtering power (W)	Pre-sputtering time (min)	Sputtering time (min)
Parameter	0.5	60	200	3	30

9.5.1.4 Process Design for Three-layer SiO$_2$/Al/SiO$_2$ Film

The structure of three-layer SiO$_2$/Al/SiO$_2$ film is shown in Fig. 9-17. As shown in the figure, the dielectric layer (SiO$_2$) was coated firstly, and the reflective layer (Al) was coated subsequently, and the protective film (transmissive material: SiO$_2$) was coated finally, to fabricate the three-layer SiO$_2$/Al/SiO$_2$ film.

9.5.1.5 Preparation of Thin Film

Prior to deposition, aramid fabrics (15 cm×15 cm) were ultrasonicly cleaned in acetone solution for 30 min and then washed with deionized water for 3~5 times followed by drying in oven at 80 ℃ to remove moisture. The deposition of Al film, SiO$_2$/Al film and SiO$_2$/Al/SiO$_2$ film on fabrics by magnetron sputtering can refer to **Section 9.3.1.3**. The corresponding deposition parameters are illustrated in Tab. 9-25, Tab. 9-26 and Tab. 9-27, respectively.

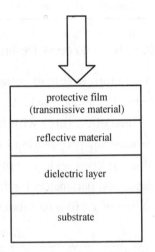

Fig. 9-17 Structure of three-layer SiO$_2$/Al/SiO$_2$ film

9.5.2 Property Test and Characterization of Film

9.5.2.1 Morphological Structure

Metal spraying by ion sputter was performed on the surface of the coated fabrics, subsequently, the surface microtopography of the films coated on fabric surface was characterized by a scanning electron microscope (SEM) (S-3400N, Japan) under the circumstances: resolution ratio=4 nm, magnification=300 000, and acceleration voltage=5~15 kV.

9.5.2.2 Crystal Structure

Crystal structure of the coated fabrics was studied by X-ray diffractometer (X'Pert Pro) under the circumstances: Kα ray from Cu target, Ni filtering, tube voltage=40 kV, tube current=40 mA, measuring wavelength λ=0.154 18 nm, scanning range 2θ=5°~90°, scanning speed=0.8 s/step.

9.5.2.3 Reflectivity

(1) UV-Vis-IR Reflectivity

Reflectivity (0.18~2.5 μm) of the films coated on fabric surface was investigated by UV-Vis-IR spectrophotometer.

(2) FTIR Reflectivity

Reflectivity (4000~400 cm^{-1} or 2.5~25 μm) of the films coated on fabric surface was investigated by FTIR spectrophotometer under the circumstances: attenuated total reflection, sampler: OMNI-Sampler™, scanning times=32, data interval=1.929 cm^{-1}, resolution ratio=8 cm^{-1}.

9.5.2.4 Anti-thermal Radiation Properties

Anti-thermal radiation properties of the coated fabrics are studied by the methods as described in **Section 9.2.2**.

9.5.3 Results and Discussion

9.5.3.1 Morphological Structure

SEM images with different magnifications of Al film, SiO_2/Al film and SiO_2/Al/SiO_2 film coated on fabric surface are shown in Fig.9-18. As can be seen from Fig.9-18(a) Ⅰ and (b) Ⅰ, the coated fabric surface is smooth. However, some of large particles are observed on the fabric surface when coated with SiO_2/Al/SiO_2 film, as shown in Fig. 9-18 (c) Ⅰ. As illustrated in Fig.9-18 (a) Ⅱ, small spherical Al particles are deposited on the fabric surface to form a uniform thin film. In this case, excellent mirror effect could be achieved due to the uniform and smooth membrane structure, thus satisfying the requirements of anti-heat-radiation. As can be observed in Fig.9-18 (b) Ⅱ, the particle size of the SiO_2/Al film is large and the particles present in geometric shape with superior smooth surface. Compared with the SiO_2/Al film, the SiO_2/Al/SiO_2 film [Fig. 9-18 (c) Ⅱ] exhibits a denser surface.

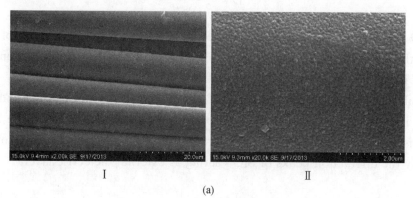

(a)

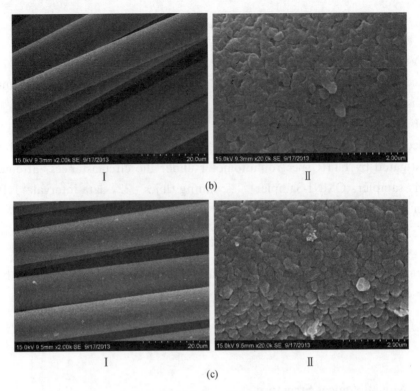

Fig. 9-18 SEM images of the coated fabrics with different magnifications: (a) Al film; (b) SiO$_2$/Al film; (c) SiO$_2$/Al/SiO$_2$ film. (I—2000; II—10 000)

9.5.3.2 Crystal Structure

XRD patterns of Al film, SiO$_2$/Al film and SiO$_2$/Al/SiO$_2$ film are shown in Fig. 9-19. Characteristic peaks are observed at 38.51°, 44.75°, 65.11° and 78.29°. According to the standards of pure aluminum (JCPDS cards 652869), 78.29° corresponds to (311) crystal face of Al. Therefore, it can be concluded that the metal Al is contained in all three membrane-structure. The average grain diameter of Al film can be calculated on basis of XRD patterns and the Scherrer formula. In addition, the Scherrer formula is expressed as follows:

$$D = \frac{0.89\lambda}{\beta \cos\theta}$$

where, D refers to the average grain diameter, λ refers to the wavelength of X-ray, β refers to the full width at half maximum (FWHM) of the strongest diffraction peak and expressed in radian, θ refers to the diffraction angle.

By calculation, the average diameter of grain in Al film, SiO$_2$/Al film and SiO$_2$/Al/SiO$_2$ film is about 28, 28 and 29 nm, respectively (calculation based on the first three characteristic peaks). In addition, the characteristic peaks within the range of 15°~35° are mainly the diffraction peaks of SiO$_2$ (acting as external dielectric layer).

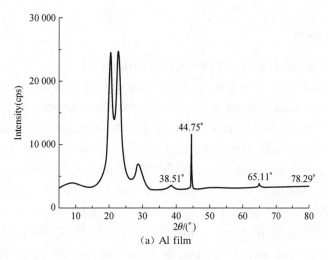

(a) Al film

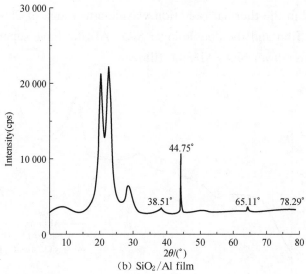

(b) SiO$_2$/Al film

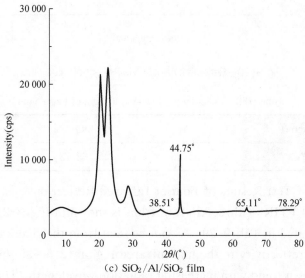

(c) SiO$_2$/Al/SiO$_2$ film

Fig. 9-19 XRD patterns of the coated films

9.5.3.3 Reflectivity

(1) Analysis of Test Results by UV-Vis-IR Spectrophotometer

Reflectivity of UV-Vis-IR light of three kinds of films is shown in Fig. 9-20. The corresponding data are illustrated in Tab. 9-28. As demonstrated in Fig. 9-20, the single-layer Al film and the double-layer SiO_2/Al film have the similar reflectivity, whereas the three-layer $SiO_2/Al/SiO_2$ film has a relatively lower reflectivity. It can be concluded that SiO_2 particles (acting as dielectric layer) located in the outermost layer can absorb thermal radiation, especially obvious in the UV and visible light region, which results in a lower reflectivity of multi-layer film than single-layer film. As illustrated in Tab. 9-28, the reflectivity of films within the range of 250~2500 nm could be achieved based on the integral calculation. It can be observed that all three films have certain heat reflection within the thermal radiation wavelength range of 250~2500 nm, where the single-layer Al film and the double-layer SiO_2/Al film have superior heat reflection than that of the three-layer $SiO_2/Al/SiO_2$ film.

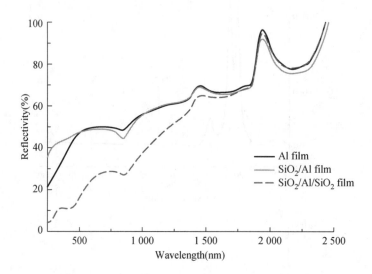

Fig. 9-20 Reflectivity of UV-Vis-IR light of three films

Tab. 9-28 Reflectivity of UV-Vis-IR light of three films

Membrane structure	Al	SiO_2/Al	$SiO_2/Al/SiO_2$
Reflectivity(%)	64.26	63.73	53.93

(2) Analysis of Test Results of Fourier Infrared Reflection

Reflectivity of FTIR of three kinds of films is shown in Fig. 9-21. The corresponding data are illustrated in Tab. 9-29. As can be seen from the figure and table, all three films have similar reflectivity in the mid-infrared region (2.5~25 μm), and the reflectivity presents a down trend with increasing radiation wavelength. The reflectivity of the three-layer $SiO_2/Al/SiO_2$ film within the range of 4000~1100 cm^{-1} (2.5~9.09 μm) is a little higher than those of other two films, and decreases to some extent (still a little higher than that of the single-layer Al film) within the range of 1100~400 cm^{-1} (9.09~25 μm).

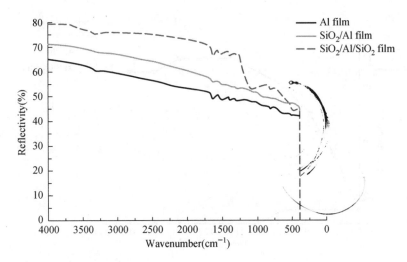

Fig. 9-21 Reflectivity of FTIR of three films

Tab. 9-29 Reflectivity of FTIR of three films

Membrane structure	Al	SiO_2/Al	$SiO_2/Al/SiO_2$
Reflectivity (%)	56.68	63.05	71.41

Combined with the data acquired by FTIR spectrophotometer and UV-Vis-IR spectrophotometer, the total reflectivity of three membrane structures within the range of 0.25~25 μm could be calculated. As illustrated in Tab. 9-30, the three-layer $SiO_2/Al/SiO_2$ film has the highest reflectivity, and then is the double-layer SiO_2/Al film, finally is the single-layer Al film. It can be clearly seen that the reflectivity increases with the increasing number of film layers.

Tab. 9-30 Total reflectivity of three films

Membrane structure	Al	SiO_2/Al	$SiO_2/Al/SiO_2$
Reflectivity (%)	51.59	57.38	64.97

9.5.3.4 Anti-thermal Radiation Properties

Temperature differences between back and face of fabrics are shown in Fig. 9-22. As can be seen from the figure, the temperature differences of the magnetron-sputtered fabrics are higher than that of the fabric substrate. The results indicate that the heat-insulating properties and anti-thermal radiation properties of the fabrics can be strengthened after coating with Al SiO_2/Al or $SiO_2/Al/SiO_2$ films. In addition, the greatest temperature differences between two sides of the single-layer fabric and the three-layer fabric reach 12 ℃, and these two fabrics exhibit superior anti-thermal radiation properties. The temperature difference between two sides of the double-layer fabric is about 5~6 ℃, and no temperature difference exists for the fabric substrate.

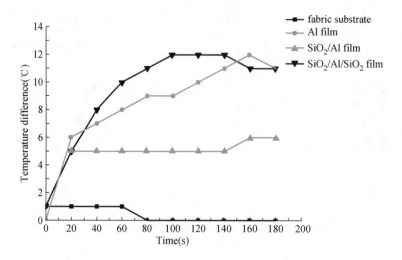

Fig. 9-22 Temperature differences between back and face of fabrics

9.5.4 Summary

In this study, Al film, SiO_2/Al film and $SiO_2/Al/SiO_2$ film are coated on aramid and basalt fabrics by magnetron sputtering. The experimental results show that compact nano-scale membrane structure consisting of Al and SiO_2 particles are formed on fabric surface after magnetron sputtering. In addition, the Al film has the smoothest surface with finest metal particles, whereas both the SiO_2/Al and the $SiO_2/Al/SiO_2$ films have rough surface with large particles. Moreover, Al's characteristic peak can be observed in XRD patterns of all the coated fabrics, and the particle diameter of Al is about 28~29 nm. In addition, the $SiO_2/Al/SiO_2$ film has the best thermal properties. The reflectivity within the range of 0.25~25 μm of $SiO_2/Al/SiO_2$, SiO_2/Al and Al films are 64.97%, 57.38% and 51.59%, respectively. The anti-thermal radiation properties of the fabrics can be improved after coating with Al, SiO_2/Al and $SiO_2/Al/SiO_2$ films.

9.6 Preparation and Characterization of Ag/ TiO₂ Film Coated PSA Fabric via Magnetron Sputtering

In this study, Ag/TiO_2 film is coated on PSA fabric substrate by magnetron sputtering. Morphological structure, crystal structure, FTIR reflectivity and heat-insulating properties of the coated fabrics are systematically studied. The optimal sputtering processing parameters for effective Ag/TiO$_2$ film is also investigated.

9.6.1 Experimental

9.6.1.1 Experimental Materials

Materials used in this study include acetone (purity≥99.5%), absolute alcohol (purity≥99.7%), silver target with diameter of ϕ80 mm×6 mm (purity: 99.99%), TiO$_2$ target with diameter of ϕ80 mm×6 mm (purity: 99.99%), and argon gas (volume fraction: 99.99%). The specification of PSA fabric is shown in Tab.9-31.

Tab.9-31 Specification of PSA fabric

Material	Warp density [piece/(10 cm)]	Weft density [piece/(10 cm)]	Surface density (g/m²)
PSA fabric	312	237	173.73

9.6.1.2 Design of Ag/TiO$_2$ Membrane Structure

It should be noticed that Ag is an easy-oxidized metal substance, and it can be oxidized into Ag$_2$O when being exposed to air. In this case, the thermal reflection properties of the silver film will be weakened. Therefore, dielectric layer with excellent light transmittance, such as SiO$_2$, ZnO or TiO$_2$ films, can be introduced to prevent Ag from being oxidized. The valence band of TiO$_2$ consists of $2p$ band of oxygen and its conduction band mainly consists of $3d$ band of titanium with great band width (approximately 3 eV), and it is difficult for TiO$_2$ to absorb visible light. Therefore, the structure of the coated fabric can be designed as shown in Fig.9-23 and TiO$_2$ is selected as the dielectric layer. In this case, the visible light could pass through the TiO$_2$ layer and arrive at the Ag layer.

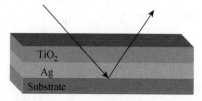

Fig.9-23 Structure of the coated fabric

9.6.1.3 Introduction of Orthogonal Experiment

$L_{27}(3^{13})$ orthogonal experiment is designed based on the sputtering parameters, including deposition pressure (A), sputtering time (B), sputtering power (C), argon flow rate (D), substrate temperature (E) of Ag film, and deposition pressure(F), sputtering time (G), sputtering power (H), argon flow rate (I), substrate temperature (J) of TiO$_2$ film, as shown in Tab.9-32 and 9-33.

Tab. 9-32 Level table of factors

	Ag					TiO$_2$				
A(Pa)	B(min)	C(W)	D(mL/min)	E(℃)	F(Pa)	G(min)	H(W)	I(mL/min)	J(℃)
0.9	30	20	30	100	0.7	20	100	10	300
0.7	20	60	10	200	0.9	40	200	20	100
0.5	10	40	20	300	0.5	30	300	30	200

Tab. 9-33 Orthogonal experimental program

Experimental No.	A	B	C	D	E	F	G	H	I	J
1	0.9	30	20	30	100	0.7	20	100	10	300
2	0.9	30	20	30	200	0.9	40	200	20	100
3	0.9	30	20	30	300	0.5	30	300	30	200
4	0.9	20	60	10	100	0.7	40	200	20	200
5	0.9	20	60	10	200	0.9	30	300	30	300
6	0.9	20	60	10	300	0.5	20	100	10	100
7	0.9	10	40	20	100	0.7	30	300	30	100
8	0.9	10	40	20	200	0.9	20	100	10	200
9	0.9	10	40	20	300	0.5	40	200	20	300
10	0.7	30	60	20	100	0.9	20	200	30	200
11	0.7	30	60	20	200	0.5	40	300	10	300
12	0.7	30	60	20	300	0.7	30	100	20	100
13	0.7	20	40	30	100	0.9	40	300	10	100
14	0.7	20	40	30	200	0.5	30	100	20	200
15	0.7	20	40	30	300	0.7	20	200	30	300
16	0.7	10	20	10	100	0.9	30	100	20	300
17	0.7	10	20	10	200	0.5	20	200	30	100
18	0.7	10	20	10	300	0.7	40	300	10	200
19	0.5	30	40	10	100	0.5	20	300	20	100
20	0.5	30	40	10	200	0.7	40	100	30	200
21	0.5	30	40	10	300	0.9	30	200	10	300
22	0.5	20	20	20	100	0.5	40	100	30	300
23	0.5	20	20	20	200	0.7	30	200	10	100
24	0.5	20	20	20	300	0.9	20	300	20	200
25	0.5	10	60	30	100	0.5	30	200	10	200
26	0.5	10	60	30	200	0.7	20	300	20	300
27	0.5	10	60	30	300	0.9	40	100	30	100

9.6.2 Characterization

9.6.2.1 Morphological Structure

The experimental equipment and methods are the same as those described in **Section 9.5.2.1**.

9.6.2.2 Crystal Structure

The experimental equipment and methods are the same as those described in **Section 9.5.2.2**.

9.6.2.3 Reflectivity

The experimental equipment and methods are the same as those described in **Section 9.5.2.3**.

9.6.2.4 Heat-insulating Properties

(1) Composition of Self-developed Heat-insulating System

The heat-insulating properties of the coated fabric were investigated by a self-developed heat-insluting testing system(Fig.9-24), which consists of data acquisition instrument (HORIZON), temperature transmitter (SBWR KO-300 ℃), thermocouple (TC-KBB2×0.5L=1500 mm), far-infrared radiation element (120 mm×120 mm), electronic temperature controller, and notebook computer.

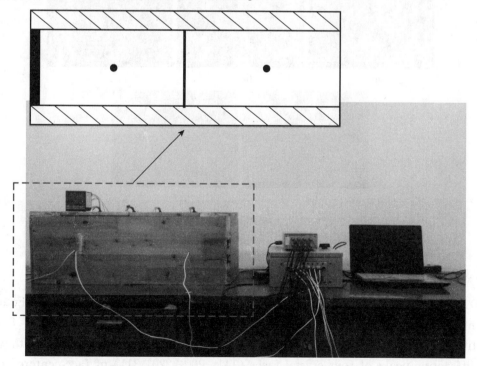

Fig.9-24 Self-developed heat-insluting testing system

(2) Test Method for Heat-insulating System

a. The coated surface of sample is faced to the heat source and placed in the middle part of the gap of heat-insulated cotton, and kept away from the far-infrared radiation element about 35 cm.

b. The far-infrared radiation element was enabled by connecting the power supply, and the internal temperature of the cabinet was initialized by an intelligent temperature controller, with distances between the former/latter point for measuring temperature and the heat source of 17.5 cm/60 cm.

c. Real-time display of temperature acquired by thermocouple was achieved by dedicated software, data acquisition instrument, temperature transmitter; in addition, the temperatures before/after each time node were recorded.

9.6.3 Results and Discussion

9.6.3.1 Morphological Structure

SEM images of the coated PSA fabric with different magnifications are shown in Fig.9-25. A thin film with smooth, flat, continuous and dense surface was observed on fibers. As illustrated in Fig.9-25 (d) and (e), particle agglomerations are discovered on film, and most of particles exhibit spherical shape.

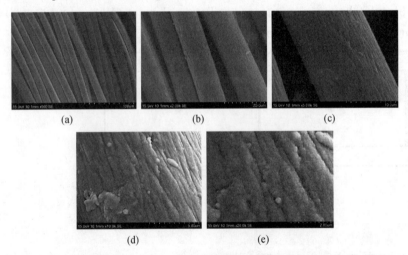

Fig.9-25 SEM images of the coated PSA fabric with different magnifications:
(a) 500; (b) 2000; (c) 5000; (d) 10 000; (e) 20 000

9.6.3.2 Crystal Structure

XRD pattern of Ag/TiO_2 film coated on PSA fabric is illustrated in Fig.9-26. Four apparent characteristic diffraction peaks are found at 38.0°, 44.2°, 64.3° and 77.5°, respectively. Compared with the standards of pure Ag (JCPDS cards 04-0783), they are the diffraction peaks of four crystal faces (111, 200, 220, 311) of face-centered cubic structure elemental silver, and it indicates that the nano-scale silver crystal of cubic crystal system is deposited on the fabric substrate. The average diameter of Ag grain in the Ag/TiO$_2$ film is calculated as 25 nm based on the Scherrer formula. This fine nano-scale membrane structure can contribute to fabricate continuous and compact metal film on fabric surface and form mirror effect to increase the thermal reflectivity of fabric.

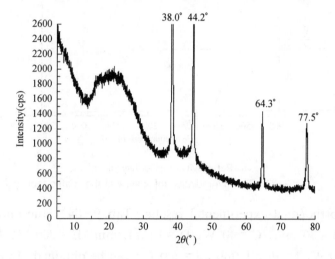

Fig. 9-26 XRD pattern of Ag/TiO$_2$ film coated on PSA fabric

9.6.3.3 Fourier Infrared Reflectivity

The thermal-radiation reflectivity of the coated fabrics is shown in Fig. 9-27. As demonstrated in the figure, the experimental No. 21 exhibits the greatest reflectivity of 81.6%, with specified parameters (A=0.5 Pa, B=30 min, C=40 W, D=10 mL/min, E=300 ℃; F=0.9 Pa, G=30 min, H=200 W, I=10 mL/min, J=300 ℃). As shown in Fig. 9-27 (b), the overall testing wavenumber range presents a high reflectivity, and the reflectivity presents a gentle downtrend from 4000 to 400 cm^{-1}.

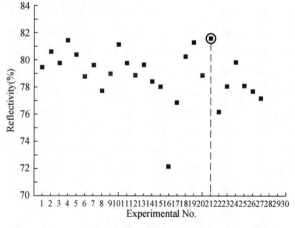

(a) Reflectivity of samples

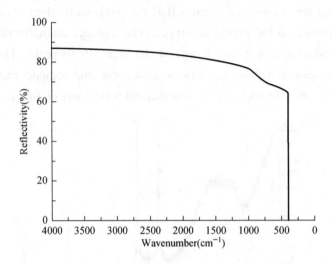

(b) Reflectivity of experimental No. 21

Fig. 9-27 Thermal-radiation reflectivity of the coated fabrics

From the orthogonal experimental results (Tab. 9-34), the optimum parameters (A = 0.9 Pa, B = 30 min, C = 40 W, D = 10 mL/min, E = 300 ℃; F = 0.7 Pa, G = 40 min, H = 300 W, I = 10 mL/min, J = 200 ℃) can be obtained. In addition, the primary and secondary sequence of affecting the reflectivity is as follows: B>H>A>J>C>G>I>E>F>D.

Tab. 9-34 Analysis table of orthogonal experiment

Experimental No.	A	B	C	D	E	F	G	H	I	J	Reflectivity(%)
1	0.9	30	20	30	100	0.7	20	100	10	300	79.44
2	0.9	30	20	30	200	0.9	40	200	20	100	80.62
3	0.9	30	20	30	300	0.5	30	300	30	200	79.75
4	0.9	20	60	10	100	0.7	40	200	20	200	81.42
5	0.9	20	60	10	200	0.9	30	300	30	300	80.41
6	0.9	20	60	10	300	0.5	20	100	10	100	78.77
7	0.9	10	40	20	100	0.7	30	300	30	100	79.64
8	0.9	10	40	20	200	0.9	20	100	10	200	77.74
9	0.9	10	40	20	300	0.5	40	200	20	300	79.01
10	0.7	30	60	20	100	0.9	20	200	30	200	81.12
11	0.7	30	60	20	200	0.5	40	300	10	300	79.77
12	0.7	30	60	20	300	0.7	30	100	20	100	78.86
13	0.7	20	40	30	100	0.9	40	300	10	100	79.64
14	0.7	20	40	30	200	0.5	30	100	20	200	78.39
15	0.7	20	40	30	300	0.7	20	200	30	300	78.02
16	0.7	10	20	10	100	0.9	30	100	20	300	72.12
17	0.7	10	20	10	200	0.5	20	200	30	100	76.85

(continued)

Experimental No.	A	B	C	D	E	F	G	H	I	J	Reflectivity(%)
18	0.7	10	20	10	300	0.7	40	300	10	200	80.23
19	0.5	30	40	10	100	0.5	20	300	20	100	81.30
20	0.5	30	40	10	200	0.7	40	100	30	200	78.88
21	0.5	30	40	10	300	0.9	30	200	10	300	81.60
22	0.5	20	20	20	100	0.5	40	100	30	300	76.15
23	0.5	20	20	20	200	0.7	30	200	10	100	78.03
24	0.5	20	20	20	300	0.9	20	300	20	200	79.81
25	0.5	10	60	30	100	0.5	30	200	10	200	78.10
26	0.5	10	60	30	200	0.7	20	300	20	300	77.68
27	0.5	10	60	30	300	0.9	40	100	30	100	77.14
K_1	717	721	703	709	709	712	711	698	713	704	
K_2	705	711	713	712	708	710	713	715	709	711	
K_3	709	699	714	710	713	708	707	718	708	715	
R	11.8	22.8	11.2	2.8	4.8	4.1	6.0	20.7	5.4	11.2	

9.6.3.4 Heat-insulating Properties

The fabric to be tested is placed in the self-developed heat-insulating testing system, subsequently, the temperature of former/latter point is recorded respectively and the temperature difference is calculated after the heat source is turned on to achieve the balance of former/latter point for measuring temperature (approximately 5~6 h). The relevant data are shown in Tab.9-35. It can be observed that the finial temperature of former/latter point of fabric substrate maintains at 78 ℃. Most of the temperature differences of former/latter point of double-film-coated fabrics increase, and the sample corresponding to experimental No. 27 presents the greatest temperature difference of 90 ℃. However, the temperature differences of some samples present a downtrend, that is because the TiO_2 film exhibits high thermal conductivity and superior thermal conductivity when being introduced. In this case, it can be deducted that the heat-insulating properties of the coated fabrics can be improved. The thermal reflectivity of the samples increases when the heat reflection effect is enhanced and it can overcome the effect of high thermal conductivity, whereas it decreases when the heat reflection effect fails to overcome the effect of high thermal conductivity. Because of the excellent heat-insulating properties and thermal reflection properties of the sample of experimental No. 21, the sputtering parameters for preparation of this sample are selected as the optimum processing.

Tab. 9-33 Resultant temperature differences of samples

Experimental No.	Voltage of former point(V)	Temperature of former point(℃)	Voltage of latter point(V)	Temperature of latter point(℃)	Temperature difference(℃)
Fabric substrate	2.93	144.75	1.89	66.75	78.00
1	2.85	138.75	1.85	63.75	75.00
2	2.91	143.25	1.89	66.75	76.50
3	2.86	139.50	1.87	65.25	74.25
4	2.88	141.00	1.86	64.50	76.50
5	2.92	144.00	1.87	65.25	78.75
6	2.87	140.25	1.85	63.75	76.50
7	2.89	141.75	1.84	63.00	78.75
8	2.95	146.25	1.83	62.25	84.00
9	2.97	147.75	1.84	63.00	84.75
10	2.92	144.00	1.85	63.75	80.25
11	2.93	144.75	1.83	62.25	82.50
12	2.95	146.25	1.81	60.75	85.50
13	2.90	142.50	1.83	62.25	80.25
14	2.93	144.75	1.81	60.75	84.00
15	2.93	144.75	1.82	61.50	83.25
16	2.96	147.00	1.79	59.25	87.75
17	2.96	147.00	1.80	60.00	87.00
18	2.96	147.00	1.82	61.50	85.50
19	2.98	148.50	1.82	61.50	87.00
20	2.99	149.25	1.82	61.50	87.75
21	2.99	149.25	1.82	61.50	87.75
22	3.00	150.00	1.82	61.50	88.50
23	3.00	150.00	1.82	61.50	88.50
24	2.99	149.25	1.81	60.75	88.50
25	2.99	149.25	1.81	60.75	88.50
26	2.98	148.50	1.81	60.75	87.75
27	2.99	149.25	1.79	59.25	90.00

9.6.4 Summary

In this part, Ag/TiO$_2$ films are coated on PSA fabric by magnetron sputtering. The experimental results show that the coated film on PSA fiber is smooth with excellent mirror effect and the average grain diameter of Ag reflective layer is 25 nm. Based on the orthogonal experimental results, the optimum parameters (A = 0.9 Pa, B = 30 min, C = 40 W, D = 10 mL/min, E = 300 ℃; F = 0.7 Pa, G = 40 min, H = 300 W, I = 10 mL/min,

J=200 ℃) can be obtained. In addition, the primary and secondary sequence affecting the reflectivity is given as: B>H>A>J>C>G>I>E>F>D. In addition, the highest reflectivity within the range of 4000~400 cm^{-1} can reach 81.6%, and the temperature difference of former/latter point in the heat insulation experiment can reach 90 ℃, which is 12 ℃ higher than the temperature difference before coating Ag.

References

[1] Wang X D, Liu X X, Lin L T. Investigation on fireproof fabrics' anti-thermal radiation and heat insulation properties. Shanghai Textile Science & Technology, 2012, 40(1): 1-4.

[2] Liu P. The development and trend of the industrial robot in our country. Robot Technique and Application, 2012(5): 20-22.

[3] Zhuang M Y, Liu X X, Wang T T. Research progress of robot protective fabric. Cotton Textile Technology, 2014(5): 78-82.

[4] Li J, Shi L H, Zhang W Y. High temperature resistant protective clothing and its development trend. China Personal Protective Equipment, 2005(1): 16-18.

[5] Liu Y. Preparation and property of high temperature resistance composite materials. Donghua University, 2011.

[6] Yang Z, Qing N. Current state for research and development of thermal insulating materials. New Chemical Materials, 2011, 39(5): 21-24.

[7] Hua T. Analysis of thermal protective performance of thermal protective clothing. Technical Textiles, 2002(8): 28-31.

[8] Chen J J, Yu W D. Research in thin flexible multilayer thermal insulation materials. Materials Review, 2008, 22(6): 41-44.

[9] Li W D. The main varieties of high temperature resistant fiber and its performances. China Fiber Inspection, 2011(9): 77-79.

[10] Cai Y D, Zhang S G. Production process of silver powder coated fabric. Beijing Textile Journal, 2002(3): 21-24.

[11] Lin X L, Zhou Y, Huang J W. Progress in surface modification of aramid fibers. Journal of Tianjin Polytechnic University, 2011, 30(3): 11-18.

[12] Wang H X, Xie G Y, Chen Q W. Surface modification methods of aramid fiber. Technical Textiles, 2012(2): 32-34.

[13] Tao A, Rumin W, Zhou W Y. Effect of grafting alkoxysilane on the surface properties of Kevlar fiber. Polymer Composites, 2007, 28(3): 412-416.

[14] Sproul W D, Legg K O. Opportunities for Innovation: Advanced Surface Engineering. Technomic Publishing Co., 1994.

[15] Kelly P J, Arnell R D. Magnetron sputtering: A review of recent developments and applications. Vacuum, 2000, 56(3): 159-172.

[16] Li Y, Wang H B, Gao W D. Structure and properties of nano-copper thin films deposited on carbon fiber farbic by magnetron sputtering. Journal of Textile Research, 2012, 33(9): 10-14.

[17] Chu C L, Bi S M, Bao J Y. Preparation and performance of magnetron sputtering nanometer TiO_2 antibacterial PBT/PET fabrics. New Chemical Materials, 2012, 40(5): 31-33.

[18] Ehiasarian A, Pulgarin C, John K W. Inactivation of bacteria under visible light and in the dark by

Cu films. Advantages of Cu-HIPIMS-sputtered films. Environmental Science and Pollution Research International,2012,19:3791-3797.

[19] Yang Q B,Yu W D. Thermal property of soybean protein fibers. Journal of Textile Research,2005(2):53-55.

[20] Zeng Q Y,Zheng X F. Review on current methods of SiO_2 thin film preparation. Vacuum,2009(4):36-40.

Chapter 10 Development of Functional PSA Nano-composites

Zhuoming Chen*, Binjie Xin, Shixin Jin and Wenjie Chen

PSA fiber is a new kind of high temperature resistant material and it has excellent heat resistance, flame retardancy and thermal stability[1]. Therefore, it can be used to prepare protective products for applications in aerospace, high-temperature environment and civil fields with flame retardant requirements. PSA fibers and related products are also widely used in military industry.

10.1 Preparation of PSA Composite Fibers by Wet-spinning

It is reported that raw PSA has high electrical resistance and poor ultraviolet resistance, which cause some difficulties in its manufacturing procedures and limit its application in the development of functional textiles. Therefore, it is an important work to improve the electrical conductivity and ultraviolet resistance of PSA fibers. It has been proved that CNT has excellent electrical conductivity, mechanical properties and thermal conductivity[2-3]; and TiO_2 has outstanding scattering and absorption of ultraviolet. In this section, PSA composite fibers, including PSA/CNT, PSA/TiO_2 and PSA/(CNT/TiO_2) fibers, are prepared by wet spinning method to enhance the electrical conductivity or ultraviolet resistance of PSA composites. The corresponding composite membranes are also fabricated by spin-coating.

Corresponding Author:
Zhuoming Chen
School of Fashion Technology, Shanghai University of Engineering Science, Shanghai 201620, China
E-mail: chenzhuoming178041@163.com

10.1.1 Experimental

10.1.1.1 Preparation of Materials

PSA polymer was used as spinning solution with an intrinsic viscosity of 2.0~2.5 dL/g and relative molecular mass of 462. Multi-walled carbon nanotubes (S-MWNT-1020, short for CNT) were blended as functional particles for improving the electrical conductivity of PSA fiber. CNTs, with diameter of 10~20 nm and length of 1~2 μm, were used after being treated with a mixed solvent of 70% nitric acid (20 mL) and 98% sulfuric acid (60 mL) for 2 h. The purity of the treated CNTs was about 93%. Rutile titanium dioxide (short for nano-TiO_2) with diameter of 30~50 nm and rutile content of about 99% was blended as functional particles for improving the ultraviolet resistance of PSA and used without further purification. Dimethylacetamide (DMAC) was selected as a dissolvent in this study. The above materials were supplied by Shanghai Tanlon Fiber Co., Ltd.

A certain amount of functional particles, including the treated CNT, TiO_2 and CNT/TiO_2, were pre-dispersed in DMAC using ultrasonic vibration for 30 min and then added into the PSA solution, respectively. PSA composite spinning solutions with various mass fractions of nano-particles can be prepared after mechanical stirring for 1 h and ultrasonic vibration for 2 h. The experimental data are shown in Tab. 10-1.

Tab. 10-1 Sample list of PSA composites

Samples with different mass fraction of functional particles (CNT or TiO_2)	Mass of PSA(g)	Mass of CNT(g)	Mass of TiO_2(g)	Volume of DMAC(mL)
PSA	100	0	0	0
PSA/1%CNT	100	0.121 2	—	1
PSA/3%CNT	100	0.371 1	—	3
PSA/5%CNT	100	0.631 6	—	5
PSA/7%CNT	100	0.903 2	—	7
PSA/1%TiO_2	100	—	501.42	1
PSA/3%TiO_2	100	—	498.23	3
PSA/5%TiO_2	100	—	501.82	5
PSA/7%TiO_2	100	—	497.28	7
PSA/1%(CNT/TiO_2)	100	0.060 6	0.060 6	1
PSA/3%(CNT/TiO_2)	100	0.185 6	0.185 6	3
PSA/5%(CNT/TiO_2)	100	0.315 8	0.315 8	5
PSA/7%(CNT/TiO_2)	100	0.451 6	0.451 6	7

10.1.1.2 Preparation of PSA Composite Fibers by Wet Spinning

PSA fiber and PSA composite fibers were prepared by a small-scale single-screw wet spinning apparatus as shown in Fig.10-1.

The spinning solution was added into the barrel and then flow to the spinning nozzle with a pore size of (0.18 ± 0.03) mm. As shown in Fig.10-1, the spinning trickling is pressed out from the spinning nozzle under the nitrogen pressure and then into the water bath. The nascent fibers were formed after the solvent precipitating into the coagulation bath and then dried in electrical blast oven for 24 h to finish the thermosetting process and remove the residual solvent.

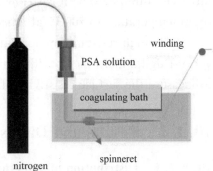

Fig. 10-1 Schematic diagram of wet spinning process

10.1.1.3 Preparation of PSA Composite Membranes

PSA composite membranes were prepared by using a SJT-B digital spin-coating instrument. An appropriate amount of spinning solution was used to spread on the substrate of the instrument into a thin liquid film at rotation speed of 2000 r/min for 5 s and 4000 r/min for 20 s. Then the membranes was put into water to remove the solvent. The PSA composite membranes with different contents of functional particles can be obtained after drying in electrical blast oven for 24 h. The thickness of the PSA composite membranes was about 1 mm. The PSA composite membranes were prepared for electrical conductivity test.

10.1.1.4 Characterizations

S-3400N scanning electron microscope (SEM) with a resolution at nano-scale was used to characterize the morphological structure of the PSA composite fibers. The machine was operated at 10~15 kV.

American AVATAR 370 fourier transform infrared spectroscopy (FTIR) was used to investigate the molecular structure and chemical composition of the PSA composite fibers. The spectra data were recorded from 4000 to 500 cm^{-1}, with a 4 cm^{-1} resolution over 32 scans and the step size of about 1.929 cm^{-1}.

K780 FirmV_06 X-ray diffraction (XRD) was used to characterize the crystalline structure of the PSA composite fibers using CuKα radiation (λ = 0.154 nm) at a voltage of 40 kV and a current of 40 mA. The spectra were obtained at 2θ of 5°~90° with a scanning speed of 0.8 s/step.

YG006 electronic single fiber strength tester was used to investigate the mechanical properties of the PSA composite fibers. The sample gauge length was 10 mm. The elongation speed was 20 mm/min. The measurements for each sample were conducted for 10 times and the average value was used for results analysis.

Germany STA PT-1000 thermal gravimetric analyzer (TGA) was used to investigate the thermal stability of the PSA composite fibers. TGA experiment was carried out in nitrogen atmosphere with gas flow rate of 80~100 mL/min. Samples were heated from room temperature to 700 ℃ at a heating rate of 20 ℃/min.

ZC36 high resistance meter with a measuring range of $10^8 \sim 10^{17}$ Ω and UT70A universal Ddgital multi-meter with a higher limit of measuring range about 10^8 Ω were used to measure the surface resistivity of the PSA composite membranes.

10.1.2 Results and Discussions

10.1.2.1 Distribution of Functional Particles in PSA Composites

As shown in Fig.10-2, a small mass fraction of CNT (1% or 3%) is dispersed uniformly in PSA matrix. The size of most nano-particles is about 30~50 nm, however, as light aggregation is observed, the size is about 100 nm. When the mass fraction of CNT is increased to 5%, the distribution of nano-particles becomes inhomogeneous. With the mass fraction of CNT increasing to 7%, the agglomeration is more serious and it is hard for CNT to disperse evenly throughout the PSA matrix, and the size of aggregated particles is about 150~200 nm.

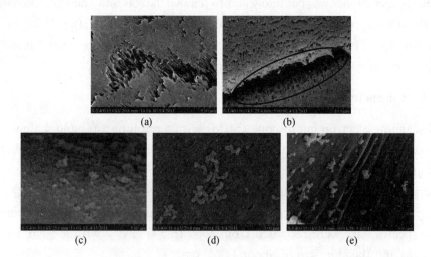

Fig. 10-2　SEM images of partial cross-sectional of PSA/CNT composite fibers: (a) PSA; (b) PSA/1%CNT; (c) PSA/3%CNT; (d) PSA/5%CNT; (e) PSA/7%CNT

As demonstrated in Fig.10-3, 1% TiO_2 (mass fraction) is dispersed evenly throughout PSA matrix and the size of nano-particles is about 50~60 nm. With the mass fraction of TiO_2 is increased to 3%, a slight aggregation can be observed. When the mass fraction of TiO_2 is increased to 5% and 7%, its dispersion in PSA becomes inhomogeneous because of the large specific surface and high surface polarity, and the aggregation size is about 100~300 nm. It is difficult for TiO_2 with high mass fractions to distribute uniformly in the blending system.

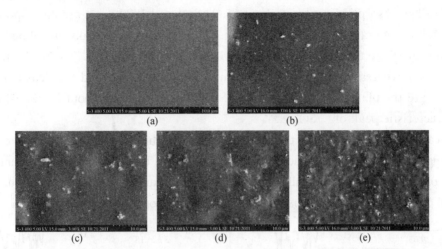

Fig. 10-3 SEM images of PSA/nano-TiO_2 composite membranes: (a) PSA; (b) PSA/1%TiO_2; (c) PSA/3%TiO_2; (d) PSA/5%TiO_2; (e) PSA/7%TiO_2.

As illustrated in Fig. 10-4, 1%(CNT/TiO_2) (mass fraction) is distributed homogeneously in PSA matrix and its size is about 30~50 nm. When the mass fraction of (CNT/TiO_2) is increased to 3%, a slight aggregation is observed and its size is about 100 nm. With an increased mass fraction of (CNT/TiO_2) from 5% to 7%, the distribution of functional particles in PSA matrix becomes inhomogeneous and it is hard for nano-particles with high mass fraction to disperse evenly in the blending system, the size of the aggregations is about 150~300 nm.

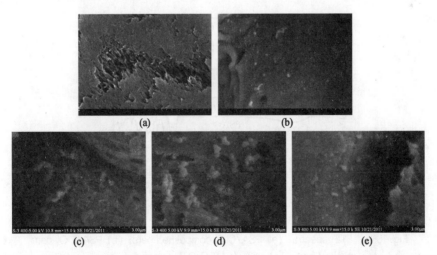

Fig. 10-4 SEM images of partial cross-sectional of fibers: (a) PSA; (b) PSA/1%(CNT/TiO_2); (c) PSA/3%(CNT/TiO_2); (d) PSA/5%(CNT/TiO_2); (e) PSA/7%(CNT/TiO_2)

10.1.2.2 Fourier Transform Infrared Spectral

FTIR spectra of PSA composite fibers are shown in Fig. 10-5. The absorption peaks exhibiting at about 3300~3339 cm^{-1} attribute to the stretching vibration of N—H in amido bond. The absorption peaks at about 1660 cm^{-1} indicate the stretching vibration of —C=O. The absorption peaks at about 1590 cm^{-1} and 1530 cm^{-1} are corresponding to the stretching vibration of benzene ring. The absorption peaks at about 1500~1300 cm^{-1} attribute to the plane bending vibration caused by C—H. The peaks at 1300~1000 cm^{-1} are corresponding to the skeleton vibration of C—C and the peaks at 1000~650 cm^{-1} are the plane bending vibration of C—H. The peak at about 1 149. 94 cm^{-1} is the characteristic peak of —SO$_2$—.

As can be seen in Fig. 10-5, the characteristic peaks shift from 3 338. 99 cm^{-1} to a short-wave range after addition of nano-particles and that can be attributed to the quantum size effect[2] of nano-particles. Generally, it is no significant changes to the molecular structure and chemical composition of PSA fibers with addition of functional nano-particles.

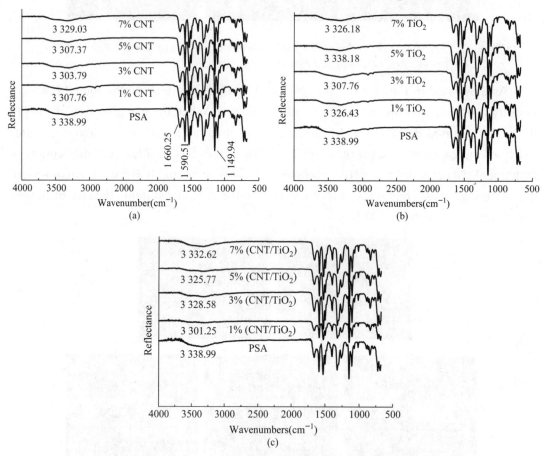

Fig. 10-5 FTIR spectra: (a) PSA/CNT composite fibers; (b) PSA/TiO$_2$ composite fibers; (c) PSA/(CNT/TiO$_2$) composite fibers

10. 1. 2. 3 Crystallization Structure

Fig. 10-6 shows XRD spectra of PSA composite fibers blending with different functional nano-particles. It can be found in Fig. 10-6 (a) and (b), all samples show diffraction peaks at 11.85° and 21.25°. The obviously sharp diffraction peaks at 11.85° of PSA composites compared with that of PSA fiber indicate there are crystalline structures in PSA composite fibers[4]. It indicates that the crystallization of PSA can be improved with blending of TiO_2, which can act as a nucleation agent. What's more, the shape of the diffraction peaks at about 21.25° of PSA composites broadens obviously with increasing mass fraction of nano-particles and this proves that the size of crystal region becomes smaller[5]. As depicted in Fig. 10-6 (b), the PSA composite fibers have diffraction peaks at 27.54°, 36.15°, 41.35° and 54.40°, this is because of the blending of nano-TiO_2[6-7]. As shown in Fig. 10-6 (c), all samples have diffraction peaks at about 20° and there is a shift excursion of PSA composites compared with PSA fiber. In addition, the sharpness of the diffraction peaks at about 20° is enhanced gradually with the increasing mass fraction of CNT/TiO_2. It indicates that the crystallization of PSA/CNT/TiO_2 composites is improved because CNT/TiO_2 can act as a nucleation agent. Moreover, diffraction peaks of rutile titanium dioxide are also observed at 27.45°, 36.15°, 41.35° and 54.40° because of TiO_2.

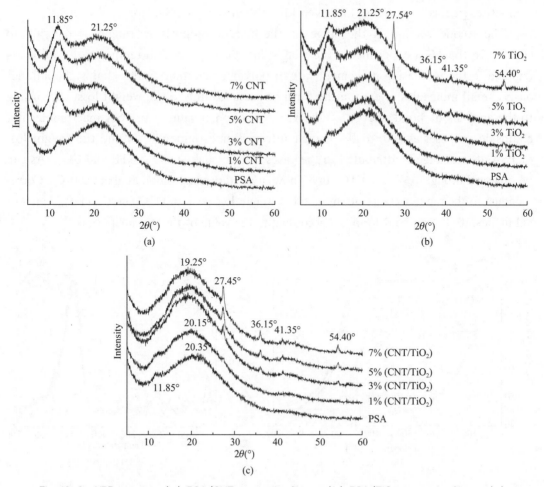

Fig. 10-6 XRD spectra: (a) PSA/CNT composite fibers; (b) PSA/TiO_2 composite fibers; (c) PSA/CNT/TiO_2 composite fibers

10.1.2.4 Thermal Stability

Fig. 10-7 presents TG and DTG curves of PSA composites. The key parameters of these curves are summarized in Tab. 10-2. As shown in Fig. 10-7, the thermal decomposition of all samples can be divided into three sections.

The first section is the stage of small weight loss from room temperature to 400 ℃. At this stage, the volatilization of additives and the bound water among molecules lead to the reduction of material weight from room temperature to about 100 ℃. With the temperature continuously increased to 400 ℃, the decomposition of oligomers with a small molecular weight leads to the weight loss of composites. As shown in Fig. 10-7 (a), the changes of weight loss of each PSA/CNT composite present a similar trend, but PSA fiber has a significantly high weight loss compared with the other samples. It suggests that the thermal stability of PSA blending with CNT can be improved markedly. Similar results can be also observed in Fig. 10-7 (b) and (c). However, as depicted in Tab. 10-2, as the mass fraction of CNT/TiO_2 is increased to 5% and 7%, T_{10wt} of PSA composites begins to decrease and even lower than that of PSA fiber.

The second section is the stage of thermal decomposition ranging from 400 to 600 ℃. In the TGA experiment, the weight loss during the high temperature can be attributed to the increasing movement rate of polymer macromolecular chains. Simultaneously, small molecules release in the form of gases leading to the weight loss. According to the analysis of bond energy[8], the C−N section of amide in PSA macromolecular chains decomposes at 500~600 ℃ under nitrogen environment[9-10], and the weight loss of PSA at this stage is attributed to the gases release such as SO_2, NH_3 and CO_2. As can be observed in Fig. 10-7, each TG curve shows a rapid decomposition at about 500 ℃. Corresponding to the rapid decomposition there is a weight loss peak in DTG curve and T_{max} (presented in Tab. 10-2) can be determined according to the value of the maximum peak[11].

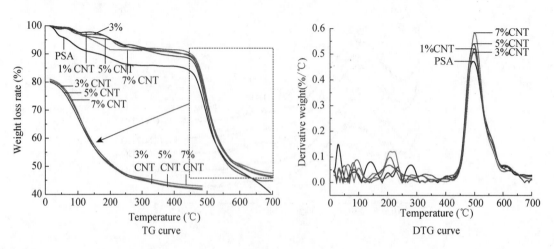

(a) PSA/CNT composites

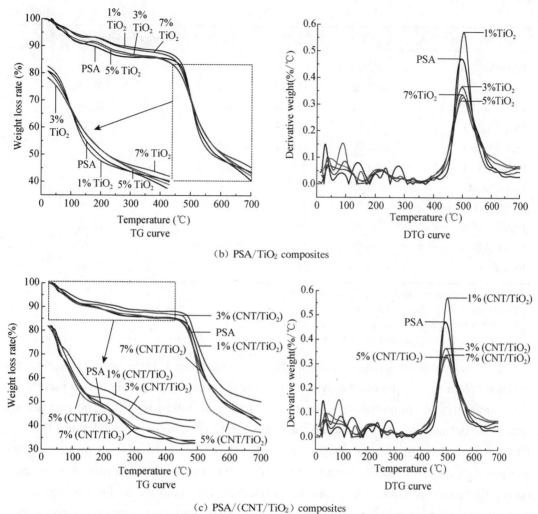

Fig. 10-7 DSC curves of PSA composites with different mass fraction of functional particles

The third section is the high-temperature phase of carbon formation ranging from 600 to 700 ℃. At this stage, most of polymers are carbonized and the rising temperature has less impaction on the weight loss of the residues. As depicted in Fig. 10-7, the residual mass at the terminal temperature of most of PSA composites tends to a steady state, while the weight loss of PSA has been decreasing. As illustrated in Tab. 10-2, the residual mass of most of PSA composites at 700 ℃ is higher than that of PSA fiber indicating that the thermal stability of PSA composite fibers blending with functional nano-particles can be improved significantly.

Tab. 10-2 Parameters of different PSA composite fibers during the thermal decomposition

Sample	T_0(℃)	$T_{10\%}$(℃)	T_{max}(℃)	Residual weight rate at 700 ℃ (%)
PSA	460.90	170.19	495.41	40.52
PSA/1%CNT	467.06	352.36	496.42	44.74

(continued)

Sample	T_0(℃)	$T_{10\%}$(℃)	T_{max}(℃)	Residual weight rate at 700 ℃ (%)
PSA/3%CNT	467.79	425.80	498.81	46.06
PSA/5%CNT	469.93	425.79	497.74	46.58
PSA/7%CNT	468.47	450.38	499.75	47.48
PSA/1%TiO_2	472.55	273.13	501.42	43.75
PSA/3%TiO_2	462.52	213.17	498.23	43.34
PSA/5%TiO_2	458.10	201.01	501.82	42.32
PSA/7%TiO_2	460.05	252.82	497.28	45.44
PSA/1%(CNT/TiO_2)	470.56	251.15	481.46	50.38
PSA/3%(CNT/TiO_2)	460.76	212.26	476.08	42.81
PSA/5%(CNT/TiO_2)	466.37	153.08	497.62	37.61
PSA/7%(CNT/TiO_2)	462.38	159.21	488.89	42.62

The test with the coefficient of variation less than 2%;
T_0—Initial decomposition temperature;
$T_{10\%}$—Temperature corresponding to weight loss rate of 10%;
T_{max}—Temperature at the maximum decomposition rate.

The thermal decomposition rate of the PSA composites is influenced by thermal resistance (R) and thermal conductivity (λ) of the blending system. These two factors are determined by the comprehensive function of PSA and nano-particles. The thermal resistance is defined as a ratio between the temperature of sample's two sides and the heat flow through the unit area of sample vertically. It is one of the indexes to measure the thermal properties of materials and it can reflect the heat resistant ability. On the contrary, the thermal conductivity can be defined as the quantity of heat which passes through a plate of particular area and thickness in the unit time when its opposite face differ in temperature by one Kelvin.

In the TGA experiment, heat concentrates on the surface of PSA, and then PSA decomposes at some points when temperature increases to its decomposition temperature. Since the thermal conductivity of PSA is low [0.084 W/(m·K)][12] and the sequence of reaching the decomposition temperature of PSA is: the surface layer→the middle layer→inner. Therefore, PSA decomposes layer by layer when it's heated at a high temperature.

Taking the functional particle CNT for an example, it has excellent thermal stability and thermal conductivity [6000 W/(m·K)][13], the two ends and the area around the tubes of dispersed CNT can play a role of physical cross linking points in the polymer matrix. In addition, according to the Fourier's law, the surface temperature of PSA is higher than the temperature of its adjacent internal layer during the heating process. Therefore, the temperature gradient field can be constituted to transfer the external heat into the polymer layer by layer. As a result, the thermal decomposition behavior of the PSA/CNT composites can be postponed.

Under these circumstances, the simplified heat transferring model[14] (as shown in Fig. 10-8) filled with particles and the corresponding thermal conductivity formula [as shown in Eq. (10-1)][15] can be employed to calculate the theoretical thermal conductivity and thermal resistance of the PSA composites.

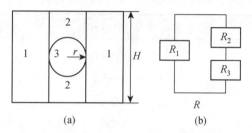

Fig. 10-8 Simplified heat transferring model of PSA composites: (a) The transferring paths of heat; (b) The series and parallel models of composites. (1—polymer; 2—the interface of the two phases; 3—nano-particles; r—the radius of nano-particles; H—the side of the cube; R—thermal resistance)

$$\lambda_c = \lambda_m \left[1 - \pi\left(\frac{3\varphi_f}{4\pi}\right)^{\frac{2}{3}}\right] + \frac{\pi\lambda_f\lambda_m}{\lambda_f\left[\left(\frac{3\varphi_f}{4\pi}\right)^{-\frac{2}{3}} - 2\left(\frac{3\varphi_f}{4\pi}\right)^{-\frac{1}{3}}\right] + 3\lambda_m\left(\frac{3\varphi_f}{4\pi}\right)^{-\frac{1}{3}}} \quad (10\text{-}1)$$

According to the Fourier's law, the transferring paths of heat [as shown in Fig. 10-8 (a)] in the PSA/CNT composites are: PSA (1)→the interface of the two phases between PSA and CNT (2)→CNT (3); or CNT (3)→the interface of the two phases between PSA and CNT (2)→PSA (1). Fig. 10-8 (b) is the R of the composites constituted by the series R between the interface of the two phases (R_2) and CNT (R_3) and the parallel R of the peripheral PSA (R_1).

The relationship between nano-particles' volume fraction (φ_f) and its mass fraction (w_f) can be obtained according to following formulas[16]:

$$\varphi_f = \frac{w_f}{w_f(1-x)+x}$$

$$x = \frac{\rho_f}{\rho_m} \quad (10\text{-}2)$$

In above formulas, λ_c, λ_m and λ_f is the thermal conductivity of composites, polymer matrix and filling particles respectively; φ_f is the volume fraction of filling particles in the composites; ρ_f is the density of filling particles (1.40 g/cm³); ρ_m is the density of polymer matrix (1.42 g/cm³).

The thermal resistance is relevant to the thickness (d) and thermal conductivity (λ) of the materials as shown in Eq. (10-3). The lower value of thermal resistance is, the better thermal conductivity performance of the materials will be.

$$R = \frac{d}{\lambda} \tag{10-3}$$

The thermal conductivity and thermal resistance (the reciprocal of thermal conductivity) of PSA/CNT composites with same thickness of 1 mm are presented in Tab. 10-3.

Tab. 10-3 The theoretical thermal conductivity and thermal resistance of PSA/CNT composites

Sample	$\lambda/[W/(m \cdot K)]$	$R/[(m^2 \cdot K)/W]$
PSA	0.084 0	0.011 9
PSA/1%CNT	0.085 0	0.011 7
PSA/3%CNT	0.090 0	0.011 1
PSA/5%CNT	0.096 0	0.010 4
PSA/7%CNT	0.102 0	0.009 8

As illustrated in Tab. 10-3, the thermal conductivity of PSA/CNT composites increases with increasing mass fraction of CNT, and it is higher than that of PSA fiber. However, the thermal resistance of PSA/CNT composites presents a decreasing trend as the mass fraction of nano-particles increased. In this case, the heat can pass from PSA to CNT through their interface with increasing temperature. Moreover, CNT has excellent thermal properties and it is difficult to decompose at a high temperature. Therefore, CNT can slow down the thermal decomposition rate and increase the residual mass of the PSA composites. Consequently, the blending of CNT in PSA matrix can improve the thermal stability of the PSA composites.

10.1.2.5 Functional Properties

PSA fiber is a new kind of flame retardant material. It has excellent heat resistance and flame retardancy, as well as excellent thermal stability, etc. However, raw PSA has high electrical resistance and poor ultraviolet resistance, which can cause some difficulties in its manufacturing procedures and limit its application in the development of functional textile products. Therefore, it is an important work to improve the electrical conductivity and ultraviolet resistance of PSA fibers. It has been proved that CNT has excellent electrical conductivity, mechanical properties and thermal conductivity, and TiO_2 has outstanding scattering and absorption of ultraviolet. It is feasible to blend CNT, TiO_2 or CNT/TiO_2 into PSA to enhance the electrical conductivity or ultraviolet resistance of PSA composites.

(1) Electrical Conductivity

The content of conductive particles in composite is one of the important parameters which can influence the electrical conductivity of the composite. Because CNT shows a high aspect ratio, it is easily for CNTs to establish electrical conductive network throughout the PSA matrix. Moreover, CNT is described to have large numbers of freely movable electrons. Therefore, the electrical conductivity of PSA composites can be improved obviously by blending of CNT.

A sketch[17] presented in Fig. 10-9 is employed to simulate the formation of electrical conductive network in the PSA/CNT composites. It is difficult for CNT at a low mass fraction (1%) to form a completely conductive network. With the mass fraction of CNT continuously increased to the percolation threshold (3%), the distance among the particles begins to decrease. The local conductive network in PSA matrix can be established basically when the maximum distance is less than 10 nm. In this case, the electrical resistivity of the PSA composites declines sharply and the conductivity can be improved. With the further increased mass fraction of CNT, more and more conductive particles connect directly and the completely conductive network can be developed (5% or 7%).

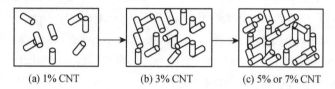

Fig. 10-9　Sketch of formation of electrical conductive network in PSA matrix

Similarly, the electrical conductivity of the PSA composites can be improved significantly by blending of CNT/TiO_2, and it can be increased gradually with the increased mass fraction of CNT/TiO_2 because of the gradually developed conductive network as shown in Fig. 10-10.

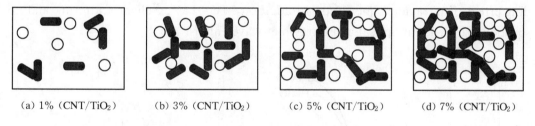

Fig. 10-10　Sketch of formation of electrical conductive network in PSA matrix

As can be seen in Tab. 10-4, the surface resistivity of PSA composites shows a downward trend with the mass fraction of CNT or CNT/TiO_2 increased, indicating an improved electrical conductivity (the reciprocal of resistivity) of the blending system. However, the improving degree of electrical conductivity decreases as the mass fraction of CNT continuously increased from 3% to 7% and the mass fraction of CNT/TiO_2 increased from 5% to 7%. This suggests that it is difficult for CNT with high mass fraction to distribute evenly in the PSA matrix and impairs the outstanding electrical properties of CNT correspondingly.

Tab. 10-4 Surface resistivity of PSA composites

Samples	Average of surface resistivity(Ω)	Samples	Average of surface resistivity(Ω)
PSA	3.10×10^{12}	PSA/1%(CNT/TiO$_2$)	8.38×10^{11}
PSA/1%CNT	1.13×10^{11}	PSA/3%(CNT/TiO$_2$)	1.84×10^{10}
PSA/3%CNT	5.96×10^{6}	PSA/5%(CNT/TiO$_2$)	3.78×10^{6}
PSA/5%CNT	4.70×10^{5}	PSA/7%(CNT/TiO$_2$)	5.98×10^{5}
PSA/7%CNT	2.10×10^{4}	—	—

(2) Ultraviolet Resistance

As exhibited in Fig. 10-11, the ultraviolet transmittance of specimens ranging from 390 to 400 nm decreases gradually with the mass fractions of TiO$_2$ increased. This suggests that TiO$_2$ can improve the ultraviolet resistance of PSA composites significantly. That is because the refraction index (*RI*) of TiO$_2$ is extremely high (2.73) and it has excellent ultraviolet scattering properties[18]. In addition, electrons in TiO$_2$ are transited from the valence band to the conduction band under the ultraviolet radiation; therefore, TiO$_2$ has outstanding ultraviolet absorption properties.

As shown in Fig. 10-12, the UV transmittance of PSA/CNT/TiO$_2$ composites in the range of 390~400 nm almost closes to zero, while the UV transmittance of PSA is high. It is suggested that the blending of CNT/TiO$_2$ can improve the scattering and absorption performance of ultraviolet of PSA/CNT/TiO$_2$ composites. Then, the ultraviolet resistance of PSA composites can be improved obviously. That is because the PSA composites blending with CNT/TiO$_2$ show a gray-black color, and then the PSA composites' shielding effect of ultraviolet improved. In addition, TiO$_2$ has excellent scattering and absorption properties at optical and ultraviolet wavelengths.

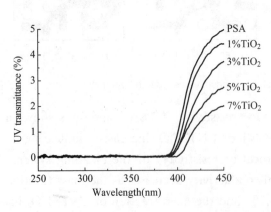

Fig. 10-11 UV transmittance of PSA/TiO$_2$ composites

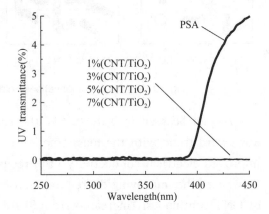

Fig. 10-12 UV transmittance spectra of PSA/(CNT/TiO$_2$) composites

10.1.3 Summary

PSA/CNT, PSA/TiO$_2$, and PSA/CNT/TiO$_2$ composite fibers and membranes with different mass fraction of functional particles were prepared. The experimental results of their properties can be summarized as follows.

(1) Functional particles at low mass fractions (1% or 3%) can be homogeneously distributed in the PSA matrix; however, it is hard for the particles with high mass fractions (5% or 7%) to distribute evenly in the blending system;

(2) There was no significant influence of the blended functional particles on the molecular structure and chemical composition of PSA fiber;

(3) The crystallinity of PSA composite fibers can be improved by blending with functional particles with low mass fractions, that is because they can act as nucleation agent;

(4) The breaking tenacity and initial modulus of PSA composite fibers can be improved obviously by blending with functional particles with low mass fractions; while the elongations at break of PSA composites decrease with the mass fraction of functional particles increasing from 1% to 7%;

(5) The thermal stability of PSA composites can be significantly enhanced by the blending of functional particles;

(6) The electrical conductivity of PSA composites can be improved by the blending of CNT and the average surface resistivity of the system decreases as the mass fraction of functional particles increases from 1% to 7%;

(7) The ultraviolet resistance of PSA composites can be increased by the blending of functional particles, the UV transmittance of PSA/TiO$_2$ composite decreases as the mass fraction of functional particles increasing from 1% to 7%, while the UV transmittance of PSA/(CNT/TiO$_2$) compsite was close to zero after the blending of CNT/TiO$_2$.

10.2 Preparation of PSA Fiber by Electro-spinning

Electro-spinning is one kind of advanced fiber spinning techniques used for the production of superfine fiber with a diameter from nanometer to micrometer. Compared with the conventional fiber, the electrospun fiber can be at least one or two orders smaller at the diameter magnitude. Because of the large specific surface areas of the electrospun fiber, it can be used for the production of filtration film[19], sensors[20], coating[21] and battery isolation film[22]. It can be also used for biomedical applications[23-24]. In recent years, more than 100 different types of polymer fibers have been successfully manufactured by electro-spinning, including organic synthetic polymers[28], synthetic biodegradable polymers[26], cellulose natural polymers and their derivatives[27].

10.2.1 Experimental

10.2.1.1 Preparation of PSA Fiber

PSA spinning solutions with different mass fraction of PSA were prepared by adding a certain amount of DMAC solvent in PSA solution and stirred for 30 min. The specifications of PSA spinning solutions are illustrated in Tab. 10-5. PSA fibers were fabricated by electro-spinning as shown in Fig. 10-13. The syringe is located perpendicularly to a metal collector. The collector is wrapped around a rotating drum, which can be controlled by a stepping motor. During the preparation, the rotation velocity of grounded collector was 42 r/min and the flow rate was about 0.1 mL/h. The effects of electro-spinning parameters, including the voltage (16, 20, 24 and 28 kV) and the tip-target distance (10, 15, 20 and 25 cm) on the morphological structure, crystallinity and thermal stability of the electrospun fibers were investigated. The experimental data are shown in Tab. 10-6.

Tab. 10-5 Specifications of PSA spinning solutions

Mass of PSA(g)	Mass of DMAC(g)	Mass fraction of PSA spinning Solution(%)
50	0	12
50	10	10
50	25	8
50	50	6

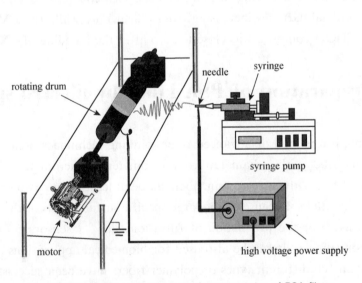

Fig. 10-13 Electro-spinning setup for preparation of PSA fibers

Chapter 10
Development of Functional PSA Nano-composites

Tab. 10-6 Experimental data

No.	Mass fraction of PSA spinning solution(%)	Applied voltage(kV)	Tip-target distance(cm)
1	6	20	15
2	8		
3	10		
4	12		
5	12	16	15
6		20	
7		24	
8		28	
9	12	28	10
10			15
11			20
12			25

10.2.1.2 Characterizations

The S-3400N scanning electron microscope (SEM) with a resolution of 4 nm was used to characterize the morphological structure of PSA fiber, and the machine was operated at 5~15 kV. A small specimen of PSA fiber mat was placed on the SEM sample holder and sputter-coated with gold membrane with a thickness of about 10 nm. The diameter of PSA fiber was measured by using Image-Pro-Plus software from the SEM pictures at a magnification of 5000.

The K780 FirmV_06 X-ray diffraction (XRD) was used to characterize the crystalline structure of composite fibers using CuKα radiation ($\lambda = 0.154$ nm) at a voltage of 40 kV and a current of 40 mA. The spectra was obtained at 2θ of 5°~90° with a scanning speed of 0.8/step.

The Germany STA PT-1000 thermal gravimetric analyzer (TGA) was used to investigate the thermal stability of composite fibers. The TGA experiment was carried out under nitrogen atmosphere with gas flow rate of 80~100 mL/min. Samples were heated from room temperature to 700 ℃ at a heating rate of 20 ℃/min.

10.2.2 Results and Discussion

10.2.2.1 Morphological Structure

(1) Effects of Mass Fraction of PSA Spinning Solution on Fiber Morphology

SEM images of electrospun PSA fibers prepared with different mass fraction of PSA spinning solution are shown in Fig. 10-14. As can be observed in Fig. 10-14 (a), many beaded fibers were prepared when the mass fraction of PSA spining solution is 6%. The beads changed from spherical shape to spindle-like shape with the mass fraction of PSA spinning solution increased from 6% to 8%. The beaded fibers decreased and uniform fibers were prepared eventually when the mass fraction of PSA spinning solution is further increased from 8% to 12%. The generation of the beaded fibers can be attributed to the low mass fraction of PSA spinning solution, and then the solution viscosity is below than the critical value and the chain entanglements are not sufficiently enough to resist the external forces, which can cause a premature break-up among chains. When the mass fraction of PSA spinning solution is increased to a relatively high value, the viscoelastic force can resist the rapid change in fiber shape, which leading to an uniform fiber formation[28-29]. However, as observed in Fig. 10-14, the average diameter of fiber increased with increase of mass fraction and viscosity of spinning solution, and that is because the resistance of pulling fiber increased simultaneously. That indicates it is impossible to produce fiber when the mass fraction of PSA spinning solution is too high. According to the results in Fig. 10-14, the suitable mass fraction of PSA spinning solution is 12%[30-31].

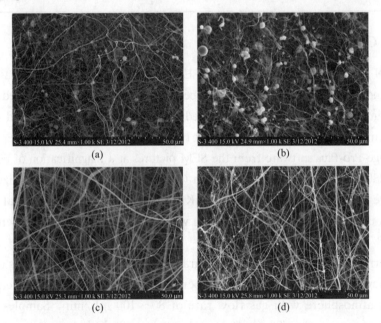

Fig. 10-14 SEM images of electrospun PSA fibers prepared with different mass fraction of PSA spinning solution: (a) 6%; (b) 8%; (c) 10%; (d) 12%

(2) Effects of Voltage on Fiber Morphology

SEM images and fiber diameter of electrospun PSA fibers prepared with different voltages are shown in Fig. 10-15. As demonstrated in this figure, the average diameter of fiber decreased from 472 to 133 nm as the electro-spinning voltage gradually increased from 16 to 28 kV. That is because the net charge density on the surface of jets increased with the raising voltage, and then leads to the improvement of whipping and split ability of jets. Moreover, the jet velocity can be increased when the applied voltage is increased and the jets can be solidified quickly on the collector without enough draft, which resulted in the preparation of thick fibers. In addition, stable jets can be formed in electric field when the PSA solution conductivity is low, which can narrow the distribution of fiber diameter[32-33]. According to the results in Fig. 10-15, the suitable voltage is 28 kV.

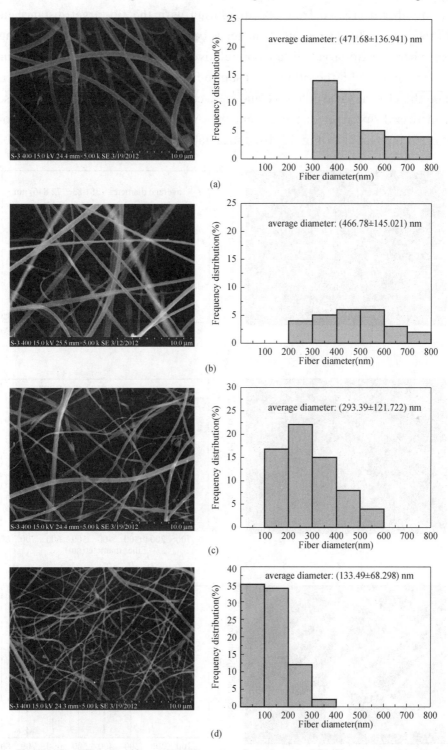

Fig. 10-15　SEM images and fiber diameter of electrospun PSA fibers prepared with different voltage: (a) 16 kV; (b) 20 kV; (c) 24 kV; (d) 28 kV

(3) Effects of Tip-target Distance on Fiber Morphology

SEM images and fiber diameter of electrospun PSA fibers prepared with different tip-target distances are shown in Fig. 10-16. As demonstrated in the figure, the average diameter of fiber decreased from 382 to 133 nm with the tip-target distance increased from 10 to 15 cm. The decreasing diameter is because of the enhanced stretching ability for fiber when the tip-target distance is relatively large. The fiber diameter increased from 133 to 508 nm with the tip-target distance further increased from 15 to 25 cm. That is because the electric field intensity and the accelerated velocity of jet decrease as the further increased tip-target distance, and then weaken the stretching ability for fiber. According to the results in Fig. 10-16, the suitable tip-target distance is 15 cm.

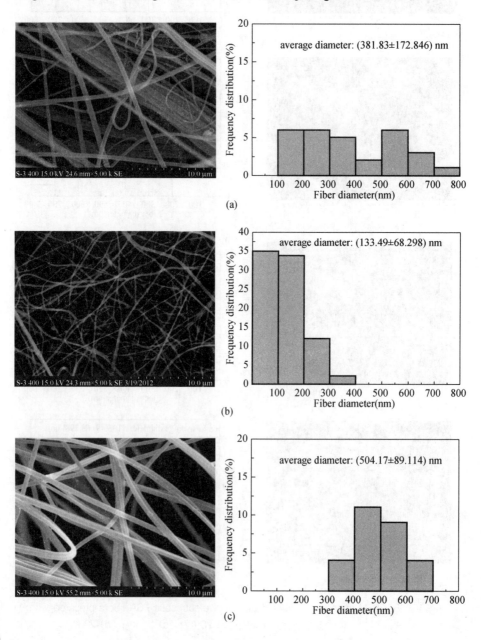

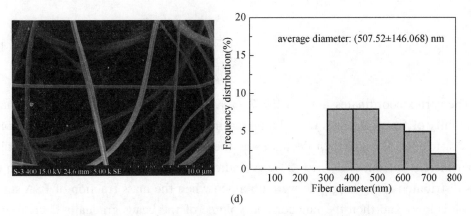

Fig. 10-16 SEM images and fiber diameter of electrospun PSA fibers prepared with different tip-target distances: (a) 10 cm; (b) 15 cm; (c) 20 cm; (d) 25 cm

10.2.2.2 Crystal Structure

(1) Effects of Mass Fraction of PSA Spinning Solution on Fiber Crystallinity

XRD patterns of PSA fibers prepared with different mass fraction of PSA spinning solution and the relationship between the mass fraction of PSA spinning solution and the crystallinity of fiber are shown in Fig. 10-17.

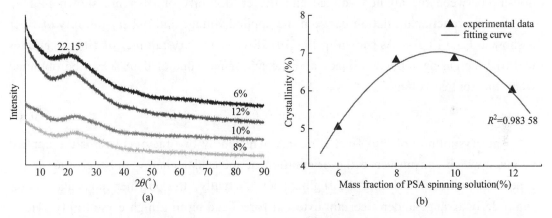

Fig. 10-17 (a) XRD patterns of PSA fibers prepared with different mass fractions of PSA spinning solution; (b) Relationship between the mass fraction of PSA spinning solution and the crystallinity of fiber

The functional relation between the mass fraction of PSA spinning solution and the crystallinity of fiber is shown in Eq. (10-4).

$$y = ax^2 + bx + c \qquad (10\text{-}4)$$

In Eq. (10-4), y and x represent the crystallinity of fiber and the mass fraction PSA spinning solution, respectively. The value of a, b and c are $-0.164\,38$, $3.106\,25$ and $-7.637\,5$, respectively. The vertex coordinates of regression model can be calculated according to Eq (10-5).

$$\begin{cases} x = -\dfrac{b}{2a} \\ y = \dfrac{4ac - b^2}{4a} \end{cases} \quad (10-5)$$

The vertex coordinates are (9.45, 7.04). Therefore, it can be deduced that the crystallinity of fiber increases with the increasing mass fraction of PSA spinning solution from 6% to 9.45%, and then decreases as the mass fraction of PSA spinning solution further increased from 9.45% to 12%. As indicated in Section 5.2.2.1, fiber with many beads distributed along longitude were obtained when the mass fraction of PSA spinning solution is low, and then the number and volume of the beads gradually decreased with increasing mass fraction of PSA spinning solution. Therefore, the crystallinity of fiber can be improved. However, increasing the mass fraction of PSA spinning solution will increase the resistance of pulling fiber, and then decrease the force of guiding the molecular chain orientation. Therefore, the crystallinity of fiber decreases when the mass fraction of PSA spinning solution is beyond 9.45%.

(2) Effects of Voltage on Fiber Crystallinity

XRD patterns of PSA fibers prepared with different applied voltages and the relationship between the applied voltage and the crystallinity of fiber are shown in Fig. 10-18. The functional relation between the applied voltage and the crystallinity of fiber is show in Eq. (10-6). As shown in Fig. 10-18 (b), the crystallinity of fiber decreases with the increasing applied voltage, and the minimum value of 6.08% can be obtained when the applied voltage is 28 kV.

$$y = 0.0006875\, x^2 - 0.03722\, x + 6.5834 \quad (10-6)$$

The crystallinity of PSA fiber is mainly affected by two factors. One is the applied voltage, which can influence the orientation of molecular chain, and the other is the time of crystallization during the flight of jet. Actually, the polymer molecular chains can orderly arrange under a certain external force leading to a high crystallinity. However, the original ordered arrangement of PSA molecule can be destroyed by applied voltage when it exceeds a critical value, resulting in decreased crystallinity of fiber. In addition, increasing the applied voltage can increase the flight velocity of jet and shorten the flight time of jet from spinneret to collector, which means that there is no sufficient time for PSA fiber to crystallize during the preparation, and the crystallinity of fiber would be low. However, it was found that the quadratic coefficient of Eq. (10-6) was very low. In addition, as shown in Fig. 10-18 (b), the difference among crystallinity is also small. Therefore, it is deduced that the applied voltage had no significant effect on the crystallinity of electrospun PSA fiber.

(3) Effects of Tip-target Distance on Fiber Crystallinity

XRD patterns of PSA fibers prepared with different tip-target distances and the relationship between the tip-target distance and the crystallinity of fiber are shown in Fig. 10-19. As demonstrated in Fig. 10-19 (b), the crystallinity of fiber increased when the tip-target distance increased from 10 to 20 cm. The maximum value of crystallinity of fiber is 7.36%. However, the crystallinity of fiber decreased with the further increasing tip-target distance from 20 to 25 cm. The results in Fig. 10-19 indicate that the crystallinity of fiber can be improved when the tip-target distance is ranging from 10 to 20 cm and the jet has a sufficient time to crystallize during the flight. However, the crystallinity of fiber decreases when the tip-target distance exceeds a certain value, that is because the reduced tip-target distance can decrease the electric intensity, and then further decrease the stretching force on fiber and applied electric force, which can induce an orderly arrangement of molecular chain.

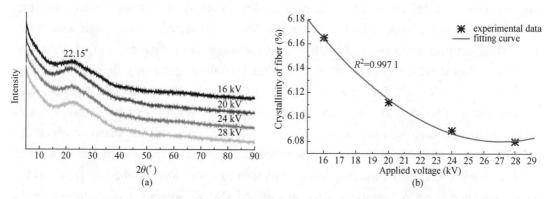

Fig. 10-18 (a) XRD patterns of PSA fibers prepared with different applied voltages;
(b) Relationship between the applied voltage and the crystallinity of fiber

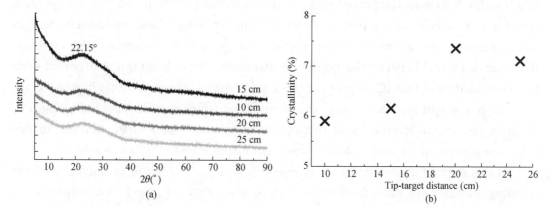

Fig. 10-19 (a) XRD patterns of PSA fibers prepared with different tip-target distances;
(b) Relationship between the tip-target distance and the crystallinity of fiber

10.2.2.3 Thermal Stability

Thermal behavior is an important property of heat-resistant fiber and influences the post-processing properties and application of fiber. Thermal analysis can designate any technique involves the measurement of a physical quantity while the temperature is changed or maintained. The relationship between the weight loss of electrospun PSA samples and temperature (or time) controlled by the system program is investigated by using thermogravimetric analysis (TGA)[34]. Thermogravimetric (TG) curve is used to represent the relationship between the weight loss of samples prepared under different electro-spinning conditions and temperature (or time), while the derivative thermal gravimetric analysis (DTG) curve is applied to descript the relationship between the weight loss of samples and temperature.

TG and DTG curves of PSA fibers prepared with different conditions are shown in Fig. 10-20. As illustrated in the figure, the thermal decomposition of all samples can be divided into three sections.

The first section is the stage of small weight loss from room temperature to 400 ℃. At this stage, the TG curves of all samples declined sharply upon temperature rose from room temperature to about 140 ℃. Correspondingly, all samples had a high weight loss rate when temperature was at about 100 ℃, because of the volatilization of DMAC solvent as well as some water in samples. And then the curves gradually changed to be gentle until at about 330 ℃, during which the weight loss is attributed to both the continuing volatilization of residual DMAC solvent and some tightly-bound water in samples as well as the additives, such as oiling agent. The TG curve of each sample between about 330 and 390 ℃ was almost a platform. As depicted in Fig. 10-20, the weight loss of samples decreased with increased mass fraction of PSA spinning solution, applied voltage and tip-target distance. This is because a large amount of DMAC solvent was contained in the spinning solution with a low mass fraction of PSA when compared with the spinning solution with a high mass fraction of PSA. In this case, samples prepared with the spinning solution of a lower mass fraction of PSA had a higher weight loss. In addition, the jet changed much faster as the increasing applied voltage, which can shorten the vaporization-time of DMAC between the needle tip and collector, indicating the prepared samples contained more DMAC solvent. In this case, samples prepared with a higher voltage had a higher weight loss. Moreover, the short tip-target distance reduces the flying time of jet in the space, leading to an incomplete vaporization of DMAC solvent. In this case, samples prepared with a shorter tip-target distance had a higher weight loss.

At this stage [Fig. 10-20 (a)], the weight loss of S_1, S_2, S_3, S_4 manufactured from spinning solution with mass fraction of PSA of 6%, 8%, 10% and 12% respectively, was 38.82%, 31.5%, 29.5% and 29% respectively. For the same principle, the TGA analysis [Fig. 10-20(b)] of samples prepared under different applied voltage shows the weight loss of samples is affected by applied voltage that the weight loss of samples gradually increased, 28.7%, 29%, 29.5% and 32% respectively, with the applied voltage ranging from 16 to 28 kV in the initial stage.

The second section is a stage of thermal decomposition ranging from 400 to 600 ℃. At this stage, the weight loss of samples was attributed to the increasing movement rate of polymer macromolecular chains. As the temperature increased, PSA macromolecular chains were ruptured. Simultaneously, some substances constituted small molecules accompanied by gas were produced, which leaded to the weight loss of samples. The analysis of the bond energy[35] showed those released gas may be SO_2, NH_3 or CO_2. As a result, duing to the increasing of produced gas, the weight loss rate of samples was gradually increased.

Meanwhile, the third section is a high-temperature phase of carbon formation ranging from 600 to 700 ℃ and most of polymers are carbonized at this stage. As depicted in Fig. 10-20 (a) and (b), the weight loss at the terminal temperature of PSA fibers increased with raising mass fraction of PSA spinning solution from 6% to 12%, while decreased with increasing applied voltage from 16 to 28 kV. The increasing weight loss at 700 ℃ can indicate an improving thermal stability of PSA fibers. As observed in Fig. 10-20 (c), the tip-target distances had no significant influence on the weight loss of samples.

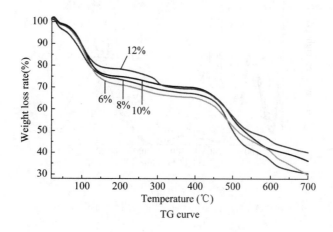

TG curve

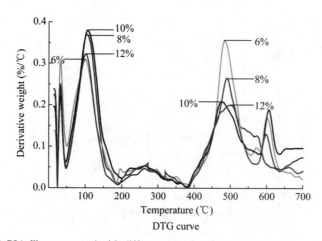

DTG curve

(a) PSA fibers prepared with different mass fraction of PSA spinning solution

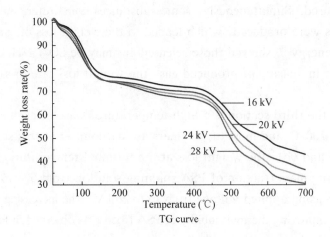

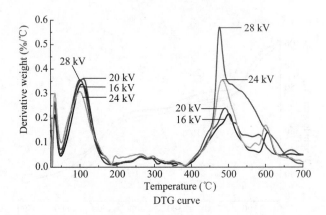

(b) PSA fibers prepared with different applied voltages

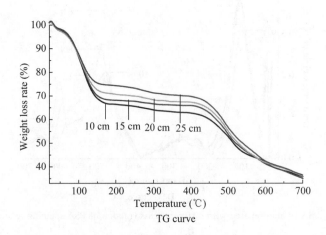

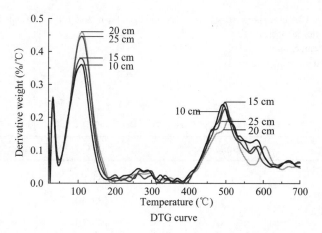

(c) PSA fibers prepared with different tip-target distances

Fig. 10-20 DSC curves of PSA fibers prepared with different conditions

Some key thermal degradation indexes are summarized in Tabs. 10-7 and 10-8. T_i is defined as the initial decomposition temperature when the change of accumulation mass can be detected by thermal-balance. The crossing point obtained from the tangent of descending curve in the first section and the extension cord of baseline can be represented by T_0, while the crossing point obtained from the tangent of the descending curve in the third section and the extension cord of the horizontal line can be represented by T_e[36]. The thermal stability of materials is usually characterized by T_i, T_0 as well as T_e.

Tab. 10-7 Parameters of TG curves of PSA fibers during the thermal decomposition

Samples	Sample No.	T_i(℃)	T_0(℃)	T_e(℃)	T_f(℃)	$T_f - T_i$ (℃)	α(%)
PSA fibers prepared with different mass fraction of spinning solution	S_1	399.00	434.07	565.68	627.99	228.99	70.10
	S_2	406.99	436.60	570.70	615.01	208.02	69.48
	S_3	409.54	438.46	573.89	610.32	200.78	63.80
	S_4	418.35	442.26	576.21	618.06	199.71	59.83
PSA fibers prepared with different applied voltages	S_5	419.03	443.67	584.74	611.65	192.62	56.98
	S_4	418.35	442.26	576.21	618.06	199.71	59.83
	S_6	393.59	440.39	575.29	609.37	215.78	63.52
	S_7	392.57	438.94	572.53	613.97	221.41	69.48
PSA fibers prepared with different tip-target distances	S_8	397.74	440.71	594.23	614.21	216.47	66.19
	S_7	392.57	438.94	572.53	613.97	221.41	69.48
	S_9	417.67	442.92	595.13	623.62	205.95	64.00
	S_{10}	414.94	439.78	581.52	623.06	208.13	64.99

S_1, S_2, S_3 and S_4—The PSA fibers prepared with different mass fraction of spinning solution of 6%, 8%, 10% and 12%, respectively;

S_5, S_4, S_6 and S_7—The PSA fibers prepared with different applied voltages of 16, 20, 24 and 28 kV, respectively;

S_8, S_7, S_9 and S_{10}—The PSA fibers prepared with different tip-target distances of 10, 15, 20 and 25 cm, respectively;
T_i—Initial decomposition temperature;
T_0—Epitaxial initial temperature;
T_e—Epitaxial termination temperature;
T_f—Terminal temperature;
α—Weight loss rate.

Tab. 10-8 Parameters of DTG curves of PSA fibers during the thermal decomposition

Sample	Sample No.	T_p(°C)	$d\alpha/dt$
PSA fibers prepared with different mass fractions of spinning solution	S_1	484.67	0.347
	S_2	488.40	0.256
	S_3	495.19	0.202
	S_4	505.85	0.198
	S_5	506.55	0.183
PSA fibers prepared with different applied voltages	S_4	505.85	0.198
	S_6	495.67	0.234
	S_7	478.77	0.278
	S_8	501.07	0.238
PSA fibers prepared with different tip-target distances	S_7	478.77	0.278
	S_9	515.95	0.190
	S_{10}	502.97	0.218

T_p—Temperature at the maximum decomposition rate;
$d\alpha/dt$—The maximum decomposition rate.

As shown in Tab. 10-7, the changing trend of T_i is similar to T_0 and the temperature increased with the raising mass fraction of spinning solution, which indicates the thermal stability of samples is gradually improved. Besides, the difference between T_f and T_i also showed opposite trend, $S_1 > S_2 > S_3 > S_4$, suggesting the internal structure of manufactured PSA fiber by the spinning solution with low mass fraction needed to be completely transformed in a very wide temperature range and the stability of these samples' internal crystalline structure was correspondingly poor. This is because when the spinning solution's concentration is low, fibers with bead distributed along longitude were manufactured. Two different heat transferring models of beaded fibers are shown in Fig. 10-21.

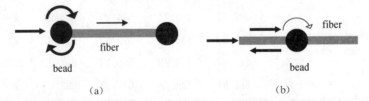

Fig. 10-21 Two different heat transferring models of beaded fibers

The thermal conductivity of this system has significant effect on the thermal decomposition behavior of samples. The internal structure of beads is different from that of fiber, which leads to a quite different conductivity. The heat cannot be evenly delivered. On the contrary, the heat concentrated on either bead or fiber, and then one of them decomposes at one point when the temperature increased to its decomposition temperature. Therefore, the samples electrospun with low mass fraction of spinning solution showed lower thermal stability and higher weight loss ratio.

Moreover, the crystallinity of fiber also has effect on the thermal stability of samples. As illustrated in Tab. 10-8, T_p of samples increased with raising mass fraction of spinning solution, which also indicated an improving thermal stability of samples.

As illustrated in Tabs. 10-7 and 10-8, decreasing T_i and T_0 as well as increasing value of $d\alpha/dt$ with raising applied voltage can suggest a degradative thermal stability of samples. This can be ascribed to two factors. One is the decreasing crystallinity of fiber with increasing applied voltage because the non-crystalline structure has poor heat resistance. Therefore, the samples with a low crystallinity show poor thermal stability. In addition, the average diameter of fiber decreased and the specific surface area of fiber greatly increased when the applied voltage increased[37], which allows fibers to absorb more heat at the same temperature. Therefore, the thermal stability of PSA fiber decreases gradually as the applied voltage increased.

As can be observed in Tabs. 10-7 and 10-8, S_7 has the minimum value of T_i and T_0 as well as the maximum value of $d\alpha/dt$. These data can indicate that the thermal stability of S_7 is the poorest. That can be attributed to the smallest diameter of fiber when it is prepared with the tip-target distance of 15 cm, which can increase the absorption of heat. In addition, S_7 showed a relatively low crystallinity when it is fabricated with the tip-target distance of 15 cm leading to a poor thermal stability.

10.2.3 Summary

PSA fibers have been prepared by electro-spinning with different mass fractions of spinning solution, applied voltages and tip-target distances. The effects of these electro-spinning parameters on the morphology, crystallinity and thermal stability of PSA fibers have been investigated. The experimental results can be summarized as follows.

(1) The morphology of PSA fibers can be changed from spherical beaded shape to spindle-like beaded shape with the mass fraction of spinning solution increased from 6% to 8%. The beaded fibers decreased and uniform fibers were obtained when the mass fraction of spinning solution is further increased from 8% to 12%. The average diameter of PSA fibers decreased and the diameter distribution was narrowed when the applied voltage increased from 16 to 28 kV. The average diameter of PSA fibers decreased when the tip-target distance increased from 10 to 15 cm, and then increased with the tip-target distance further increased from 15 to 25 cm.

(2) The optimal value of the mass fraction of spinning solution, applied voltage and tip-target distance corresponding to the highest crystallinity of PSA fibers are 10%, 16 cm and 20 cm, respectively.

(3) The thermal stability of PSA fibers can be influenced by the interaction of morphology and crystallinity of fibers. The beaded structure, the small diameter and poor crystallinity of fibers can decrease the thermal stability of fibers. The optimal value of the mass fraction of spinning solution, applied voltage and tip-target distance corresponding to the highest crystallinity of PSA fibers are 12%, 16 cm and 20 cm, respectively.

References

[1] Wang X F, Wang F H, Ren J R, et al. The performance and application of PSA fibres. In: Yu MF., The symposium of development and application of high-temperature PSA fibres in 2009. Shanghai, 2009: 3-7.

[2] Zhang L D, Mou J M. The structures of nanomaterials. Beijing: Science Press, 2001: 420-476.

[3] Avouris J P, Appenzeller M R, et al. Carbon nanotube electronics. In: IEEE, 2003, 91(11): 1772-1784.

[4] Yang W T. The characterization and testing of polymer. Beijing: China Light Industry Press, 2008, 72-80.

[5] Meng Q H, Hu J L, Zhu Y. Shape-Memory Polyurethane/Multiwalled Carbon Nanotube Fibers. Journal of Applied Polymer Science, 2007, 106: 837-848.

[6] Chen J S, Liu M C, Zhang L, et al. Application of nano TiO_2 towards polluted water treatment combined with electro-photochemical method. Water Research, 2003, 37(16): 3815-3820.

[7] Xia H S, Wang Q. Ultrasonic Irradiation: A Novel Approach To Prepare Conductive Polyaniline/Nanocrystalline Titanium Oxide Composites. Chemistry of Materials, 2002, 14(5): 2158-2165.

[8] Zhang X Y, Cheng Y, Zhao J B. Polymer chemistry. Beijing: China Light Industry Press, 2000: 285.

[9] Broadbelt L J, Chu A, Klein M T. Thermal stability and degradation of aromatic polyamides. Part 2 Structure-reactivity relationships in the pyrolysis and hydrolysis of benzamides. PolymDegrad Stab, 1994, 45(1): 57-70.

[10] Broadbelt L J, Chu A, Klein M T. Thermal stability and degradation of aromatic polyamides. Part 2. Structure-reactivity relationships in the pyrolysis and hydrolysis of benzamides. PolymDegrad Stab, 1994, 44(2): 137-144.

[11] Yang W T. Characterization and testing of polymer materials. Beijing: China Light Industry Press, 2008: 144.

[12] Wang D. Assembly of Lamellar Magnesium Hydroxide Nano Crystal onto Modified Polysulfonamide Fabric. Shanghai: Donghua University, 2009.

[13] Zhu H W, Lu D H, Xu C L. Carbon nanotubes. Beijing: Mechanical Industry Press, 2003, 176.

[14] Liang J Z, Qiu Y L. Preview on the thermal conductivity of composites of aluminum oxide/silicone rubber. China Rubber Industry, 2009, 56(8): 476-479.

[15] Liang J Z, Liu G S. Heat Transfer Models and Finite Element Simulation in Inorganic Particle-Filied Polymer Composites. Special Purpose Rubber Products, 2006, 27(5): 36-38.

[16] Liang J Z, Li R K. Measurement of dispersion of glass beads in PP matrix. Journal of Reinforced Plastics Composites, 2001, 20(8): 630-638.

[17] Jiang F D. Study on the preparation and structure and performance of the Cool polyurethane/CNT nano composites. Beijing: Beijing University of Chemical Technology, 2009.

[18] Liu J X, Tang Z Y, Zhang D R, et al. Preparation and performance of Polysulfonamide nanocomposites and it's fiber. Technical Textiles, 2007, (2): 14-20.

[19] Veleirinho B, Lopes-Da-Sil J A. Application of electrospun poly(ethylene terephthalate) nanofiber mat to apple juice clarification. Process Biochemistry, 2009, (44): 353-356.

[20] Shi W, Lu W, Jiang L. The fabrication of photosensitive self-assembly Au nanoparticles embedded in silica nanofibers by electro-spinning. Journal of Colloid Interface Science, 2009, (340): 291-297.

[21] Sundarrajan S, Venkatesan A, Ramakrishna S. Fabrication of nanostructured selfdetoxifying nanofiber membranes that contain active polymeric functional groups. Macromolecular Rapid Communications, 2009, (30): 1769-1774.

[22] Zheng Y. Preparation and research of poly(ethylene-co-cinyl alcohol). Harbin: MasterDissertation, Harbin University of Science and Technology, 2007.

[23] Hong Y L, Chen X S, Jing X B. Fabrication and drug delivery of ultrathin mesoporous bioactive glass hollow fibers. Advanced Functional Materials 2010, (20): 1503-1510.

[24] Rockwood D N, Akins R J, Parrag I C. Culture on electrospun polyurethanescaffolds decreases atrial natriuretic peptide expresson by cardiomyocytes in vitro. Biomaterials, 2008, (29): 4783-4791.

[25] Gopal R, Kaur S, Feng C Y. Electrospun nanofibrous polysulfone membranes as pre-filters: Particulate removal, Journal of Membrane Science, 2004, (41): 30-36.

[26] Tan S H, Inai R, Kotaki M. Systematic parameter study for ultra-fine fiber fabrication via electrospinning process. Polymer, 2005, (46): 6128-6134.

[27] Kim C, Frey M, Marquez M. Preparation of submicron-scale, elecrospuncellulosefibers via direct dissolution. Journal of Polymer Science: Part B: PolymerPhysics, 2005, (43): 1673-1683.

[28] Jeun J P, Lim Y M, Nho Y C. Study on morphology of electrospun poly(caprolactone) nanofibers. Journal of IndustrialEngineering Chemistry, 2005, (11): 573-578.

[29] Gupta P, Elkins C, Long T. Electro-spinning of linear homopolymers of polly(-methyl methacrylate): Exploring relationships between fiber formation, viscosity, molecular weight and concentration in a good solvent. Polymer, 2005, (46): 4799-4810.

[30] Demir M M, Yil I, Yilgor E. Electro-spinning of polyurethane fibers. Polymer 2002, (43): 3303-3309.

[31] Deitzel J M, Kleinmeyer J, Harris D. The effect of processing variables on themorphology of electrospun nanofibers and textiles. Polymer, 2001, (78): 1149-1151.

[32] Hohman M M, Shin M, Rutledge G. Electro-spinning and electrically forced jets. I. Stability theory Physics Fluids, 2001, (12): 2201-2220.

[33] Hayati I, Bailey A, Tadros T F. Investigations into the mechanism of electrohydrodynamic spraying of liquids: II. Mechanism of stable jet formation and electrical forces acting on a liquid cone. Journal of Colloid Interface Science, 1987, (117): 1138-1142.

[34] Yang W T. Characterization and testing of polymer materials. Beijing, China: ChinaLight Industry Press, 2008.

[35] Zhang X B, Cheng J. High polymer chemistry. Beijing, China: China Light Industry Press, 2000.

[36] Zhang M Z, Liu B J, Gu X Y. Study method of polymer materials. Beijing, China: China Light Industry Press, 2000.

[37] Gibson P, Schreuder-gibsonH, Rivin D. Transport properties of porous membranes based on electrospun nanofibers. Colloids Surfaces A Physicochemical Engineering Aspects, 2001, (187-188): 469-481.

Chapter 11 Functionalization of Cellulose Based Materials

Binjie Xin*, Shan He and Yanjuan Cao

11.1 Introduction

Recently, the applications of biodegradable natural cellulose have attracted much attention because of the "white pollution" problem caused by the non-biodegradable polymer materials. So far, environmental-friendly material has been considered as one of the key development orientations in material field. It has been reported by authoritative sources that hundreds of billions of natural biological have been synthesized into cellulose every year globally, which goes far beyond the oil resources[1]. As the regenerated cellulose membranes, cellophane and copper cellophane are mainly used as dialysis membranes[2] by different preparation methods (i.e., viscose process or cuprammonium process). The microfiltration membrane and ultrafiltration membrane made of regenerated cellulose have been much more popularly used for the development of protein contamination resistance and renal dialysis, and selected as the key medical products. Some developed countries have transferred the production bases to underdeveloped countries because of serious environment pollution, which has been caused by the preparation (viscose process or cuprammonium process) of conventional regenerated cellulose fiber, and the annual output of viscose fiber has been decreased. Recently, some researchers presented their studies on the development of newly-developed solvent system and preparation

---

Corresponding Author:
BinjieXin
School of Fashion Technology, Shanghai University of Engineering Science, Shanghai 201620, China
E-mail: xinbj@sues.edu.cn

technology for cellulose fiber. The polyacrylonitrile (PAN), which possesses excellent chemical resistance, light fastness, heat resistance and microbial properties, could be dissolved in organic solvents, such as DMAC, DMF, NMP, etc. The PAN/cellulose composite nano-fiber has the potential to be applied for the development of drug-loading release, biological dressing, tissue engineering, precision filtration, etc.

Membrane separation technique (e.g., microfiltration, ultrafiltration, nanofiltration and reverse osmosis) has been used for the development of environmental protection, electronics, energy source, chemical engineering and biotechnology, which covers most of the industrial sectors[3-5]. As can be seen from Tab.11-1, membrane properties are determined by their pore size directly, what's more, the higher the requirement of separation technique is, the smaller the pore size will be. So far, except for conventional industrial applications, the applications of cellulose in nanomedicine, nanocomposite and new energy, have been investigated by overseas and domestic scientists. The application of cellulose is enhanced by combination with chemistry, physics, material science, biology and bionics[5].

Tab. 11-1 Scope of applications for various separation membranes[4]

Name	Separation mechanism	Separated object	Pore size (nm)
Particle filter	Volume	Solid ion	>10 000
Microfiltration	Volume	Solid ion with diameter of 0.05~10 μm	50~10 000
Ultrafiltration	Volume	1000~1 000 000 daltons, colloid	2~50
Nanofiltration	Solution-diffusion	Organics with ionic/molecular weight less than 100	<0.5
Reverse osmosis	Solution-diffusion	Organics with ionic/molecular weight less than 100	<0.5
Pervaporation	Solution-diffusion	Organics with ionic/molecular weight less than 100	<0.5

Preparation of nano-material with excellent functions has been frontier of cellulose science, by arranging the aggregates of cellulose molecule and its supermolecular at nano-scale[6-7]. The nano-cellulose has a large number of excellent properties[8], and electro-spinning has been considered as one of the most feasible methods toprepare nano-fiber[9]. The electrospun fiber has been much more popularly used for the development in quite a lot of fields, such as biomedicine, filtration, protective clothing and sensor, because of its extremely-high specific surface area and porosity. For example, the filtration efficiency of filter material made of nano-fiber is much better than that made of ordinary filter material. Many researchers[10] have attempted to apply the nano-fiber in preparation of multi-function membrane, for the purpose of bringing new opportunities to develop membrane technology.

Membrane fouling may be caused by the absorption of particulate and colloidal particle in the feed solution or solute macromolecule, and it is an unavoidable difficulty for membrane applications. The experimental results[11,12] showed that as the carrier of microbe, fiber materials may result in membrane fouling since the microorganism may grow on the surface of cellulose membrane, which limits the application of cellulose membrane. In this case, the enhancement of antibacterial properties and development of membrane material have been the research priorities of engineering science.

The antimicrobial agents can be classified into natural antibacterial agents, inorganic antibacterial agents and organic antibacterial agents, among which, inorganic antibacterial agents remain invincible in the antibacterial markets because of its unique features, such as broad-spectrum antibacterial, non-toxic, durable, safe and wide scope of applications[13-14]. At present, the most-popularly used inorganic antibacterial agent is the silver antibacterial agent, which is characterized by sufficiently-strong antibacterial activity, low dosage, no drug resistance and superior antibacterial effect of silver[15].

The numbers of retrieved literatures from the year of 2000 to now are illustrated in Fig. 11-1, by inputting the keywords of "cellulose electrostatic spinning" or "copper antimicrobial". It can be found that quite a lot of researches have been done for the development of electro-spinning of cellulose fiber and antibacterial properties of copper ion; whereas very few works have been reported to investigate the antibacterial properties of electrospun cellulose fiber based on combination of them. In recent years, electro-spinning of cellulose fiber has aroused wide attention among researchers. Although many studies about electro-spinning have been reported and the electro-spinning of cellulose derivative have also been studied, the researches on the electro-spinning of cellulose fiber have so far been barely conducted. The numbers of literature for electro-spinning and cellulose electro-spinning is depicted in Fig. 11-2, retrieved on the basis of Wanfang database.

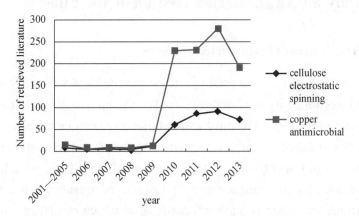

Fig. 11-1 Numbers of retrieved literature for cellulose electrostatic spinning and copper antibacterial in Web of Science

Recently, development and application of functional composite membrane made of antibacterial cellulose have become a research hotspot in nano-fiber field, including recovery and regeneration of natural cellulose, functional finishing of fiber based on physical and chemical methods, preparation of superfine nano-fiber, etc. The research on

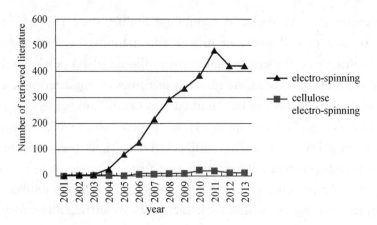

Fig. 11-2 Numbers of retrived literature for electro-spinning and cellulose electro-spinning in Wangfang database

development of natural cellulose fiber is very important to satisfy the industrial applications, which could be electrospun to be functional membrane at nano scale, and the prepared fiber based products could be used for antibacterial protection. It is proposed in this study that PAN/cellulose composite nano-fiber possessing antibacterial properties could be fabricated by using electro-spinning, which can endow the fiber nano-meter characteristics. In this case, the applications of cellulose at nano-scale in ultrafiltration and antibacterial fields could be expanded and high-performance products with superfine structure and antibacterial effect could be developed.

11.1.1 Study on Antibacterial Treatment for Fiber

11.1.1.1 Classification of Antibacterial Agent

Quite a lot of technologies can be used to endow the material with antibacterial properties, and the most commonly-used methods are functional finishing or utilization of fiber with antibacterial properties. The antibacterial materials can be classified into two categories: one is the finished fabric characterized by poor fastness and easy to fall off, where the antibacterial agents and materials are combined based on adhesion and impregnation, meanwhile, the enhancement of fastness is considered as a technical difficulty in the future; the other is materials made of antibacterial fibers, in addition, the antibacterial fiber mainly refers to the artificial antibacterial fiber, including nanometer antibacterial fiber, photocatalytic antibacterial fiber, Modal Fresh fiber[16], chitin fiber, charcoal fiber, etc. Internationally, especially Japan is in a leading position in investigation on antibacterial. However, domestic study on antibacterial textile started rather late, and the development of textile with features of high added value, multiple function, wide scope of application and industrial use[17] has been the focus of researches.

The antibacterial agents used for preparation of antibacterial fiber are composed of three series: natural, organic and inorganic. However, it is difficult for the natural antibacterial agents to be promoted due to their limited resource. The organic antibacterial agents have been applied for more than 30 years because of their excellent bactericidal power and effect. However, the organic antibacterial agents have some disadvantages, such as poor security, chemical stability, heat resistance, drug resistance and secondary pollution. The inorganic antibacterial agents, which have been the research hotspot, remain invincible in the antibacterial markets because of the advantages, such as broad-spectrum antibacterial, non-toxic, durable, safe and wide scope of applications[18-19].

The inorganic antibacterial agents consist of two categories: silver series materials (Ag, Cu) and titanium series materials. The silver antibacterial agent is characterized by sufficiently-strong antibacterial activity, low dosage, no drug resistance and superior antibacterial effect of silver[20]. Nevertheless, the promotion of fabric of silver antibacterial series is limited, which is because, it is characterized by high price and prone to be oxidized and discolored resulted in discoloration of antibacterial fabric, thus remarkably decreasing the antibacterial effect. The research on copper mainly focuses on the investigation of copper-coated fiber and copper-bearing stainless steel fiber, and the antibacterial fabric made of Cupron copper-based antibacterial fibers (KAPOOR, U.S.) is developed with the function of preventing bacterial infections[15]. The copper, which is also called the "secondary silver", belongs to the silver antibacterial inorganic agents with relatively-low price and excellent antibacterial effect[15].

11.1.1.2 Research Status of Copper-based Antibacterial Fiber

Preparation of copper-based antibacterial fiber can be achieved by the two following methods: one is to coat the copper oxide on the cotton fiber; the other is to apply the copper oxide powders into the spinning solution when blending. The fabricated fiber presents poor hand-feel, and the original color of the copper has effects on both above-mentioned preparation methods, with the solution being coloration of light blue because of copper ion. The fabric or fiber finished via chemical methods (especially copper ion complexation) is characterized by light color, soft and good hand-feel (different from copper-coated fiber). Diverse organic functional and coordinating groups, which are able to achieve the goal of complexation of metal ions under appropriate circumstances, are contained in some synthetic fibers and natural fibers[21-22]. In 1998, the relationship between coordinate structure and antibacterial properties of copper ion complex fiber was investigated by Chen Wenxing[23-35]. It was concluded that the more unstable the coordination structure was, the poorer antibacterial and washability properties will be. In addition, the deodorant and antibacterial fiber with excellent antibacterial effect and wash durability was fabricated by designing reasonable coordination structure. The antibacterial mechanism of copper ion, which is belonged to the contact-type antibacterial mechanism, is that the copper ion sufficiently attracts the sulfydryl of protease in the microbial cells, and then results in the loss of cell activity because of the destructions of enzyme activity. The antibacterial principle of copper ion is shown in Fig. 11-3.

$$\text{Enzyme} \Big\langle {}^{SH}_{SH} + Cu^{2+} \longrightarrow \text{Enzyme} - S_2Cu + 2H^+$$

Fig. 11-3 Antibacterial principle of copper ion

At present, most of copper-based antibacterial fibers have been fabricated through graft modification, namely, copper ion combines with the coordination bond of modified fiber, thus endowing the fiber with antibacterial properties[23]. To fabricate the antibacterial fiber via graft modification, the important processing is to graft the antibacterial compound groups onto the fiber surface by conducting surface treatment. In this case, the antibacterial compound groups could be combined with the finished fiber and the antibacterial fiber is achieved. In 2002, the copper-based antibacterial fiber was achieved by Yu Zhicheng and Chen Wenxing, using the polybasic carboxylic acid to conduct chemical modification on cellulose fiber[26]; in 2007, the copper-based antibacterial cotton fiber was fabricated by Wang Jun through alkalinization, epoxidation and enimization of cotton fiber and then the absorption of Cu(Ⅱ) ion[27].

Recently, a large number of researches on the development of copper-based antibacterial fiber focused on antibacterial properties of fabric achieved via complexation of copper ion and cellulose fiber (based on cellulose fiber fabric being finished by [Cu(NH$_3$)$_4$](OH)$_2$) have been done.

The investigation on effects of response time and ratio of coagulating based on complexation capacity of copper ion and complexation equilibrium was performed, in addition, the anti-UV properties and antibacterial properties of finished fabric were characterized by Qin Zhongyue from Soochow University[28-29]. It was found that the excellent antibacterial effect of bamboo pulp fiber after copper ion complexation could be achieved when the critical concentration of cuprammonia was set to be 0.05 mol/L. Moreover, the antibacterial effect of bamboo pulp fiber after copper ion complexation against the staphylococcus aureus is illustrated in Fig. 11-4(a) and the antibacterial effect against the colibacillus is illustrated in Fig. 11-4(b).

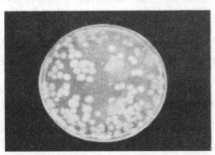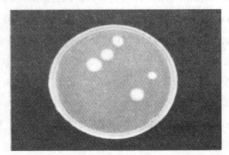

(a)

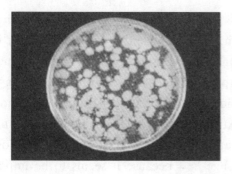 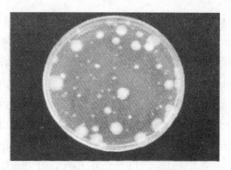

(b)

Fig. 11-4 Antibacterial effects of bamboo pulp fiber: (a) Antibacterial effect against staphylococcus aureus; (b) Antibacterial effect against colibacillus

11.1.1.3 Assessment Methods for Antibacterial Test

The newly-issued international standards for the assessment of antibacterial effect, which are sufficiently significant for the development of antibacterial fiber and antibacterial textile, are as follows: ISO 20645:2004 *Textile Fabric-Determination of Antibacterial Activity-Agar Diffusion Plate Test*, ISO 20743:2007 *Textile-Determination of Antibacterial Activity of Antibacterial Finished Products*, AATCC Test Method 100:2004 *Antibacterial Finishes on Textile Materials: Assessment*[29-31]. In addition, the methods for antibacterial test of national standards mainly consists of AGAR plate diffusion method, absorption method and oscillation method, and this study is supported by GB/T 20944.1—2007 *Textiles Evaluation for Antibacterial Activity—Part 3: Oscillation Method* [32], with principle as follows: the sample and check sample were fed into the conical flask filled with experimental bacteria solution with a certain concentration, after that, the conical flask was oscillated for a certain time at specified temperature for the purpose of testing the variation of concentration of viable bacteria before/after oscillation, in this case, the inhibition rate was calculated and the antibacterial effect of the sample was evaluated[33].

The Gram positive bacteria could be selected as the representative: staphylococcusaureus could be considered as the representative of Gram positive bacteria because the staphylococcus aureus belonging to the non-spore bacteria is considered as the pathogenic bacteria with strongest resistance; the Gram negative bacteria could be selected as the representative as well: colibacillus is usually considered as the representative of Gram negative bacteria because it can exhibit a wide distributed scope of ion[33].

11.1.2 Research Status of Electro-spinning of Natural Cellulose

11.1.2.1 Structure, Properties and Application of Cellulose

As can be seen from Fig. 11-5, most of free hydroxyls of natural cellulose have formed hydrogen bonds, only certain free hydroxyls can be found on the surface of amorphous region rather than that of crystalline region. The sufficiently strong hydrogen bonds inside/between the cellulose molecules enhance linear integrity and rigidity of molecular chain of cellulose remarkably, which provides the cellulose with complex aggregation structure, high crystallinity, stable chemical and physical properties, and high glass-transition temperature. So far, it is extremely difficult for some natural synthetic cellulose to satisfy the requirements of utilization, where the dissolution of cellulose has been considered as the most difficult challenge[34], because it seems to be impossible for the cellulose to be dissolved in common solvents, thus limiting the development and application of cellulose (serving as the most abundant natural products). Therefore, the investigation on the dissolution mechanism of cellulose and development of solvent system is rather important, quite a lot of researches have been done for the development of appropriate solvent system since the macromolecular structure of cellulose was confirmed.

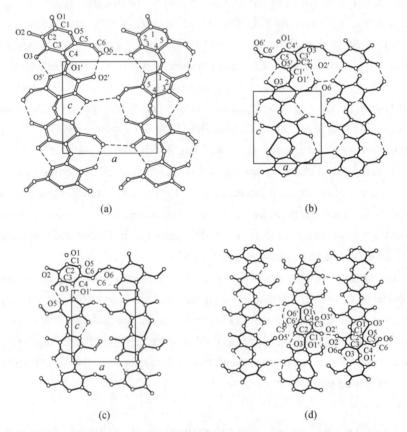

Fig. 11-5 Hydrogen bonds inside/between the cellulose molecular chain molecules[21]: (a) Cellulose I (side 020); (b) Cellulose I (side 110); (c) Cellulose II (side 020); (d) Cellulose II (side 110)

Usually, the cellulose solvents could be classified into derivatization solvent (i.e., derivatization reaction occurs during cellulose dissolution via viscose process and cuprammonium process) and non-derivative solvent (no derivative is generated during cellulose dissolution, i.e., direct solvent: N-NMMO/H_2O system, LiCl/DMAC system, NaOH/urea system and ionic liquid[35-36]). Since the regenerated cellulose has been developed to successfully fabricate diverse products, such as related cellulose and derivative material, the investigation on directly dissolving cellulose in solvent to prepare transparent organic solution with stable chemical and physical properties[36] has become the most concerned topic. In this case, the relevant solvent and dissolution method has been a hotspot in the cellulose industrial and basic study.

11.1.2.2 Conventional Preparation Technology for Cellulose Fiber

The viscose process and cuprammonium process are selected as the main conventional technologies for preparing regenerated cellulose. The conduction of viscose process is extremely complicated and the generated sulfide could results in serious environment pollution: firstly, such cellulose as bamboo pulp or cotton pulp was treated by concentrated alkaline for a certain period, after that, it was shattered, matured and sulfated; finally, it was dried and molded after the sulfuric acid was stripped from the cellulose (coagulating bath of wet spinning)[37-38]. In addition, certain noxious gases could be generated during viscose process, such as HS_2 and CS_2.

The other conventional method for preparation of generated cellulose is cuprammonium process, where the complex solvent of copper ammonia is selected as the earliest solvent. The cuprammonium process presents following disadvantages[38]: (1) complicated process and low production efficiency; (2) high recovery costs of chemical substances (such as ammonia and sulfate generated in this experiment) with serious environment pollution. Therefore, the amount of viscose fiber and copper ammonia fiber fabricated by conventional technologies has significantly decreased. In addition, how to select appropriate preparation technology which is relevant to resource consumptions, effects on ecological environment and product handling has caused wide attention[38].

11.1.2.3 Research Status of Activation Process of Natural Cellulose

The dissolution of cellulose fiber, which mainly consists of two links: activation and dissolution of cellulose, has been always considered as a classic topic. Low chemical reactivity of cellulose fiber is caused by following features, strong hydrogen-bond interaction inside/between the molecules and high crystallinity of locally-structured cellulose fiber[39]. The uniformity and efficiency of cellulose reaction as well as the physical and chemical properties of fabricated fiber are determined by the activation process directly.

As early as in 1979, McCormick[40-41] proposed that cellulose could be dissolved in the LiCl/DMAC solvent system and activation processes for diverse cellulose samples, which were put forward by some researchers based on the previously-known theory[42-44]:

diverse cellulose samples were activated in liquid ammonia, water, NaOH solution or hot DMAC, after that, the solvent displacement was conducted for the purpose of investigation on its solubility. It was also found that the solution remained stable properties and no degradation occurred with a viscosity loss of 2%~3% after the solution was stewed in room temperature (about 30 ℃) for 30 days. The investigation on dissolution of cellulose in the LiCl/DMAC solvent system is mainly based on the research performed by McCormick, including the measurement of relative molecular mass of cellulose via size exclusion chromatography (SEC)[45].

It was found by Zhou Xiaodong[46] from Qingdao University that the cellulose treated by ultrasonic activation exhibited excellent properties, with three methods being selected to conduct the cellulose activation. It was proposed by Wang Yuanlong[44] that the cellulose activation was an chemical treatment with potential application in the future and the steam explosion was considered as an effective physical technology while it has higher requirements on the equipment, based on conclusion of seven physical and chemical methods for cellulose pre-treatment, investigation on effects of mechanism of diverse methods on reaction properties of cellulose, conclusion of advantages and disadvantages of activation. The ultrasonic activation and alkali swelling to the microcrystalline cellulose were performed by Wang Xianling[47], comparing the effects of two methods on supermolecular structure and physical characteristics of microcrystalline cellulose. The experimental results showed that the two activation methods could strengthen the selective oxidation of cellulose remarkably, the content of aldehyde groups increased from 71.3% to 85.0% and 88.8% respectively after activation.

At present, quite a lot of domestic researches on cellulose electro-spinning or preparations of regenerated cellulose fiber have been done, focused on the activation in DMAC and NaOH. The cellulose-based stimuli-responsive material was fabricated by Doctor Luo Hongsheng[48-50] from The Hong Kong Polytechnic University, and it has strictly-controlled dissolution condition for cellulose during process: N_2 protection when using high temperature condition, stirring at room temperature for over 10 h. The activation was improved and optimized, and the stirring time was shortened in this study based on the previously-known researches. It was found that moisture-control was a key factor which affects the cellulose dissolution, and appropriate measures could be taken for the drug disposal and dissolution. In this case, the organic cellulose solution with excellent dissolution effect could be achieved in shorter time, which contributes to the conduction of subsequent experiments. For example, regenerated cellulose membranes can be prepared by phase inversion and cellulose/PAN blended with nano-fiber which is fabricated by electro-spinning.

11.1.2.4 Cellulose Dissolution System Applicable to Electro-spinning

The properties of spinning solution have significant effects on the feasibility of electro-spinning during the electro-spinning process. At the end of the 20th century, the cellulose could be dissolved in NMMO/H₂O solvent system directly for the purpose of commercial production. The non-derivative solvent system of cellulose continues to emerge with the further investigation, whereas very few solvents can be used for the development of electro-spinning, mainly including NMMO/H_2O, LiCl/DMAC and room-temperature ionic system[10].

(1) LiCl/DMAC Solvent System

Variation of electric charge between Li atom and Cl atom could be found since Li—O bond and Li—N bond are formed because of dissolution of LiCl in DMAC. Cl atom is with more negative charges and exhibits strong electronegativity, and it is prone to form hydrogen bond with cellulose molecule. Mcormick[51] proposed that hydroxyl proton on cellulose macromolecule connects to Cl⁻ through hydrogen bond, and Cl⁻ connects to Li⁺ (DMAC), in this case, DMAC solvent penetrates gradually onto the cellulose surface due to the interaction of electric charges, thus the cellulose dissolves in the end, as illustrated in Fig. 11-6.

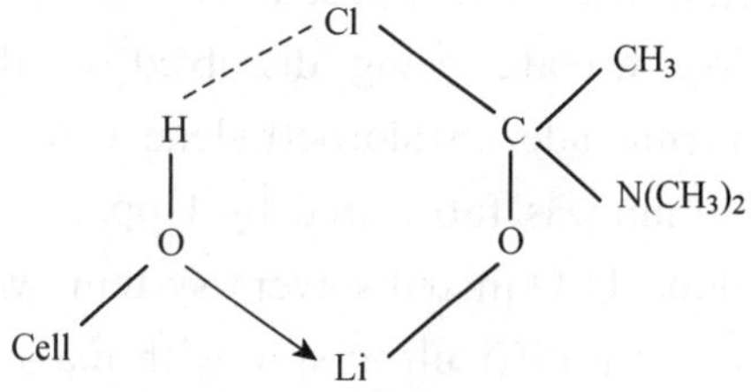

Fig. 11-6 Dissolution mechanism of cellulose in LiCl/DMAC

(2) NMMO/H_2O Solvent System

NMMO, whose dissolution to cellulose is considered as the direct dissolution (dissolution is achieved based on destructions of hydrogen bond between molecules and no cellulose derivative is generated during dissolution), is an aliphatic ring-shaped tertiary amine oxide[52]. The cellulose-NMMO complex is generated by the two lone pairs of electrons on oxygen atom in strongly-polar functional group N—O in NMMO macromolecule and hydrogen nuclei on two hydroxyl groups, the complexation transmits from the non-crystalline region to crystalline region, thus destroying the original hydrogen bond and aggregation structure between cellulose macromolecules[53], and the dissolution mechanism is illustrated in Fig. 11-7.

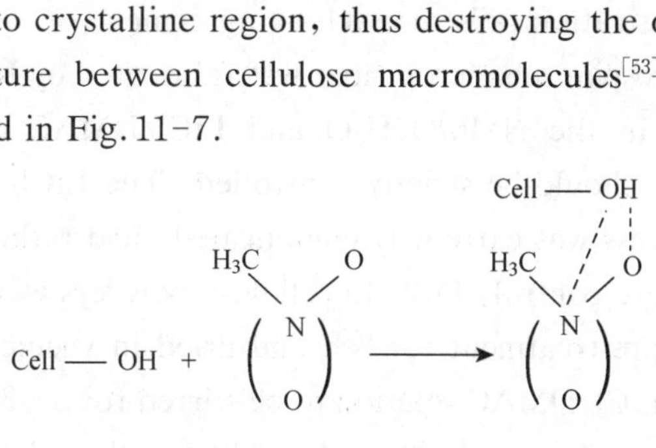

Fig. 11-7 Dissolution mechanism of cellulose in NMMO/H_2O

11.1.2.5 Research Status of Electro-spinning of Cellulose

Following aspects of research on cellulose electro-spinning are focused: (1) solvent system; (2) polymerization degree of cellulose; (3) electro-spinning technology;

(4) electrospun fiber structure at micro/nano-scale.

The superfine cellulose acetate fiber was fabricated by Jaeger[54] dissolving the cellulose in the acetone solution, with nano-fiber diameter of 16 nm～20 mm. The electrospun ethyl-cyanoethyl cellulose fiber was prepared in the tetrahydrofuran solution by Zhao Shengli[41], with nano-fiber diameter of 250～750 nm. The organic solution with mass fraction of 8% was achieved via such cellulose as cotton and absorbent cotton being dissolved in the ethylenediamine/thiocyanate solvent system, in addition, the electrospun superfine cellulose nano-fiber with excellent properties was fabricated when the high-voltage power supply was set to be 30 kV. The superfine and high-affinity membrane with fiber diameter of 200 nm～1 mm, which can be used for filtration of water and biochemical products, was obtained by Ma[55] through electro-spinning, with cellulose acetate being dissolved in the solvent system consisting of acetone/dimethylformamide/trifluoroethylene (3 : 1 : 1). The electrospun cellulose with diameter of 80 nm was fabricated by Uppal[56] dissolving cellulose in the NMMO/N-methylpyrrolidone/H_2O mixed solvent system, with electro-spinning voltage set to be 28 kV, however, it is difficult to spin with mass production and beads can be found in the process.

In 2003, the activation of pulp cellulose in the LiCl/DMAC solvent system and stability after dissolution were investigated by Dupont[57]. It was proposed that salt concentration, sample source and pre-treatment were considered as the key parameters, in addition, it was concluded that the final process for the pulp cellulose was to activate the pulp cellulose in the water/methanol/DMAC solvent system through exchange reaction and no aggregation stratification could be found after the stewing for certain months, with the pulp cellulose being dissolved at the LiCl concentration of 8% at 4 ℃. It has been reported that the solvent displacement for dissolution in the direct solvent is conducted during fiber activation, as the principle for solvent displacement before cellulose dissolution and after activation fails to reach an agreement.

In 2006, the electrospun submicron fiber was fabricated by Kim[58], with cotton cellulose being dissolved in the NMMO/H_2O and LiCl/DMAC direct solvent systems, whose moisture content should be strictly controlled. The LiCl/DMAC solvent system, whose preparatory process was extremely complicated, had rather high requirements on moisture and temperature control: DP1140 cellulose powders were applied into water at room temperature for pre-treatment for 8 h and dried in vacuum at 60 ℃, after that, they were fed into the LiCl/DMAC solution to be stirred for 2～3 h at 60 ℃ and continuously stirred for 12 h at room temperature. In addition, the solubility of 3% for DP1140 powder in the LiCl(8%)/DMAC solvent could be achieved and the activation of cellulose could be used for reference in this study. In the NMMO/H_2O solvent system, the cellulose should be dried in vacuum for 8 h at 80 ℃ and stirred at room temperature to achieve the goal of dissolution; furthermore, the antioxidant of 0.5%～1.0% should be applied. In this case, the solubility of 3% for DP1140 powder could be achieved; while the solubility of 9% for DP210 in the NMMO/H_2O solvent system could be obtained. It had high requirements on electro-spinning temperature of fiber and the collector, where the rotary collector should be heated. Activation of cellulose during pre-treatment and improvement of electro-spinning device could be used for references in this study.

Cellulose nano-fiber was achieved by Wan Hejun[59] from Zhejiang Sci-tech University, with brown cotton fiber being dissolved in the LiCl/DMAC solvent system and fiber being treated by methanol and water at 0 ℃ (serving as coagulating bath) respectively for the purpose of removing the residual solvent. However, as can be seen from the SEM images, severe adhesion could be found between fibers. The preparation and stability of cellulose LiCl/DMAC solution was investigated by Mamat[25, 60], the experimental results showed that the higher the mass fraction of cellulose was, the longer the time of heated stirring as well as stirring at room temperature for preparing this solution was, the greater the polymerization degree of cellulose was, the poorer the dissolution effect was. It was possible for the cellulose to be dissolved completely when the mass faction of LiCl was set to be greater than 6%, in addition, the regenerated cellulose membrane with preferable strength could be fabricated at 20 ℃. The cellulose fiber was prepared by Li[61], with cotton fiber being dissolved in the LiCl/DMAC solvent, the heating device being developed to enhance the solvent volatilization. The dissolution behaviors of bacterial cellulose in diverse solvents were investigated by Liu[62], with NMMO, LiCl/DMAC, LiCl/DMAC, AMIMCl/NMMO being selected as the solvents. The experimental results showed that the bacterial cellulose exhibited different dissolution abilities in different solvents and the sequence of dissolution ability of bacterial cellulose in the same condition was as follows: NMMO>AMIMCl>AMIMCl/NMMO>LiCl/DMAC.

11.1.2.6 Properties, Structure and Application of PAN

PAN (popularly called acrylic), which is considered as the homopolymer or copolymer of acrylonitrile, is characterized by high hardness and rigidity[63], excellent compatibility of substances with homopolarity, soluble in such organic solvent as DMAC and DMF. PAN fiber exhibits quite a lot of advantages, such as good weather resistance and resistant to sunlight, stable performance to chemical reagent, in addition, it has been used for the ultrapure water production and wastewater treatment & reuse because of its excellent anti-fouling ability and water flux[64]. In recent years, the scope of applications of PAN in ultrafiltration materials, composite materials and functional textiles have been widened, by nano-fiber with great specific surface area being made of PAN.

11.1.2.7 Research Status of Electro-spinning of PAN

There are quite a lot of researches mainly focused on variation of electro-spinning parameters[65], however, very few researches have been reported on the blended electro-spinning of PAN and natural cellulose. Recently, some researchers presented their works on the development of PAN/cellulose blending system focused on blending of modified cellulose and PAN, by using the method of traditional wet spinning or membrane stewing, usually the diameter of spun fiber is at micron-scale[66]. It was reported that[67] Tianjin Polytechnic University adopted the dry-wet spinning method to prepare PAN/cellulose blended fiber inion solution (i.e., [BMIM]CI solution) and investigated the spinnability and compatibility of the blended fiber. The multi-layer ultra-filtration/nano-filtration membrane was fabricated and the effects of cellulose acetate concentration and membrane formation technology on its ultra-filtration properties were investigated by Wei Jielin[55] from Donghua University, with the cellulose acetate membrane being selected as the functional separating layer and the chemical fiber mesh as the supporting of the substrate layer. In 2012, the composite filtration membrane made of cellulose triacetate (CTA) was fabricated by Xu[68] on the basis of this, the prepared cellulose membrane via phase inversion by both of them was composed of the electrospun PAN nano-fiber membrane in the form of upper layer and lower layer, which indirectly avoided the difficulties arising from the electro-spinning of natural cellulose. The PAN/hydroxyethyl cellulose blended nano-fiber was spun by Zhang[69], it was found that the blending ratio and concentration of spinning solution have effects on the surface properties of the fiber, the hydrophily of nano-fiber membrane increases with the increasing of mass fraction of hydroxyethyl cellulose.

Quite a lot of researches focused on modified cellulose (e.g., cellulose acetate) of electro-spinning in direct solvent have been done; whereas very few researches focused on un-modified natural cellulose have been conducted. The preparation of electrospun nano-fiber, whose technology remains in initial experimental stage and is limited in laboratory, mainly focuses on the investigation of activation of cellulose, parameter affecting cellulose solvent and improvement on collector used for electro-spinning. There are a large number of researches on the electro-spinning of PAN, especially the research on preparation of high-strength carbon fiber by applying carbon nanotube for the purpose of enhancing strength; however, very few researches have been done on the investigation of blending of PAN and cellulose for electro-spinning, which is because, the hygroscopicity of acrylic can be improved and optimized via wet spinning. Meanwhile, it can be concluded that electro-spinning could be conducted based on the blending of cotton fiber and PAN, whereas the electro-spinning parameters should be strictly controlled.

The preparation technology of electro-spinning for cellulose fiber in direct solvent and related functional modification (especially the modification of antibacterial function) for the purpose of endowing cellulose nano-fiber with novel functions are considered as the research focuses in this study.

11.2 Activation and Dissolution of Natural Cellulose and Preparation of Regenerated Cellulose Membrane

11.2.1 Application of Regenerated Cellulose Membrane

Two preparation methods of cellulose membrane are achieved based on the derivative generated by chemical modification[1, 70, 71]: one is that the cellulose derivative is generated by cellulose, for example, cellulose acetate membrane and ethyl cellulose membrane with excellent separation and permeation performance have been the important products in membrane formation industry, and the cellulose diacetate has been selected as an important material for preparation of reverse osmosis membrane, ultrafiltration membrane and microfiltration membrane; the other is that the regenerated cellulose membrane can be fabricated on the basis of changing the natural cellulose into soluble cellulose derivative[72]. And cellophane and regenerated copper ammonia cellulose membrane with superior hydrophily and anti-fouling abilities are fabricated via this method, whereas their production technology is cumbersome with great limitations which has been discussed in detail in **Chapter 1**[73].

For quite some time, investigation on preparation method retaining and utilizing the natural properties of cellulose has been the research focus[74]. Direct dissolution of cellulose contributes to developing new cellulose membrane, which is characterized by certain advantages at least[1, 75], such as high strength and integrity, good hydrophily, excellent biocompatibility, stable thermal stability, stable chemical properties.

Phase inversion, a membrane preparation process in which the regenerated cellulose membrane is fabricated after solvent removal on the basis of mass transfer of solvent and non-solvent of membrane casting solution in coagulating bath, has been selected as the commonly-used membrane preparation technology[74, 76]. In addition, these three procedures should be conducted to achieve the goal of membrane preparation via phase inversion: preparing membrane casting solution, wiping in flowing state on the substrate material with smooth surface, removing the solvent in the coagulating bath for the purpose of preparing regenerated cellulose membrane.

11.2.1.1 Application in Separation Membrane

At present, quite a lot of researches have been done by Akozo Nobel (the Europe's largest manufacturer for copper ammonia regenerated cellulose membrane, Holand) for the development of preparation of regenerated cellulose membrane with diverse micropore sizes based on NIMMO system via controlling the preparation technology. In addition, this kind of regenerated cellulose membrane could be applied in osmotic membranes for seawater desalination and hemodialysis.

11.2.1.2 Application in Food Processing and Packaging

Gas permeability plays a significant role in prolonging shelf life of product or improving food freshness in food packaging. It was proposed by Fink and Gregory that the food fermentation and spoilage arising from the residual CO_2 generated by metabolism of such respiratory as vegetable, fruit and sausage could be avoided in food packaging because of the excellent permeability of regenerated cellulose membrane[77-78]. In addition, the regenerated cellulose membrane can also be used for packaging sugar, pastry, drug and rubbish[79], which is because, it is biodegradable and causes no pollution to environment. The casing breaking can be prevented during food processing since no chemical reaction occurs and no gas is generated during the LiCl/DMAC membrane preparation[80].

11.2.1.3 Application in Agriculture

Recently, the regenerated cellulose membrane has been much more popularly used for the development of agriculture, mainly including adjusting illumination, saving irrigation water, plastic mulching, enhancing multiple cropping indexes of the land, increasing grain yield; greenhouse film which facilitates to the growth of vegetable need high temperature and solar radiation; storage of green fodder: the regenerated cellulose membrane could relieve the effects of air and humidity to fodder, fermentation with rich lactic acid based on appropriate sugar in anoxia condition may occur, which contributes to maintaining the color, scent, completeness and freshness of the fodder and providing the livestock with fresh fodder in winter, thus increasing the milk yield.

In view of the above-mentioned significance and application of regenerated cellulose membrane, diverse properties of regenerated cellulose membrane fabricated in LiCl/DMAC solvent system via phase inversion are investigated in this study.

11.2.2 Experimental

11.2.2.1 Materials

DMAC, analytical reagent, Sinopharm; LiCl, analytical reagent, Sinopharm; natural cotton fiber (white cotton with DP of 10 000~15 000); ethanol (CH_2OH), analytical reagent, Sinopharm.

11.2.2.2 Activation and Dissolution of Natural Cellulose

Activation is one of the most important preprocessing steps to ensure fully dissolving of natural cellulose[29]. The purpose of activation is to destroy the hydrogen bond between cellulose macromolecules, increase the number of reactive hydroxyl groups, and ultimately improve the reactivity of natural cellulose fibers. There are three main ideal methods for the activation of cellulose in LiCl/DMAC system generally: (1) Hot alkali activation[39] is a method which swelling fibers in NaOH with high concentration and then washed in deionized water and followed by drying process; (2) Solvent exchange method[40] refers to fibers swelling and solvent exchanging sequentially in H_2O, methanol (or ethanol) and DMAC. However, the swelling time and exchanging frequency vary in different experiments and materials, and the activation process is long time-consuming and complex in general; (3) Hot DMAC activation[81] refers to immersing the fibers into DMAC whose temperature is above 150 ℃. LiCl is added after a certain period of time. This method is simple in operation and time saving, but the degradation of cellulose could be caused due to high temperature. Based on the previous research work[48, 57, 82], an optimized solution is illustrated in **Section 11.3.3.4** after comparison between dissolution effects and mechanical properties of regenerated membranes under different activation temperature. The optimized solution is that to activate natural cotton fibers in the reflux of hot DMAC for about 60 min at 130 ℃ in N_2 atmosphere. The solution can effectively reduce the degradation degree of natural celluloses' oxidation and has better dissolution results.

Microamount moisture in LiCl/DMAC solvent system has effects on its solubility and stability[44], which is because, the moisture in the solution has effects on the dissolution of cellulose and impedes the interaction between compound ion and cellulose macromolecule. Both LiCl and DMAC are classified into the hygroscopic chemical raw materials, so the moisture in all chemical raw materials should be removed through specified process before dissolution of cellulose.

A series of organic solutions with different concentrations were prepared by dissolving the natural cotton fibers in LiCl/DMAC solvent system. This study used LiCl without drying as shown in Fig.11-8 (a), if drying in ordinary conditions to dissolve cellulose, the results were not satisfactory since the fibers could not be sufficiently dissolved and stratification phenomenon appeared, the comparison was shown in Fig.11-9. To remove the influence caused by traces of water, the treatment of LiCl was illustrated as follows: a certain amount of LiCl was dried in a vacuum oven above 130 ℃ for 5 h to remove crystal water and stored in desiccators with silica ($mSiO_2 \cdot nH_2O$). Fig.11-8 (b) and (c) showed traces of water in DMAC reagent and removed by the sodium-4A molecular sieve, and the dissolving equipment was shown in Fig.11-8 (d). The whole experiment process has been using N_2 protection to reduce the effect of trace amounts of water.

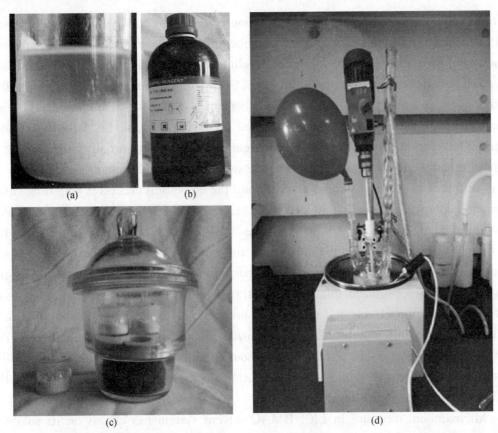

Fig. 11-8 Pretreatment of reagent and activation equipment: (a) Undissolved cellulose; (b) Moisture removal for DMAC; (c) LiCl storage; (d) Activation device

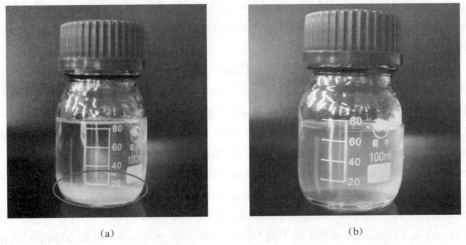

Fig. 11-9 Influence of water on cellulose/LiCl/DMAC solution system: (a) Dissolution phenomenon of cotton fiber (LiCl without vacuum drying); (b) Dissolution phenomenon of cotton fiber (LiCl with vacuum drying)

11.2.2.3 Preparation of Cellulose/LiCl/DMAC Organic Solutions

A certain amount of DMAC was fed into flask and then be heated to 130 ℃. Chips of

natural cotton fibers were applied into flask and activated for 60 min under protection of N₂ and reflux condensation. When the temperature decreased to 100 ℃, LiCl with mass fraction of 9% was fed into the above solutions and activated for 120 min at a stirring speed of 470 r/min. The solutions were then cooled down to room temperature (20～25 ℃) and continued stirring for 120 min to prepare homogeneous and stable cellulose/LiCl/DMAC organic solutions(casting solution). The bubbles in the solutions were eliminated through ultrasonic treatment. In this way, the cellulose/LiCl/DMAC solutions with different mass fraction from 0.5% to 4.5% were prepared, as shown in Fig.11-10.

Fig. 11-10　Cellulose /LiCl/DMAC solutions with different mass fraction

11.2.2.4　Membrane Forming Process

Two coating methods were selected for preparation of regenerated cellulose membranes using cellulose/LiCl/DMAC solutions. In process 1, high-speed spin technique (KW-4A spin coater) is illustrated in Fig.11-11 (a). First rotate is conducted for 10 s at 500 r/min to coat the solutions on the substrates uniformly, after the setting time the spinning speed automatically converted to 1000 r/min and kept spinning for 15 s to throw out the excess solutions and uniform the thickness. Finally, the substrates were placed into water or ethanol to extract organic solvent and obtained regenerated cellulose membranes. In process 2, low-speed scraping (AFA-Ⅱ film applicator) is shown in Fig.11-11 (b). The solutions were pushed in horizontal at 3 cm/s to obtain the membranes of uniform thickness. The process of solvent extraction was the same as process 1. The regenerated cellulose membranes prepared through two processes above were dried in air dry oven at 30 ℃ and used for characterization.

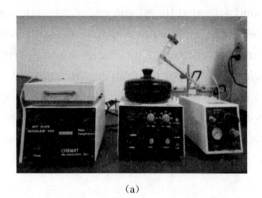 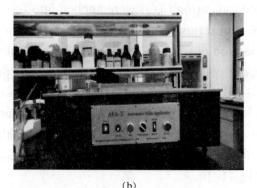

Fig. 11-11　Membrane equipment: (a) KW-4A spin coatet; (b) AFA-II film applicator

11.2.2.5　Characterization

Viscosity measurement: NDJ-1 viscometer was used to test the viscosity of cellulose solutions with different mass fractions at different temperatures.

Morphology measurement: JNOEC XS-213 optical microscope whose magnification is 400 and equipped with CCD digital color video camera and S-3400N scanning electron microscope (SEM) were used to observe the surface morphology and structure of regenerated cellulose membranes.

Mechanical properties measurement: The samples were balanced under constant temperature and humidity for 24 h and then cut into rectangular, approximately 20 mm×3 mm and to be tested by YG006 electronic single fiber strength tester. The drawing rate was set at 50 mm/min and at a distance of 10 mm. Each sample should be tested for 10 times and took the average number as final result.

Crystallinity: The crystallinity of membranes was tested by X-ray (K780FirmV_06X) with CuKα radiation ($\lambda = 0.154$ nm).

Thermal properties: The thermal properties of regenerated cellulose membranes were characterized by STA PT-1000 thermo gravimetric analysis (TGA). The samples were located in nitrogen atmosphere at the flow rate of 80~100 mL/min and heated from room temperature to 800 ℃ in 40 min.

FTIR measurement: AVATAR 370 Fourier Transform Infrared Spectroscopy (FT-IR) was used to characterize the different of molecular structure and chemical composition before and after the natural cellulose regeneration.

Static contact angle measurement: OCA15EC contact angle measuring device was used to test the wettability of regenerated cellulose membranes at a stable temperature of 20 ℃ and humidity of 60%. Test infiltration properties of regenerated cellulose membranes and characterize the regenerated cellulose membrane surface infiltration performance.

11.2.3 Results and Discussion

11.2.3.1 Viscosity of Cellulose Organic Solution

Fig. 11-12 shows the viscosity curves of cellulose organic solutions with different mass fractions at five temperature gradients of 20 ℃, 30 ℃, 35 ℃, 40 ℃, 50 ℃. Since the natural cotton fiber is a polymer, the viscosity would be changed significantly when the mass fraction is set to be greater than 1%. Under the condition of same temperature, the viscosity of solution becomes higher when the mass fraction of cellulose organic solution increases. The reason is that the force between macromolecules also increases when the mass fraction increases and the flowability of solution becomes poor. If the other conditions keep unchanged, the higher temperature would lead to a lower viscosity of solution, for the force between macromolecules would be weakened as the temperature increases, which contributes to the movement of macromolecules[29]. At room temperature (20 ℃), when the absolute viscosity of the solution whose mass fraction being 4.5% reaches above 5500 MPa · s, the solution can hardly flow. When the temperature reaches above 30 ℃, the viscosity of solution with different mass fractions obviously decreases. At the same temperature range, the solution's viscosity of high mass fraction decreases more compared with those of low mass fraction (lower than 1.5%).

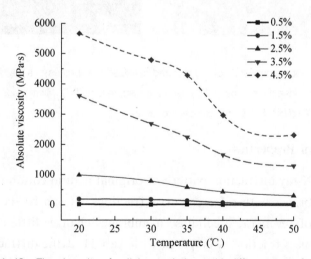

Fig. 11-12 The viscosity of cellulose solutions with different mass fractions

11.2.3.2 Surface Morphology Analysis

Fig. 11-13 shows the macroscopic and microstructure surface morphology of regenerated cellulose membranes (mass fraction: i—4.5%, ii—3.5%, iii—2.5%, iv—1.5%, the same below). When the mass fraction is lower than 1.0%, the organic solution can hardly form membranes due to its low viscosity. It has good transparency and toughness, and does not have obvious macroporous defects, which is similar to that of membranes prepared in NMMO/H$_2$O system[2], but the application value in the field of dialysis and other medical need further exploration in biological medicine.

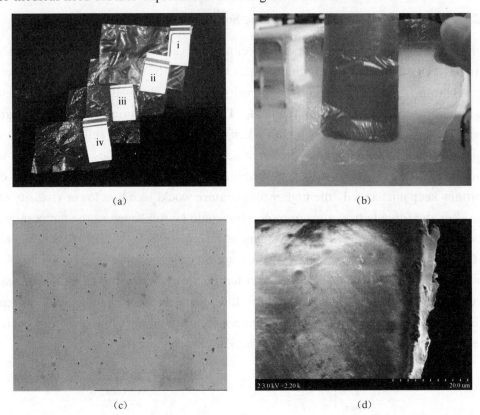

Fig. 11-13 Surface morphology of regenerated cellulose membranes: (a) A series of regenerated cellulose membranes; (b) Good transparency; (c) Microscope at magnification of 400; (d) SEM at magnification of 2200

11.2.3.3 Crystal Properties

Fig. 11-14 shows X-ray diffraction pattern of original natural cotton fiber and regenerated cellulose membranes with different mass fractions. It can be seen that intensity of diffraction peaks of regenerated cellulose membranes changes little and has no obvious relationship with mass fraction. Compared with Tab. 11-2, the diffraction peaks of natural cotton fiber at $2\theta = 15.0°, 16.9°, 23.1°, 34.6°$ exhibit characteristic peaks of cellulose I crystal form and the regenerated cellulose membranes exhibit cellulose II crystal form[33].

The diffraction peak of natural cotton fiber is very sharp while that of regenerated cellulose membranes are relatively flat and the intensity is much lower than that of cotton fiber. It can be concluded from Fig. 11-14 that the crystal form changes greatly before and after the regeneration process. Crystallinity is calculated by peak fitting named Jade software, within the error $R \leqslant 3.0$, as shown in Fig. 11-15. The crystallinity of natural cotton fiber is 74.3%, which is consistent with the report by Zhu from Donghua University, who measured the crystallinity of white cotton fiber was 70.64%[83]. The random error is less than 5%. The diffraction peaks of regenerated cellulose membranes are wide and weak, most of which are multi-non-crystalline peaks. It can be concluded from this point that the crystallinity of original natural cellulose regions is severely damaged and only a small part of the regenerated cellulose membranes recrystallizes.

Tab. 11-2 Standard spectrum peak of cellulose I and cellulose II

Types	Standard position 2θ(°)		
Cellulose I	14.8	16.6	22.7
Cellulose II	12.3	20.2	21.9

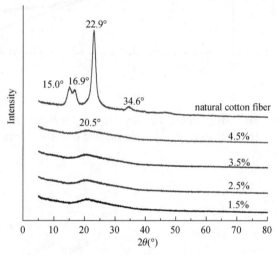

Fig. 11-14 XRD curves of natural cotton fiber and regenerated cellulose membranes with different mass fraction

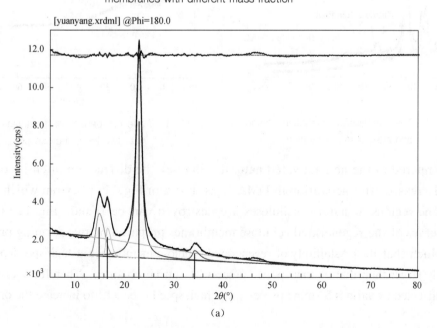

(a)

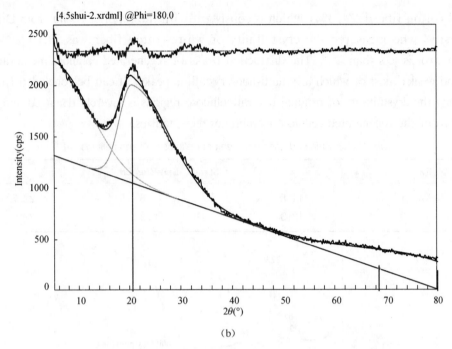

Fig. 11-15 XRD fitting curves: (a) XRD fitting curve of natural cotton fiber
(b) XRD fitting curve of regenerated cellulose membranes

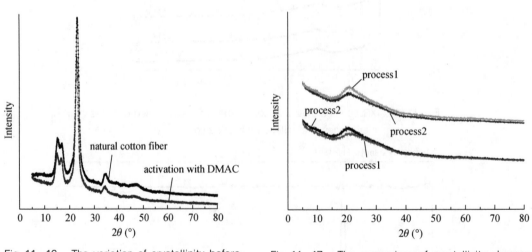

Fig. 11-16 The variation of crystallinity before and after DMAC activation

Fig. 11-17 The comparison of crystallinity changes with two forming processes

Compared to the non-activated natural cellulose, the diffraction intensity of crystal peaks decreased after activation in DMAC, as shown in Fig. 11-16, from which that the crystalline regions of natural celluloses are destroyed can be found. Fig. 11-15 shows XRD curves of the regenerated cellulose membranes prepared by two forming processes, from which that the crystallinity of regenerated cellulose membranes at high speed (process 1) is slightly larger than that at low speed (process 2) can be obtained. Therefore, crystallinity can be affected by various forming process, and high speed is benefit to increase the orientation and crystallinity of regenerated cellulose membranes. This conclusion is consistent with Fig. 11-17, the average fracture strength of regenerated cellulose membranes prepared at high speed (process 1) is higher than that at low speed (process 2).

11.2.3.4 Thermal Properties

The thermal stabilities of the regenerated cellulose membranes and natural cotton fibers were illustrated by TG and DTG curves under heating rate of 20 °C/min, as shown in Fig. 11-18, and the parameters corresponding to the process of decomposition were listed in Tab. 11-3.

Generally, it is found that the decomposition process of regenerated cellulose membranes and natural cotton fiber can be divided into three phases corresponding to different temperature ranges. Firstly, a trace weight decrease at the temperature ranging from room temperature to around 100 °C due to the decomposition of small molecular, mainly water and solvent in DMAC, corresponding to the peak appearing on DTG curves in Fig. 11-18 (b). Then the weights of all the fibers decrease slightly at temperature ranging from 100 to 400 °C, which is followed by a third range of mass decrease, corresponding to the peak around 300 °C in DTG curves; at this temperature range, the mass decrease is mainly ascribed to the decomposition of molecular chains, releasing gases like H_2O and CO_2. The third slight mass decrease ranges from 800 to 400 °C, which is mainly caused by carbon formation at high temperature.

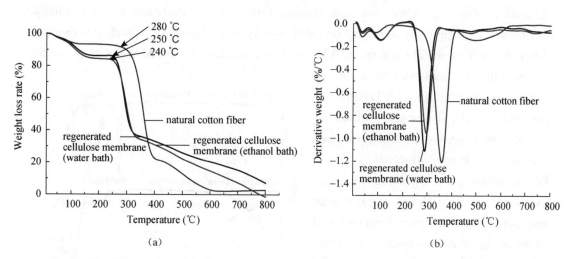

Fig. 11-18 DSC curves of regenerated cellulose membranes and natural cotton fibers: (a) TG; (b) DTG

It is shown that all the residual mass ratios of the regenerated cellulose membranes are higher than that of natural cotton fiber at 700 °C. Explanation for this phenomenon is that the crystal forms of the regenerated cellulose membranes change from cellulose I into cellulose II. Compared with cellulose I, cellulose II is not prone to reversal and more vulnerable to dehydrate or decarboxylase to H_2O, CO_2 and residues, so the residual mass ratios of the regenerated cellulose membranes are much higher[84]. The initial decomposition temperature of the regenerated cellulose membranes decreases, in comparison with natural cotton fiber, the thermal stability relatively weakened. The thermal stabilities of the regenerated cellulose membranes prepared by different coagulation baths have little difference.

Tab. 11-3 Thermal behaviors of the regenerated cellulose membrane and natural cotton fiber

Sample	T_0(℃)	$T_{10\%}$(℃)	T_{max}(℃)	α(%)
Natural cotton fiber	280	132	360	2.0
Regenerated cellulose membrane (water bath)	250	146	285	15.6
Regenerated cellulose membrane (ethanol bath)	240	139	295	7.7

T_0—The initial decomposition temperature;
$T_{10\%}$—The temperature at the weight loss rate of 10%;
T_{max}—The temperature at the maximum decomposition rate;
α—The residual weight rate at 700 ℃.

11.2.3.5 FTIR Analysis

FTIR curves of the regenerated cellulose membranes and cellulose/LiCl/DMAC solutions have been illustrated in Fig. 11-19. The characteristic absorption peak of cellulose can be seen at 3400, 2900 and 1050 cm^{-1}. Compared with regenerated cellulose membranes, the spectras of cellulose/LiCl/DMAC solutions have a slight variation. In addition to the characteristic peak of cellulose, absorption peaks of amide bond (CO-NH-stretching vibration) of DMAC also appear at 1600 cm^{-1}, 1500 cm^{-1} and 1400 cm^{-1}. The position of cellulose characteristic absorption peak changes little, but the stretching vibration intensity greatly weakens, probably due to the destruction of hydrogen bond when cellulose is dissolved. After the analysis of FTIR spectrum, it is known that cellulose/LiCl/DMAC solutions are consistent with the molecular structure of membranes prepared in LiCl/DMAC solvent system. The shapes and positions of the absorption peaks of all regenerated cellulose membranes dissolved in LiCl/DMAC solutions are similar to natural cotton fibers, which indicate no significant variation in the molecular structure and chemical composition when dissolving. When cellulose is dissolved in the LiCl/DMAC solvent system, no derivatives are produced, which belongs to direct dissolution.

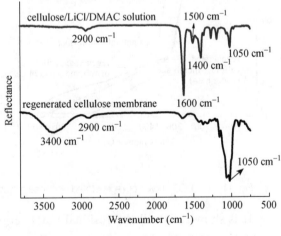

Fig. 11-19 FTIR spectra

11.2.3.6 Mechanical Properties

Quite a lot of factors have effects on strength of regenerated cellulose membrane, among which, preparation technology, activation temperature, mass fraction of cellulose and type of coagulating bath are considered as the key factors affecting the mechanical properties of membrane. A large number of researches have been done for the research on effects of coagulating bath on properties of regenerated cellulose membrane when in the NMMO system[85] and NaOH/thiourea urea system[86]. Whereas, there are very few researches have been conducted for the investigation on effects of coagulating bath in the LiCl/DMAC solvent system on properties of regenerated cellulose membrane. The initial investigation on properties of regenerated cellulose membrane fabricated in coagulating bath (water bath and ethanol bath) was conducted in this experiment.

(1) Effects of Preparation Technology on Mechanical Properties of Regenerated Cellulose Membrane

From Fig. 11-20, it can be clearly observed that the average breaking strength of regenerated cellulose membrane fabricated via process 1 is much higher than that of process 2, which is because, the orientation degree and crystallinity of regenerated cellulose membrane fabricated via process 1 increase due to the high speed during membrane preparation. As can be seen from Fig. 11-17, the diffracted intensity of characteristic peak of regenerated cellulose membrane fabricated at high speed is higher than that of low speed.

(2) Effects of Activation Temperature on Mechanical Properties of Regenerated Cellulose Membrane

Fig. 11-20 (b) shows the breaking strength of regenerated cellulose membrane (mass fraction = 4.5%, process 2) prepared at the activation temperature 130 and 150 ℃ separately, thus it can be concluded that the activation temperature has an obvious effect on the mechanical properties. Possible analysis is that high temperature and oxygen would cause degradation of cellulose mentioned in **Section 11. 1. 2. 1.** It is shown from

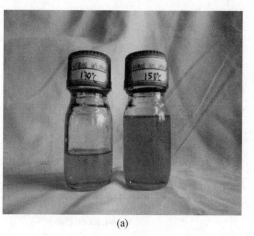
(a)

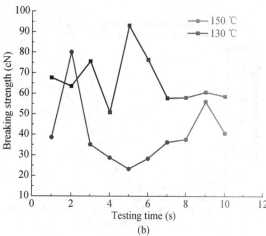
(b)

Fig. 11-20 (a) Cellulose/LiCl/DMAC solutions prepared at different activation temperature;
(b) Breaking strength of cellulose membrane at different activation temperature

Fig. 11-20 (a) that the color of cellulose/LiCl/DMAC solution activated at 150 ℃ is darker than that at 130 ℃, which is due to the cellulose degradation and oxidation at high temperature. On the premise of dissolution, the lower the activation temperature is, the smaller impact the properties of the regenerated cellulose membrane have. Nitrogen (N_2) protection is introduced into the preparation of the solution to reduce the oxidation degradation under high temperature.

(3) Effects of Mass Fraction and Coagulation Bath Type on Mechanical Properties of Regenerated Cellulose Membrane

Fig. 11-21 shows the regenerated membranes have efficient breaking strengths. With the increase of cellulose' mass fraction, the average breaking strength increases. However, when the mass fraction is above 3.5%, with the rising mass fraction, the average breaking strength becomes decreased. The reason is that as the mass fraction increases, the number of macromolecules in per unit increases, which causes large degree aggregation of macromolecules, so it can leading to the increase of density and winding of macromolecular network as well as dense structure of the membranes. With mass fraction increases, the average breaking strength of regenerated cellulose membrane has upward trend. However, when the mass fraction of cellulose membrane is above 3.5%, the average fracture strength decreases whether in water or ethanol coagulation bath. Possible explanation is that the viscosity rises rapidly as the mass fraction increases, which leads to poor fluidity. Poor fluidity results in poor uniformity of the regenerated cellulose membranes, and thus the mechanical properties of the membranes become poor. When the solution viscosity gets lower (mass fraction ≤ 0.5%), the fluidity becomes higher so the thickness of membranes can hardly be controlled and the membranes are not easily prepared. Therefore, high viscosity or low viscosity of the cellulose/LiCl/DMAC solutions is not conducive to enhancing the mechanical properties of the regenerated cellulose membranes.

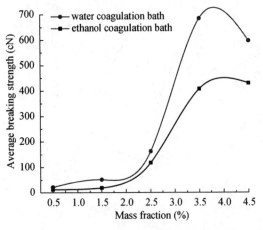
Fig. 11-21 Average breaking strength of membranes with different mass fraction and coagulation bath

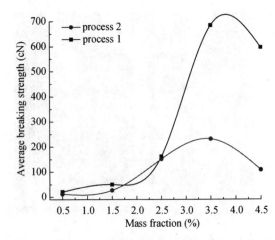
Fig. 11-22 Average breaking strength of membranes with different preparation process

(4) Effects of Coagulation Bath Type on Mechanical Properties of Regenerated Cellulose Memberane

It can be seen from Fig. 11-21, in the same preparation process and mass fraction, the average breaking strength of regenerated cellulose membrane in water coagulation bath is higher than that in ethanol coagulation bath. The reason of this phenomenon has been analyzed to as following: dense regenerated membranes are easily generated in water, while it is easy to generate porous membranes in ethanol. When the solvent and non-solvent exchange rate is larger than the phase separation of cellulose/LiCl/DMAC organic solutions, the phase separation between membrane solution and coagulation bath does not occur immediately, so the relatively dense regenerated cellulose membrane can be obtained in water coagulation bath[87]. While using ethanol coagulation bath, the porous membranes can be easily obtained.

Porosity(Fig. 11-23): The regenerated cellulose membrane is swelled in the deionized water completely and then blotted by the absorbent paper, in this case, its weight is named as W_1; after that, the membrane is dried out in a cool and well-ventilated place and placed in a hot-air oven at 50 ℃, in this case, the weight is named as W_2. The porosity of the membrane Pr could be calculated according to the following formula, with the density of cellulose at 1.528 mg/mm³.

$$Pr = \frac{(W_1 - W_2)/\rho_{H_2O}}{(W_1 - W_2)/\rho_{H_2O} + W_2/\rho_p}$$

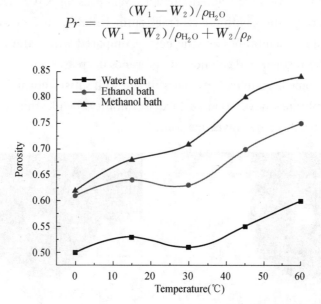

Fig. 11-23 Porosity of regenerated cellulose membranes under different solidifications

According to the different wet-abilities between different testing liquids on material surface, there are various methods about static contact angle testing. If the test liquid cannot infiltrate on the surface of regenerated cellulose membrane (contact angle $\theta \geqslant 90°$) and droplets can form a good profile, the contact angle measurements in positive direction should be applied. If not (contact angle $\theta \leqslant 90°$), the contact angle measurements in opposite direction are selected[37-38], which is also called Captive Drop Method, this is generally used for the samples which absorb the test liquid rapidly and it is difficult to quantify the wetting behavior by conventional methods. The contact angle can be read by the instrument software according to baseline profile.

Fig. 11-24 shows the contact angles of regenerated cellulose membrane and polyvinyl chloride (PVC) membrane using Captive Drop Method. As can be seen from this figure, the contact angle of regenerated cellulose membrane is about 50° which is lower than that of normal PVC membrane which is about 80°, indicating that the surface of the regenerated cellulose membrane has strong hydrophilicites and good wetting properties. A series of organic solutions with different mass fraction of cellulose are prepared by dissolving natural cotton fiber in LiCl/DMAC solvent systems after the activation of natural cotton fiber. The regenerated cellulose membranes are formed in two kinds of coagulation baths; two coating methods including high-speed spin technique (KW-4A spin coating machine) and low-speed scraping (AFA-II film applicator) were selected in this study. The results of characterization show that, on the premise of ensuring thorough dissolution, the lower activation temperature is better. Moisture content in the system must be strictly controlled, and high temperature activation could easily lead to the oxidative degradation of celluloses. However, the mechanism of activation and solutions needs to be optimized or investigated further. When using KW-4A spin coating machine, water as coagulation bath, mass fraction of cellulose at 3.5%, the properties of regenerated cellulose membranes are the best. Compared with natural cotton fibers, the crystallinity of the regenerated cellulose membranes decreases and thermal stability obviously declines to some certain extent, but still can well satisfy normal use. The regenerated cellulose membranes have good surface wettability, while the researches in related biomedical field need further investigation.

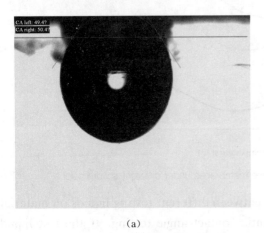 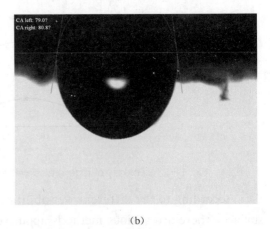

(a)   (b)

Fig. 11-24 Contact angle of different materials: (a) Regenerated cellulose membrane; (b) Ordinary PVC membrane

11.3 Preparation of Natural Cellulose/PAN Blended Nano-fiber

11.3.1 Electro-spinning Device

According to the previous investigations on natural cellulose fiber, nanometer material, membrane anti-pollution and antibacterial materials, in recent years, the LiCl/DMAC solvent system has been proposed to dissolve PAN and natural cellulose with high polymerization degree in this study. LiCl/DMAC is considered as a representative of direct solvent for cellulose and can dissolve the cellulose materials directly, furthermore, no other derivatives[80, 88-90] will be generated during the dissolution process. The electro-spinning is selected in this research to prepare the cellulose/PAN blended fiber at nano-scale, the optimization of electro-spinning parameters is conducted on the basis of previous researches, and the antibacterial treatment test is also investigated in this study, together with the researches on the morphology, thermology, contact angle and antibacterial property of nano-fiber.

The schematic diagram of electro-spinning device adopted in this research is illustrated in Fig. 11-25: position 1 refers to capillary tube/spinnerets with needle; position 2 refers to high-voltage power supply; position 3 refers to receiver. Driven by the propulsion unit, the spinning solution is extruded to the spinnerets and then sprayed into jet-flow under the action of high-voltage electric field, finally deposited in form of nano-fibers on the receiver after the solvent volatilization. The KH-2 electro-spinning machine is equipped with two propulsion units located at both left side and right side of the machine for the purpose of precisely controlling the high voltage and propulsion speed. As illustrated in Fig. 11-26: position 1 refers to propulsion unit; position 2 refers to high-voltage power supply; position 3 refers to planar receiver.

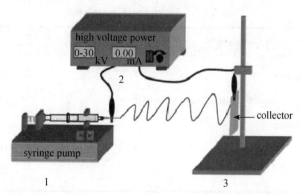

Fig. 11-25 Schematic diagram of electro-spinning principle

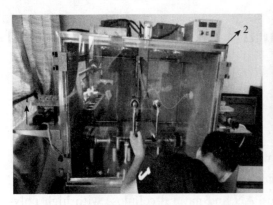

Fig. 11-26　KH-2 electro-spinning machine

11.3.2　Experimental

11.3.2.1　Material and Equipment

Experimental material and apparatus used in this study are listed in Tabs. 11-4 and 11-5, respectively.

Tab. 11-4　Experimental material

Reagent	Level
DMAC solvent	AR
LiCl	AR
Deionized water	CP
Natural cotton fiber	White cotton, with DP of 10 000~15 000
PAN	Spinning-grade

Tab. 11-5　Experimental apparatus

Equipment	Model
Ultrasonic cleaner	KQ-700
Digital-display constant-temperature oil bath pan	HH-S
Electronic balance	AL104
Vacuum oven	DZF-6050
Agitator	IKA RW20

11.3.2.2 Preparation of Cellulose/PAN Spinning Solution

The dissolution of natural cellulose fiber is conducted as follows: a certain amount of DMAC solution is fed into the three-necked flask, then it is heated by the oil bath pan until its temperature increases to 130 ℃, subsequently, the chips of natural cotton fibers are applied into the heated solution under the conditions of N_2 protection and condensation reflux, and then are activated for 60 min; in this case, the solution is not allowed to be heated continuously for the purpose of decreasing its temperature to 100 ℃, and a certain amount of LiCl anhydrous powders are added into the solution quickly so that the mixture solution with LiCl mass fraction of 9% could be configured after 120 min stirring at the speed of 470 r/min; cooling of the solution was conducted to achieve the goal of decreasing its temperature to room temperature (20~25 ℃), in this case, the stable cellulose/LiCl/DMAC spinning solution is prepared based on stirring for 120 min, finally, the ultrasonic de-foaming treatment to the spinning solution is carried out.

A certain amount of DMAC was fed into the three-necked flask, after that, PAN fibers were applied into the flask as well at room temperature and stirred for 60 min, in this case, the PAN spinning solution is prepared. Furthermore, the spinning solutions with different blending ratios are prepared byconducting ultrasonic vibration to the spinning solutions with different mass fraction of cellulose/PAN (100/0, 75/25, 50/50, 25/75 and 0/100).

11.3.2.3 Electrospun Cellulose/PAN Nano-fiber

The initialization of the electro-spinning parameters, such as the electro-spinning voltage, electro-spinning distance, and extrusion speed, can be performed by the KH-2 electro-spinning machine. The spinneret should be properly fixed on the propeller after the spinning solution is extruded into the spinneret, and then it is coated by the aluminum foil paper (around the roller). Where, in order to guarantee its surface smoothness, the aluminum foil paper should be transversely straightened. Subsequently, the tip of the spinneret is connected to the anode of high-voltage power supply, while the receiver is connected to the cathode of high-voltage power supply. The electro-spinning is achieved based on voltage regulation, and the distance between the tip of spinneret and receiver is changed by adjusting the position of spinneret pump. The drying treatment in the vacuum oven (in order to completely volatilize the surplus solvent) is conducted to series of samples of electrospun cellulose/PAN ultra-fine composite fibers for 24 h, thus diverse properties of the fiber could be investigated.

11.3.2.4 Micro-morphology Characterization of Nano-fiber

The micro-morphology of regenerated cellulose membrane is characterized by using a scanning electron microscope (SEM) (S-3400N, Hitachi, Japan) after a layer of metal film with thickness of 10 nm is coated on the surface of blended nano-fiber.

The images of samples fabricated with different parameters are digitalized by the Image-Pro-Plus image software, then 50 fibers are selected from each SEM image for the purpose of measuring their diameters so as to achieve the frequency distribution and average diameter of the fiber[91].

11.3.3 Analysis of Difficulty in Electro-spinning

Various phenomenon generated during electro-spinning are illustrated in Fig. 11-27 and corresponding troubleshooting is investigated for the purpose of solving the problem. Fig. 11-27 (a) and (b): it is difficult for the nano-fiber membrane to be stored for a long time, because the fiber or aluminum foil paper would be corroded by the residual solvent in the nano-fiber and diverse properties of nano-fiber would be affected, resultantly, it seems to be impossible to conduct complete characterization of nano-fiber after a period of time. Fig. 11-27 (c) and (d): the solutions correspond to the phenomenon illustrated in Fig. 11-27 (a) and (b), vacuum oven treatment is recommended (since the residual solvent in the nano-fiber can be evaporated by the vacuum oven at relatively-low temperature; whereas the residual solvent in the nano-fiber can be evaporated completely by the hot air oven only at high temperature, which will destroy the fiber and affect properties of the fiber severely), in addition, water bath exhibits perfect treatment effect. Fig. 11-27 (e): phenomenon occurs when the extrusion speed is too fast. Fig. 11-27 (f): the up-right phenomenon of the fiber in electric field, it seems impossible for the up-right phenomenon to be compatible in the electric field during electro-spinning process, because LiCl is a salt with relatively-high electrical conductivity, resultantly, rotary receiving instead of planar receiving is selected.

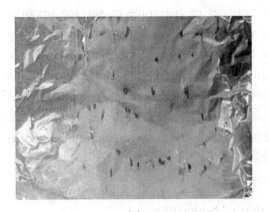
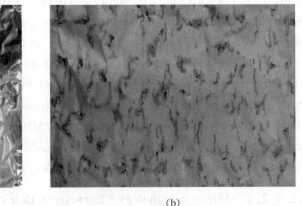

(a)   (b)

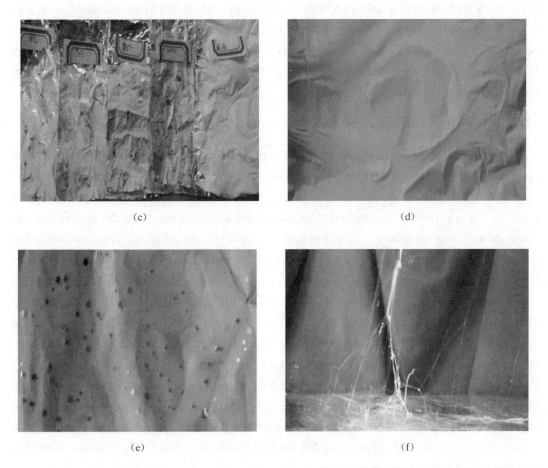

Fig. 11-27 Phenomena occurred during electro-spinning: (a) Incomplete solvent volatilization (48 h); (b) Incomplete solvent volatilization (1 week); (c) ;Comparison after 24 h vacuum drying at different temperatures (d) After water bath treatment; (e) Phenomenon arising from high extrusion speed;(f) Fiber up-right failing to be received uniformly in electric field

11.3.4 Analysis of Effects of Single-factor

The conclusion cannot be treated as the same when investigating the effects of electro-spinning parameters on fiber morphology based on single-factor analysis, such as electro-spinning voltage, electro-spinning distance and flow-speed of spinning solution. However, 'the macro conclusion is in accordance with most of the previous conclusions[6, 92]: the fiber diameter becomes finer with the increasing of electro-spinning voltage and electro-spinning distance (in specified range); the fiber diameter fails to change obviously if it exceeds the specified range, furthermore, it is difficult to form the jet and fiber, because the electric field force fails to overcome the surface tension of spinning droplet if it is lower than certain voltage. It facilitates to the solvent volatilization and deduction of the dissolution and combination for fiber, if it has a low flow speed, resultantly, the fiber is with finer diameter and excellent morphology; while the fiber diameter increases if it has a high flow speed. However, it is difficult for the fiber to be stretched, and severe fiber adhesion will occur which affects the fiber morphology if it has an extremely low flow speed. Meanwhile, corresponding effects need to be investigated particularly under different technology conditions.

11.3.4.1 Effects of Mass Fraction of Spinning Solution on Fiber Morphology

Viscosity of the spinning solution is considered as an important parameter affecting the electro-spinning, while the relative molecular mass is considered as the main factor affecting the viscosity of the spinning solution. Usually, the higher the relative molecular mass is, the severer the entanglement of macromolecule chain is, and the higher the viscosity of spinning solution is. The conclusion cannot be treated as the same, which is because, great differences could be found between the effects of different polymers on electro-spinning[9].

Molecular mass of natural cotton fiber is within the range of 160 000~250 000 (polymerization degree within the range of 10 000~15 000); while molecular mass of PAN is within the range of 70 000~100 000. Based on this, it can be concluded that the molecular mass of natural cotton fiber is much greater than that of PAN, therefore, the viscosity increasing of natural cotton fiber and PAN is different. For example, the viscosity of natural cotton fiber which mass fraction increasing from 1.0% to 4.5% will correspondingly increase from 200 to 5000 MPa·s at room temperature, while the viscosity of PAN which mass fraction increasing from 5% to 10% will correspondingly increase from 180 to 1500 MPa·s at room temperature.

The intrinsic viscosity of spinning solution is extremely low when the mass fraction of PAN spinning solution is extremely low (lower than 5%), resultantly, it is difficult to form the stable fluid and fabricate the PAN fiber. The viscosity of spinning solution is relatively low when the mass fraction of pure cellulose is lower than 1.5%, as a result, it is difficult to spin the yarn. Pure PAN exhibits superior spinnability when its mass fraction is set to be 10%; while pure cellulose presents excellent spinnability when its mass fraction is set to be 3.5%. Therefore, spinning solutions with these two mass fractions are recommended and spinning solution with mass fraction of 3.5%~10% can be selected for blended electro-spinning. Superior spinnability can be obtained when the blending ratio of cellulose/PAN is set to be 25/75. The SEM images of electrospun fibers with blending ratio of 25/75 and mass fractions of blended spinning solutions of 4%, 6%, 8% (diluting the DMAC spinning solution) are illustrated in Fig.11-28.

As illustrated in Fig.11-28 (a), breaking occurs to the molecular chain, because it is difficult for the molecular chain to resist the external force effectively as it is with low entanglement degree, thus generating quite a lot of beads in the polymer. As depicted in Fig.11-28 (b), when the mass fraction of spinning solution is higher than certain critical value, macromolecule amount in unit volume increases, the entanglement degree of molecular chain enhances, and the viscosity of spinning solution gets higher. Spinning jet has a long period of relaxation time under the action of electric field, resultantly, the breaking of local molecular chain in spinning jet is effectively restrained and the bead amount decreases. As illustrated in Fig. 11-28 (c), when the mass fraction and viscosity reach a certain value, the entanglement degree of molecular chain is extremely high, the elongation is uniform and the beads disappear. According to related researches, it can be concluded that the diameter of electrospun fiber is positively correlated with the cube of mass fraction of polymer spinning solution.

    The droplets are spurted from the spinneret and it is difficult to fabricate continuous fibers when the mass fraction of spinning solution is set to be lower than 3%. Since the mass fraction and viscosity of spinning solution are too high, it is difficult for the spinning solution to be spurted from the tip of spinneret when the mass fraction of spinning solution is set to be higher than 8%, in this case, it is difficult to fabricate the fiber and it exhibits extremely low spinning efficiency. The morphologies of fibers with mass fractions of spinning solutions within the range of 4%～8% are illustrated in following figures. As shown in Fig. 11-28 (a), quite a lot of beads are generated in the fiber mesh and the fiber diameter is extremely non-uniform with average diameter of 651 nm and diameter dispersion of 310 nm when the mass fraction of spinning solution is set to be lower than 4%. The bead amount decreases remarkably and the fiber surface becomes smoother with the increasing of the mass fraction of spinning solution (when it reaches 8%), whereas the fiber diameter increases significantly with the increasing of the mass fraction of spinning solution, the average diameter of fibers as illustrated in Fig. 11-28 (c) is 901 nm, and the reasons are in accordance with the above-analyzed reasons.

 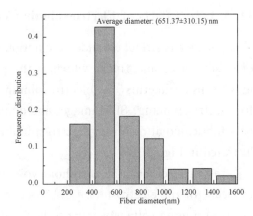

(a)

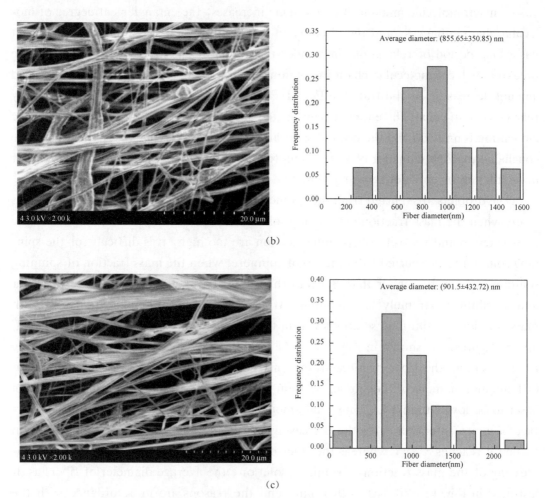

Fig. 11-28 Effects of mass fraction of spinning solution on morphology of blended fiber: (a) 4%; (b) 6%; (c) 8% (electro-spinning voltage=15 kV, electro-spinning distance=12 cm, large-sized spinneret)

11.3.4.2 Effects of Electro-spinning Voltage on Fiber Morphology

The polymer is stretched under the action of electrostatic field force to achieve the goal of being longer and finer, which is the principle of fabricating fiber via electro-spinning, considering this, the electro-spinning voltage is selected as an important parameter for electro-spinning. SEM images and diameter distribution diagrams of blended nano-fibers fabricated at different electro-spinning voltages (5, 10, 15, 20, 25, 30 kV) are illustrated in Fig.11-29.

In this research, the threshold voltage for the electrospun fiber was 8 kV. It was difficult for the electrostatic field force to stretch the spinning solution into fibers when electro-spinning voltage was set to be 5 kV. The electro-spinning voltage becomes unstable, potential risk existed and the automatic power-off protection function of electro-spinning machine enabled under the condition of high ambient humidity when the electro-spinning voltage increased to 30 kV. Combining the literatures with the experimental results, it can be concluded that the spinning jet is extremely unstable and it seems impossible for the electrostatic field force to overcome the surface tension of the droplet, resultantly, it is difficult for the spinning solution to split into fibers when the electro-spinning voltage is lower than certain voltage. The spinning procedure is conducted smoothly with high electro-spinning voltage, however, the electrostatic breakdown may occur to the electroconductive solution and spark arising from discharge may occur as well which would result in potential safety hazard and resource waste. The recommended electro-spinning voltage for this experiment was 10~20 kV.

From Fig. 11-29, it is found that the average diameter of blended fibers remarkably decreased from 323 to 214 nm with little difference of diameter dispersion when the electro-spinning voltage increased from 10 to 20 kV. In addition, the spinnable voltage range was wide when the electro-spinning voltage was within the range of 10~25 kV. The 10 nm difference occurred to the average diameter of the fibers illustrated in the SEM images and the distribution diagrams of diameter when the electro-spinning voltage was within the range of 10~15 kV. And the fiber diameter decreased below 300 nm when the applied voltage was set to be 20 kV. The previous researches showed that[111-112] the effects of electro-spinning voltage on fiber diameter mainly include following aspects: with the increasing of electro-spinning voltage, the electric charge on the surface of spinning jet increases and the cleavage ability of the jet increases as well which facilitate to decrease the fiber diameter. Whereas, when the electro-spinning voltage increases, it is difficult for the too-fast jet speed arising from the enhancement of cleavage ability to stretch and solidify the spinning solution to the receiver, in this case, the fiber diameter increases. The diameter of spun fiber is fine if the first situation predominates; on the contrary, the diameter of spun fiber is wide if the second situation predominates. In this experiment, the blended spinning solution complied with the first situation. As illustrated in Fig. 11-29(d), when the electro-spinning voltage increases from 20 to 25 kV, the fiber diameter decreased as the diameter dispersion greatly increases. Severe fi-

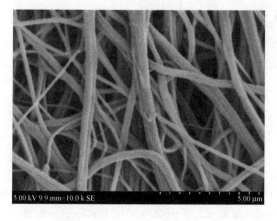
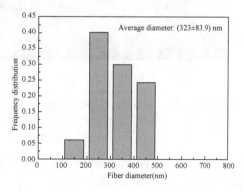

(a)

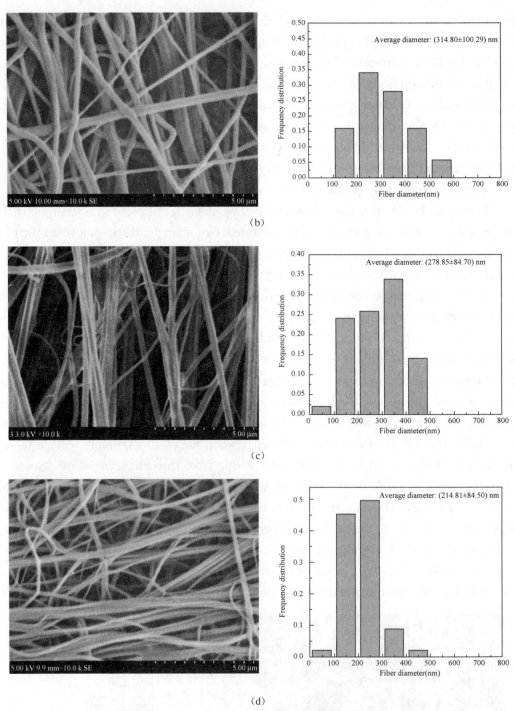

Fig. 11-29 Effects of electro-spinning voltage on morphology of blended fiber: (a) 10 kV; (b) 15 kV; (c) 20 kV; (d) 25 kV (mass fraction=8%, electro-spinning distance=16 cm, small-sized spinneret)

ber cleavage and non-uniform diameter also occurred, because the jet speed increases with the increasing of electro-spinning voltage. It is concluded that the optimum electro-spinning effect could be achieved when the electro-spinning voltage is set to be 20 kV in this experiment.

11.3.4.3 Effects of Electro-spinning Distance on Fiber Morphology

Too short or too long electro-spinning distance during the electro-spinning process has effects on the morphology of blended fiber, therefore, the electro-spinning distance is considered as a significant parameter for electro-spinning.

The SEM images and distribution diagrams of diameters of fibers fabricated with different electro-spinning distances are illustrated in Fig. 11-30. As depicted in Fig. 11-30 (a), severe fiber adhesion is generated since the electro-spinning distance is set too short, which results in short flying time for jet spraying to the receiver as well as incomplete solvent volatilization. As illustrated in Fig. 11-30 (b), the average diameter of the fibers increases a little when the electro-spinning distance increases to 10 cm, it has improved the adhesion of fiber, whereas some beads still can be found in the fiber, the diameter dispersion reaches 257.68 nm and obvious diameter differentiation occurs to the fiber diameter. As illustrated in Fig. 11-30 (c), the average diameter of the fibers decreases from 464.75 to 387.61 nm, the volumes of droplet and bead continuously decrease until they disappear, and the diameter dispersion greatly decreases with uniform distribution (seen from fiber morphology) when the electro-spinning distance increases to 15 cm. As illustrated in Fig. 11-30 (d), no obvious variation occurs to the fiber diameter, the time for the fiber spraying in the space increases and the fiber is stretched fully which facilitates to form the small-diameter fibers with the further increasing of the electro-spinning distance. Compared with Fig. 11-30 (c) and (d), it is found that the further increasing of the electro-spinning distance results in the presence of large-sized fibers (diameter increasing from 387.61 to 408.96 nm), because when the electro-spinning distance is too long, the predominant factor (i.e. electric field intensity) is weak and drafting force on the jet surface is low too, in this case, it is difficult to fabricate the nano-fiber with perfect diameter. Based on the data analysis, it can be concluded that ideal average diameter, variance and electro-spinning effect of the fibers could be achieved when the electro-spinning distance is set to be 15 cm.

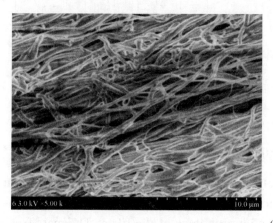 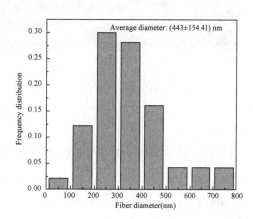

(a)

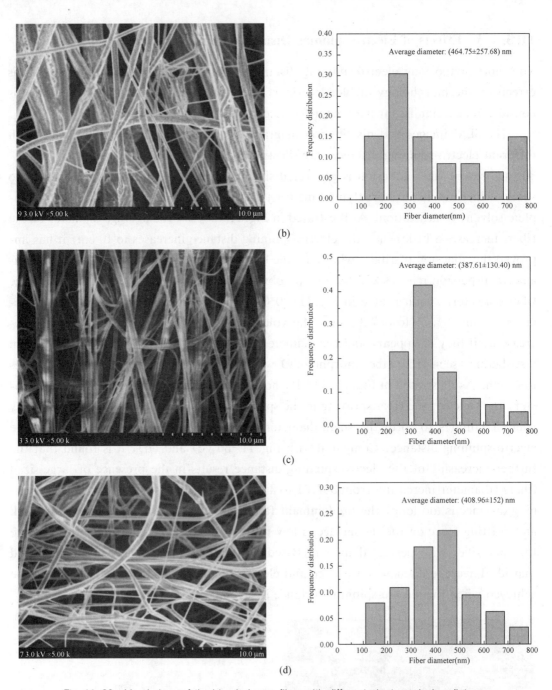

Fig. 11-30 Morphology of the blended nano-fiber with different electro-spinning distance:
(a) 6 cm; (b) 10 cm; (c) 15 cm; (d) 20 cm. (mass fraction of spinning solution=8%; electro-spinning voltage=20 kV)

11.3.4.4 Effects of Blending Ratio on Fiber Morphology

The diameter of electrospun fiber is positively correlated with the cube of concentration of spinning solution[99] and the blending ratio has effects on the viscosity of spinning solution, thus affecting the average diameter and morphology of the fiber. The mass fraction of PAN in this experiment is set to be 10% (it is difficult to fabricate the fiber if the mass fraction is lower than 5%; while the spinneret may be blocked if the mass fraction is higher than 15%). The mass fraction for cellulose spinning is set to be 3.5%, the regenerated cellulose membrane fabricated via phase inversion exhibited superior properties and the spinning solution was stable. As can be seen from Fig. 11-31 (a), (b) and (c), the diameters of fibers are fine with perfect formation, no obvious beads are generated on smooth surface, the diameter distribution is relatively concentrated with low irregularity when the mass fraction of PAN is high. As illustrated in Fig. 11-31 (d) and (e), the diameter of the fiber increases with the increasing of the mass fraction of cellulose, severe fiber adhesion and quite a lot of burrs could be found on the fiber surface although there are no obvious beads, in addition, standard deviation of sample diameter is greater than 150 and the fiber diameter dispersions are much higher than those in Fig. 11-31 (a), (b) and (c), which was because: (1) the viscosity of spinning solution increases rapidly with the increasing of mass fraction of cellulose, which is against the splitting and refinement during the electrospinning; (2) the cellulose and PAN exhibited poor compatibility with each other although both of them could be dissolved in the DMAC solution. From the experimental results, it is found that with the increasing of the mass fraction of cellulose, severe phase separation, poor fiber spinnability and branching were prone to occur when receiving the fiber, complying with the FTIR test results in **Section 11.2.3.**

 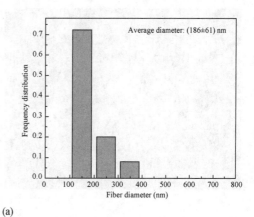

(a)

 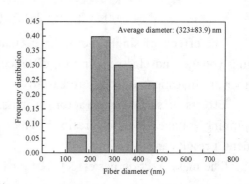

(b)

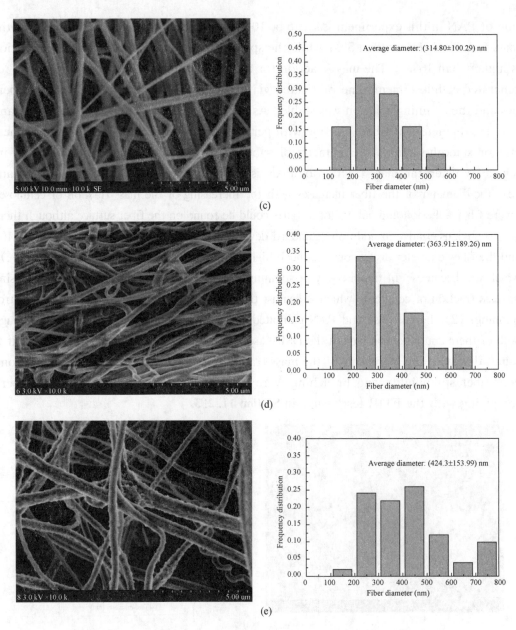

Fig. 11-31 Morphology of the blended nano-fiber with different blending ratio:
(a)0/100; (b) 25/75; (c) 50/50; (d) 75/25; (e) 100/0

The effects of single factor (e.g. mass fraction of spinning solution, electro-spinning voltage and electro-spinning distance) of electro-spinning parameters for blended fiber are investigated in this research.

Effects of such single factor as mass fraction, electro-spinning distance, electro-spinning voltage, blending ratio on fiber morphology are investigated and the experimental results are as follows.

The mass fraction has effects not only on fiber diameter but also on fiber morphology. Droplets are sprayed from the tip of the spinneret and it is difficult to fabricate continuous fibers when the mass fraction of spinning solution is set to be lower than 3%. It is difficult for the spinning solution to be spurted from the spinneret to fabricate the fibers when the mass fraction of spinning solution is set to be higher than 12%, because the mass fraction and viscosity of spinning solution are too high. Spun fiber with excellent morphology is achieved when the mass fraction of spinning solution is set to be 6% or 8%.

It results in short flying time of jet spraying to the receiver, incomplete solvent volatilization and severe fiber adhesion when the electro-spinning distance is set to be too short; while the predominant factor is the electric field intensity which causes weak electric field strength and low drafting force on the jet surface when the electro-spinning distance is set too long, in this case, it is difficult to fabricate nano-fiber with perfect diameter. Superior average diameter, variance and electro-spinning effect of the fibers are achieved when the electro-spinning distance is set to be 15 cm.

The threshold voltage for electro-spinning in this experiment is set to be 8 kV. It is difficult for the electrostatic field force to stretch the spinning solution into fiber when the voltage is set to be 5 kV. The electro-spinning procedure is smooth, however, the electrostatic breakdown may occur to the electroconductive solution, and spark arising from discharge may occur as well, which will result in potential safety hazard and resource waste when the voltage is set too high. The electro-spinning voltage suitable for this experiment should set to be 10~20 kV.

Under specified circumstances (blending ratio = 25/75, mass fraction = 8%, electro-spinning distance = 15 cm, applied voltage = 20 kV), the spun fiber is featured by more uniform diameter, lower dispersion and higher electro-spinning efficiency.

The blending ratio has effects not only on fiber morphology but also on fiberspinnability. Severe fiber adhesion occurs with the increasing of the mass fraction of cellulose, resultantly, the fabricated fiber is with unsatisfactory effect. According to the analysis on the FTIR test results, it can be concluded that the cellulose and PAN have low compatibility with each other although both of them can be dissolved in the DMAC solution. From the experimental results, it could be found that with the increasing of the mass fraction of cellulose, severe phase separation, poor fiber spinnability and branching are prone to be generated when receiving the fiber.

11.4 Characterization of Cellulose/PAN Blended Nano-fiber

Various property testing by mechanical tensile tester, TG and DTG, DSC, FTIR, optical microscope and contact angle of cellulose/PAN nano-fiber with different blending ratios are conducted to investigate the optimum parameters for preparation, in addition, hydrophily is considered as the key indicator to study the potential application of cellulose/PAN nano-fiber in medical auxiliary material field.

11.4.1 Test of Mechanical Properties

A long piece of fabric with length of 20 mm and width of 5 mm was selected (along the longitudinal direction of the fiber felt) as the sample. Weighing each sample in advance, then the sample was placed in the constant-temperature and constant-temperature laboratory (ambient temperature = (20 ± 2) ℃; ambient humidity = $(50 \pm 10)\%$) and treated for 24 h. And its tensile mechanical properties were tested by an Electronic Single-fiber Strength Tester (YG006). The relationship between the surface density and mechanical properties was investigated according to GB/T 1040.3—2006 *Plastics-Determination of tensile properties-Part 3: Test conditions for films and sheets*. In addition, the test conditions were specified (clamping length = 10 mm; drawing speed = 10 mm/min; initial tension = 0.1 cN; force measurement accuracy = 0.01 cN; 10 times of test for each sample; average value of each parameter analyzed finally).

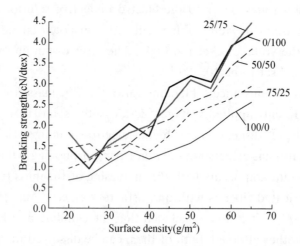

Fig. 11-32 Relationship between surface density and mechanical properties of nano-fiber with different blending ratios

Tab. 11-7 Effect analysis of single-factor for surface density on blended nano-fiber

Parameter	S_s	F	v_A	v_e	$F_{0.05}(v_A, v_e)$
Surface density	33.31	42.99	4	45	2.15
Blending ratio	9.61	27.91	9	40	2.63

S_s—Deviation square;
v_A—Interblock freedom degree;
v_e—Freedom degree of block-interior;
F—Ratio of interblock mean square and mean square of block-interior;
$F_{0.05}(v_A, v_e)$—Critical value of F distribution.

According to the unrepeated two-factor variance analysis in the above table, it is found that if $F > F_{0.05}(v_A, v_e)$ (i.e., 42.99>2.15, 27.91>2.63), both the surface density and blending ratio of nano-fiber have significant effects on the mechanical strength of blended nano-fiber membrane; whereas the effects of surface density are more significant under the condition of significant level being set to be 0.05 (confidence level = 95%). From Fig. 11-32, it can be found that the breaking strength with the variation range of 1.5~4.3 cN/dtex increases slowly with the increasing of the surface density. Based on Fig. 11-32 and Tab. 11-7, it can be concluded that the breaking strength was similar to that of 0/100 (blending ratio) and the optimum average breaking strength is achieved when the blending ratio is set to be 25/75. However, the breaking strength decreases remarkably when the mass fraction of cellulose is set to be higher than 50%, which is because, the compatibility of PAN and cellulose becomes poorer with increasing of mass fraction of cellulose. As can be observed from the SEM images that the diameter dispersions of composite nano-fiber increases significantly with severe fiber adhesion when the blending ratios are set to be 75/25 and 50/50, in this case, the breaking strength of fiber membrane decreases.

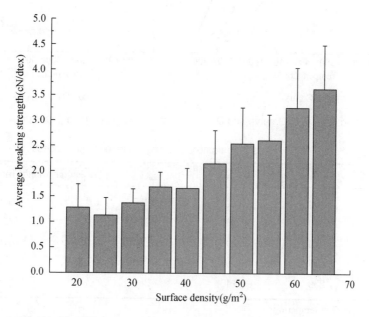

Fig. 11-33 Effects of average surface density on mechanical properties (blending ratio: 25/75)

11.4.2 Thermogravimetric Analysis (TG and DTG)

Similar to the decomposition of regenerated cellulose membrane in **Section 11.2**, the thermal decomposition of the sample illustrated in Fig. 11-34 can be divided into three sectors: the first stage is the slight weight loss stage; the second stage is the thermal decomposition stage, which mainly refers to the decomposition of low-weight oligomer;

the third stage is the carbon stabilization stage, whose weight decreases a little with the increasing of the temperature. As can be observed from Tab. 11-8, the cellulose exhibits good stability. The temperatures for maximum thermal decomposition rate are as follows: T_{max} is 315 ℃ with the blending ratio of 0/100 (i.e., pure PAN); T_{max} is 285 ℃ with the blending ratio of 100/0 (i.e., pure cellulose); T_{max} is 295 ℃ (which is about 10 ℃ higher than that of pure cellulose) with the blending ratio of 25/75 (which is superior than those with blending ratios of 50/50 and 75/25), and the thermal stability of that blending ratio is between the pure cellulose and pure PAN.

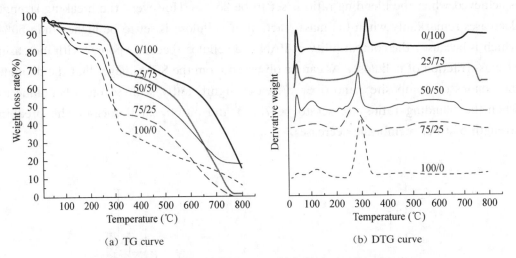

Fig. 11-34 TG curve and DTG curve of blended nano-fibre membrane

Tab. 11-8 Physical parameters during decomposition process

Blending ratio	T_0 (℃)	$T_{10\%}$ (℃)	T_{max} (℃)	Residual weight rate at 700 ℃ (%)
100/0	300	315	315	45.8
75/25	240	250	295	17.5
50/50	240	100	280	20.0
25/75	205	100	260	5.5
0/100	250	146	285	15.6

T_0—Initial decomposition temperature;
$T_{10\%}$—Corresponding temperature with weight loss rate of 10%;
T_{max}—Corresponding temperature with maximum decomposition rate.

11.4.3 Test and Analysis of Differential Scanning Calorimetry (DSC)

DSC curves of blended samples are illustrated in Fig. 11-35. The T_g of PAN sample selected in this experiment is set to be about 110 ℃ and the T_g of blended material increases with increasing of PAN component. The T_g is 125 ℃ when the blending ratio is 25/75, and it is difficult for the DSC image to measure the glass transition temperature (T_g)

of blended material with the increasing of mass fraction of the cellulose, because the T_g of cellulose is higher than its decomposition temperature. It can be expected that the T_g increases with the increasing of cellulose component in blended material. Certain molecular-level mixture can be found between PAN and cellulose with certain interlock diffusion, in this case, the T_g of PAN in blending system moves towards the high-temperature direction.

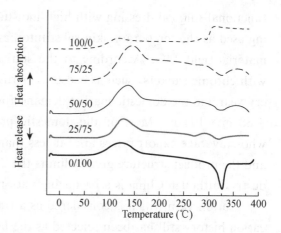

Fig. 11-35 DSC curves

11.4.4 Analysis of Fourier Transform Infrared Spectroscopy (FTIR)

Pieces of membrane with dimension of 1 cm×1 cm were selected as the sample and tests of chemical composition and molecular structure of electrospun blended nano-fiber were performed by an AVATAR 370 FR-IR Infrared Spectrophotometer.

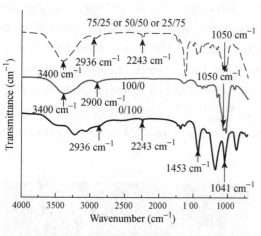

Fig. 11-36 FTIR curves

As can be observed from the FTIR curves (Fig. 11-36), the stretching vibration peaks in the pure PAN nano-fiber membrane at 2243 and 1041 cm^{-1} are the characteristic peaks of C—CN. Characteristic absorption peaks of cellulose exist at 3400, 2900 and 1050 cm^{-1} in the regenerated cellulose membrane; while the positions of characteristic peaks of various components fail to change greatly in the cellulose/PAN nano-fiber; in addition, the intensities of characteristic peaks of various components in fingerprint regions decreases and moves to the low-frequency direction. Based on the FTIR results, it can be concluded that the component of PAN/cellulose composite nano-fiber fail to change, certain compatibility in the crystalline region and certain molecular-level compatibility in the amorphous region can be found, but with low compatibility[93].

11.4.5 Test and Analysis of Dynamic Contact Angle

At present, keeping the wound environment moist and preventing bacterial infection of the wound have been considered as the two standard methods for wound therapy[94]. The functional surgical dressing with high-moisture-retention and antibacterial properties being used as the temporary skin substitute has been always the research focus in medical material field[95, 96]. According to the statistics, it is found that the domestic patients with chronic cankers, such as bedsore, burns and scalds, diabetes are over 15 million every year and the domestic market demand for high-performance surgical dressing reaches 3.88 m × 104 m. Most of the domestic products are the low-cost obsolete products, whose average export prices are far less than those of European and American countries and the product structure greatly limits the development of domestic surgical dressing industry, although China is a producing nation in surgical dressing products.

Cotton yarn which is considered as a traditional surgical dressing with a long application history still has been selected as the main material in wound care[97]. In addition, the better the hygroscopicity is, the superior the wound healing effect will be.

Since the cellulose exhibits excellent hygroscopicity due to high swelling ratio and high water-retention rate, the dynamic contact angles of nano-fiber and ordinary medical gauze are compared to investigate their hygroscopicities in this study so as to achieve the goal of investigation on potential applications of regenerated cellulose nano-fiber in medical auxiliary material field in the future.

As above mentioned in **Section 11.2**, the electrospun textile presents superior air permeability due to its porous structure. On one hand, it can relieve the stimulation of external environment to tissue and wound so as to prevent the severe body fluid loss of the wound; on the other hand, it facilitates to make the wound contact with fresh oxygen, which is helpful for cell growth and repair. Comparison of diameter and porosity of ordinary dressing and nano-fiber is illustrated in Fig. 11-37, and comparison of porosity of self-made nano-fiber membrane and ordinary non-woven fabric is illustrated in Tab. 11-9.

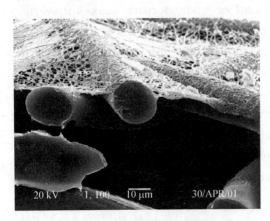
(a) A layer of nano-fiber membrane on spun-bonded base cloth[19]

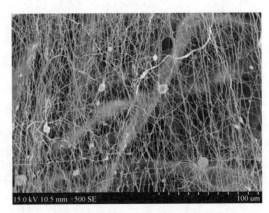
(b) Nano-fiber and non-woven fabric

Fig. 11-37 Comparison of diameter of nano-fiber and composite medical gauze

Tab. 11-9 Diameters of nano-fiber membrane and medical gauze

Sample	Minimum pore size(μm)	Maximum pore size(μm)	Average pore size(μm)
Nano-fiber membrane	1.14	1.51	1.26
Ordinary non-woven fabric	34.85	45.61	34.9

Comparison of hydrophily between nano-fiber membrane and ordinary medical gauze is conducted to achieve the goal of investigation on potential value of nano-fiber membrane in medical and health field, the test and comparison results of dynamic contact angles of electrospun nano-fiber (superior spinnability and best thermal properties available and have both advantages of natural fiber and chemical fiber when blending ratio being set to be 25/75) and ordinary medical gauze are illustrated in Fig. 11-38. It is found that the nano-fiber exhibits superior hydrophily than ordinary medical gauze.

The LiCl/DMAC solvent system is proposed to dissolve PAN and natural cellulose with high polymerization degree, furthermore, no other derivatives will be generated during the dissolution process. With the increasing of blending ratio of cellulose, severe nano-fiber adhesion and rapid increasing of diameter dispersion occur to the electrospun cellulose/PAN nano-fiber when the diameter is about 200~400 nm.

The two-component blended nano-fiber exhibits excellent thermal properties when the blending ratio is set to be 25/75, i.e., T_{max} is 295 ℃ which is about 10 ℃ higher than that of pure cellulose (T_{max} is 285 ℃ when the blending ratio is set to be 100/0).

Its own component in the PAN/cellulose blending system fails to change, certain compatibility in the crystalline region and certain molecular-level in the amorphous region could be found, but with low compatibility.

The blended nano-fiber fabricated in this experiment presents high hydrophily, it is found that the hydrophily of blended nano-fiber is superior than that of ordinary medical gauze, based on the comparison of dynamic contact angles. And the high hydrophily becomes more obvious with the increasing of the blending ratio of cellulose, however, the water absorption of nano-fiber is momentary while rather longer time is required for ordinary medical gauze.

The disadvantages are as follows: with the increasing of mass fraction of cellulose, the spinnability of blended spinning solution becomes poorer and the spinning efficiency decreases remarkably when compares with pure PAN, in addition, the fiber adhesion becomes severer and the spinneret is prone to be blocked. It is difficult for the blended nano-fiber to be stored for a long time, a little solvent which will corrode the aluminum foil paper and destroy the nano-fiber still fails to be volatilized completely although the vacuum drying has been performed. Only considering the hydrophily, the nano-fiber possess great potential value in medical auxiliary material field; however, considering the economic benefits and industrialization, it still has a long way to go.

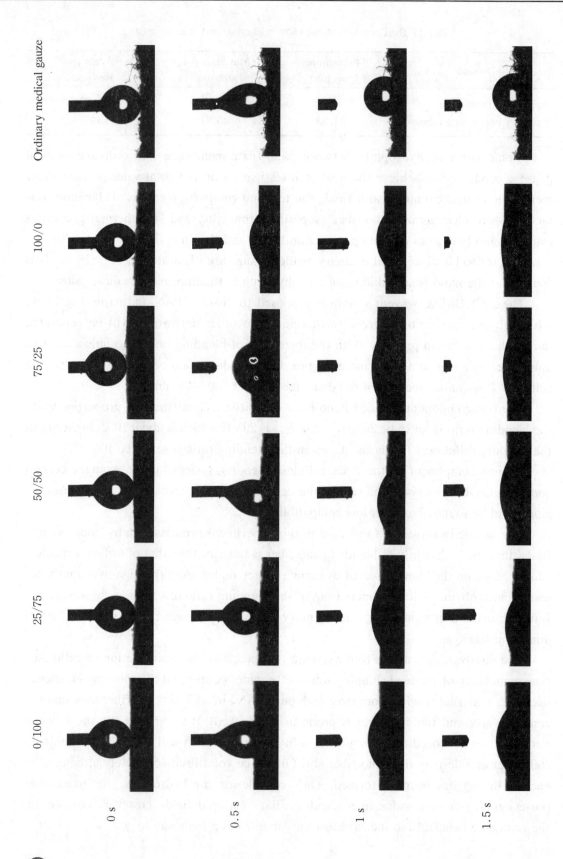

Chapter 11
Functionalization of Cellulose Based Materials

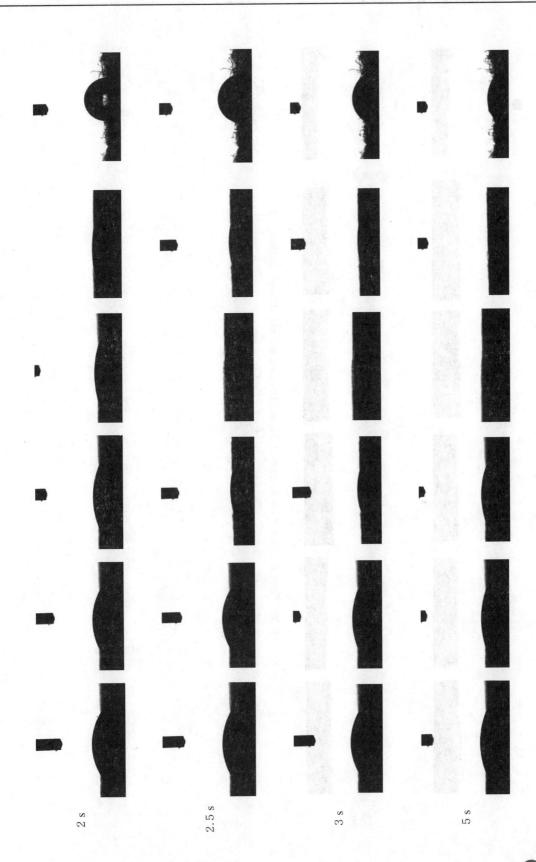

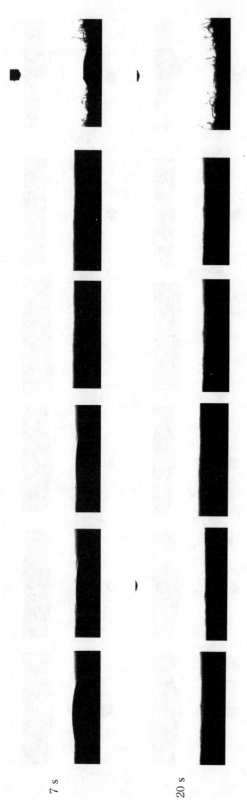

Fig. 11-38 Comparison of dynamic contact angles of nano-fibers with different blending ratios

11.5 Antibacterial Treatment of Cellulose/PAN Nano-fiber

11.5.1 Purpose of Antibacterial Treatment

Essentially, antibacterial treatment is a process of killing bacteria or impeding bacterial activity and its growth and reproduction by using physical or chemical methods. Usually, antibacterial treatment includes sterilization, disinfection, mildew, bacteriostatic, anticorrosion, where slight differences can be found among each concept, among which, bacteriostatic refers to the process of impeding bacterial activity and its growth and reproduction by using physical or chemical methods.

The concept of antibacterial involved in the preparation of antibacterial fiber mainly refers to the concept of bacteriostatic, and copper ion is selected as main antibacterial agent for the purpose of impeding the bacterial growth.

11.5.1.1 Selection of Bacteria and Strain

A large number of natural antibacterial materials could be selected from nature, however, most of the antibacterial materials are artificially prepared based on application of antibacterial agent, such as antibacterial plastics, antibacterial fiber and fabric, antibacterial ceramics and antibacterial metallic material. Usually, antibacterial is deemed to be against the bacterial which is classified into the microorganism[98]. The bacterial is characterized by multiple varieties, rapid reproduction, adaptable to diverse harsh environments, it is impossible to find a place without bacterial in nature. The bacteria could be classified into three categories: cocci, bacilli and spirilla based on external morphology, as illustrated in Fig.11-39.

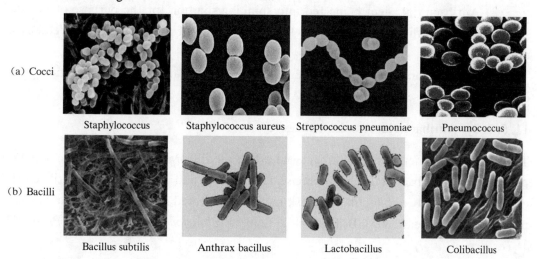

(c) Spirilla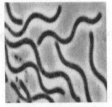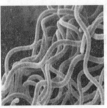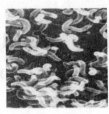

Helicobacter pylori   Spirillum volutans   Spirilla   Vibrio cholera

Fig. 11-39 Classification and morphology of bacteria[106]

The bacteria could be classified into two categories according to the influences on human body: Gram positive bacteria and Gram negative bacteria. The colibacillus, which is found to be the most widely-distributed representative of Gram negative bacteria, is one of the bacteria with the largest amount in the intestinal tract of human and some animals. Generally, the colibacillus is not pathogenic; whereas diarrhea and septicemia may be caused in specified conditions. The staphylococcus aureus, which is considered as the representative of Gram positive bacteria, is a kind of significant pathogenic bacteria. Usually, the staphylococcus aureus could be found in poisoning food of milk, meat, egg, fish and leftovers[99].

11.5.1.2 Finishing of Antibacterial Fiber

(1) Chemical method: reactive group is introduced on the fiber surface via physical or chamical modification, and the antibacterial group is grafted on the reactive group.

(2) Physical method: if there is no chemical reaction in the preparation process, the combination of antibacterial agent and fiber is determined by physical combination rather than chemical reaction, for example, the development of microporous fiber with rough surface which facilitates to let the antibacterial agent penetrate below the fiber surface during finishing; the antibacterial agent will be presented on the fiber during fiber regeneration if the antibacterial agent is applied in the coagulating bath during wet spinning.

(3) Antibacterial agent being fed into spinning solution for the purpose of blended spinning has been the main technology for development of antibacterial and deodorization fiber.

Chemical method such as antibacterial group being grafted with the reactive group on fiber surface is selected in this study.

11.5.1.3 Antibacterial Assessment

The main testing methods for antibacterial effect are illustrated in Tab. 11-10, based on properties of antibacterial fiber, precise comparison and calculation of antibacterial effect is conducted, and oscillation flask method is selected in this study which was supported by the standards of GB/T 20944.2—2007 *Evaluation of Antimicrobial Properties of Textiles*.

Tab. 11-10 Antibacterial effect of main determination methods

	Testing method	Qualitative or quantitative	Assessment basis
Halo	AATCC-90	Qualitative	Width of rejection band
	Modified AATCC-90	Qualitative	Coloration degree
Impregnation method	AATCC-100	Quantitative	Reduction rate of bacteria
	Modified AATCC-100	Quantitative	Reduction rate of bacteria
	GB/T 20944.2	Quantitative	Inhibition rate
	ISO 20743	Quantitative	Inhibition rate
	JIS L1902	Quantitative	Difference between increased/decreased value
Oscillation method	Oscillation flask method	Quantitative	Reduction rate of bacteria
	FZ/T 73023	Quantitative	Inhibition rate
	GB/T 20944.3	Quantitative	Inhibition rate
Other methods	Oscillation flask method	Quantitative	Reduction rate of bacteria
	ISO 20645	Semi-quantitative	Width of inhibition band
	GB/T 20944.1	Qualitative	Width of inhibition band
	FZ/T 73023	Quantitative	Reduction rate of bacteria
	AATCC147	Semi-quantitative	Width of inhibition band

11.5.1.4 Complexation and Antibacterial Mechanism of Cuprammonia and Cellulose

The copper ammonia fiber is generated by the reaction between cellulose and cuprammonia, the copper ion is fixed on the fabric for the purpose of antibacterial properties, the reaction mechanism is depicted in Fig. 11-40. The preparation principle of cuprammonium is expressed in equation (1), the dark-blue substance $[Cu(NH_3)_4]^{2+}$ generated by the reaction between $Cu(OH)_2$ and $NH_3 \cdot H_2O$ is selected as the active principle of the solution; (2) declares that metal coordination occurs between two hydroxyls in cellulose and Cu(II) to achieve the goal of fixing the copper ion on the fiber.

$$Cu(OH)_2 + 4NH_3 H_2O \rightleftharpoons [Cu(NH_3)_4](OH)_2 + 4H_2O$$

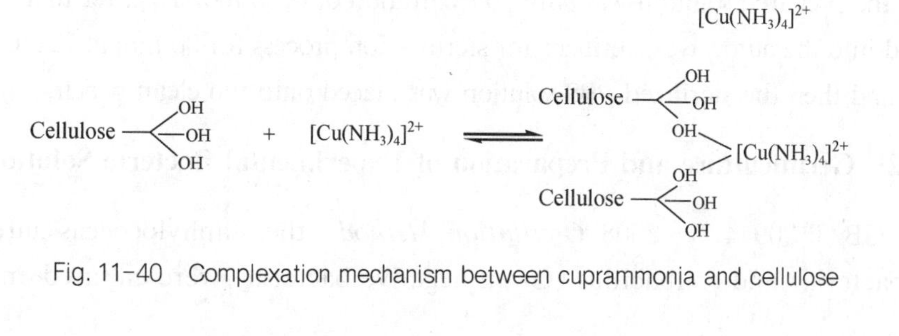

Fig. 11-40 Complexation mechanism between cuprammonia and cellulose

The antibacterial mechanism of copper ion is classified into the contact-type antibacterial mechanism whose principle was discussed previously, it is expressed as follows:

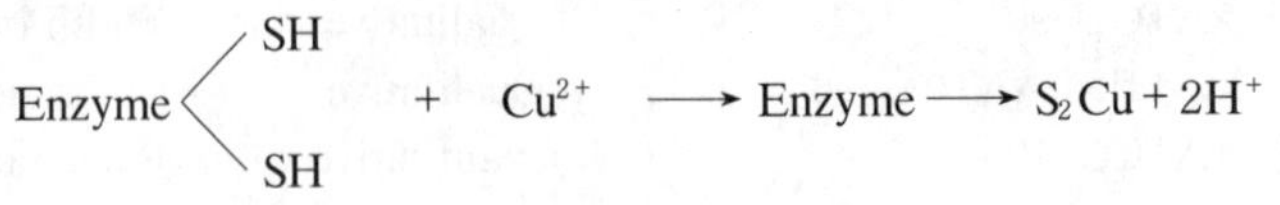

$$\text{Enzyme} \Big\langle ^{SH}_{SH} + Cu^{2+} \longrightarrow \text{Enzyme} \longrightarrow S_2Cu + 2H^+$$

Fig. 11-42   Antibacterial mechanism of copper ion

Thenano-fiber with antibacterial properties was fabricated in this research, with electrospun nano-fiber being finished by cuprammonium, based on previously-known researches.

11.5.2   Process of Antibacterial Treatment

11.5.2.1   Culture Medium and Reagent for Tests

The reagent selected for the antibacterial experiment is the analytical reagent or the reagent used for biology experiment, meanwhile, the pure water is selected as the experimental water. It has high requirements on antibacterial experiment, the overall experimental process should be conducted on the clean bench and the high-temperature sterilization is required before the operation, crossbred bacteria may grow which will interfere the experimental results and cause maximum error in the testing results if it fails to strictly observe the regulations.

Preparation for nutrient broth: beef extract (3 g), peptone (5 g), agar powder (15 g), constant volume of distilled water to 1000 mL; the PH value after the sterilization process conducted in the autoclaves sterilizer was 6.8 ± 0.2, after that, the solution was applied onto the plate for bacteria inoculation and then incubated in the oscillation incubator.

Preparation for culture medium of nutrient agar: glucose (40 g), peptone (10 g), agar powder (20 g), constant volume of distilled water to 1000 mL; the pH value after the sterilization process conducted in the autoclaves sterilizer was 6.8 ± 0.2, after that, the solution was applied onto the plate for corresponding bacteria inoculation when ambient temperature was close to room temperature.

Preparation for phosphate buffer (PBS): disodium hydrogen phosphate (2.84 g) and monopotassium phosphate (1.36 g) were fed into the distilled water (1000 mL), in this case, the prepared solution was with concentration of 0.03 mol/L, after that, the solution was placed into the autoclaves sterilizer for sterilization process for 30 min at 121 ℃ before being used, and then the sterilized PBS solution was placed onto the clean bench.

11.5.2.2   Germiculture and Preparation of Experimental Bacteria Solution

Based on GB/T 20944.3—2008 *Oscillation Method*, the staphylococcus aureus (Gram positive bacteria) and colibacillus (Gram negative bacteria) were cultured and the antibacterial properties of nano-fiber were investigated. The inoculation suspension of bacteria was prepared and the inoculation loop was used to conduct the bacterial inoculation on the agar board for 18~24 h at (37±1) ℃, after that, the inoculation loop was used to select a typical bacterial colony for the purpose of inoculation in the nutrient broth (20 mL) at (37±1) ℃, and then the oscillating culture for 18~20 h at speed of 130 r/min on the oscillation incubator was carried out, in this case, the inoculation suspension was prepared (the fresh prepared bacteria solution should be utilized in 4 h for the purpose of reactivity of inoculated bacteria), eventually, the number of bacteria was diluted to appropriate times, with 10 times dilution being selected.

11.5.2.3 Sample preparation

The antibacterial fabric and check sample were cut into chips with dimension of 5 mm×5 mm, each sample with the weight of (0.75±0.05) g was selected and packed by weighing paper, and multiple samples could be weighted according to experimental requirements, after that, the samples were placed into the autoclaves sterilizer for sterilization.

11.5.2.4 Test Principle of Oscillation Method

The samples before/after antibacterial treatment are fed into the conical flask filled with experimental bacteria solution with certain concentration respectively, after that, the oscillation culture is conducted to the samples for specified time, the concentrations of viable bacteria in the bacteria solution in the conical flask before/after oscillation are measured, and the antibacterial effect is investigated (with 10 times dilution method being selected, the solution is diluted, and then the agar culture medium is used as the plate. To obtain the concentration of viable bacteria in the conical flask, the bacterial colony on the plate is counted after the culture for 24~48 h at (37±1) ℃, thus calculating the inhibition rate.)

Concentration of viable bacteria:

$$K = ZR \qquad (11-1)$$

among which, K refers to the concentration of viable bacteria in the conical flask of each sample, CFU/mL; Z refers to the average value of bacterial colony on the plate; R refers to the dilution times.

Inhibition rate:

$$Y = \frac{W_t - Q_t}{W_t} \times 100\% \qquad (11-2)$$

among which, Y refers to the inhibition rate; W_t refers to the concentration of viable bacteria in the conical flask before antibacterial treatment, CFU/mL; Q_t refers to the concentration of viable bacteria in the conical flask after antibacterial treatment, CFU/mL.

In addition, W_t and Q_t could be achieved based on Eq.(11-1).

11.5.3 Experimental Equipment and Reagent

The equipment and material used in this study are shown in Tab. 11-11 and Tab. 11-12, respectively.

Tab. 11-11 Experimental equipment

Equipment name	Model
Constant-temperature Oscillator	THZ-82
Electronic Balance	JT601N
Biochemical Incubator	SPX-250B-Z
Medical Syringe	2 mL/5 mL/10 mL/20 mL
Clean Bench	SW-CJ
Electric Blast Drying Oven	WD-5000
Vacuum Oven	DZF-6050
Magnetic Digital-display PH Meter	PHSJ-4A
Autoclaves Sterilizer	YXQ-SG46-280S

Tab. 11-12 Experimental raw materials and reagents

Drug name	Specification
Glucose	Analytical reagent (AR)
Sodium hydroxide (sheet)	Analytical reagent (AR)
Disodium hydrogen phosphate	Analytical reagent (AR)
Monopotassium phosphate	Analytical reagent (AR)
Beef extract	Biological reagent (BR)
Tryptone	Biological reagent (BR)
Agar powder	Biological reagent (BR)
Staphylococcus aureus	ATCC6538
Escherichia coli	ATCC8739
Distilled water	4.5 L
Cupric sulfate ($CuSO_4 \cdot 5H_2O$)	Analytical reagent (AR)
Aqueous ammonia ($NH_3 \cdot H_2O$)	Analytical reagent (AR)

11.5.4 Preparation and Antibacterial Treatment of Cuprammonium

Stable complex of copper ion and cellulose/PAN can be fabricated with simple preparation process. The complexation principle of cellulose and copper ammonia is illustrated in Fig. 11-42, PAN has the cyano (—CN) with sufficiently-strong polarity, and strong negative induction effect, whose complexation ability is stronger than that of hydroxyl

(—OH) on the cellulose. The antibacterial test is performed based on the experimental results by Qin[28-29], with cellulose/PAN nano-fiber being treated by cuprammonium.

A certain amount of CuSO$_4$ and NaOH (2 times amount of CuSO$_4$) were dissolved in the deionized water respectively, after that, blue Cu(OH)$_2$ precipitation was generated based on blending of CuSO$_4$ and NaOH; subsequently, to obtain purified Cu(OH)$_2$, the vacuum filtration and then times of washing and precipitation was carried out; Cu(OH)$_2$ precipitation was dissolved in the ammonia (volume ratio of 1 : 2) and the cuprammonium[100] presented in light blue was prepared after the precipitation was dissolved completely.

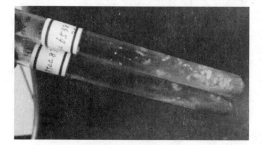
(a) Experimental strain

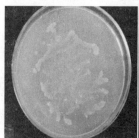

(b) Stain inoculated on the agar plate via streaking method

(c) Inoculation suspension

(d) Antibacterial nano-fiber being applied into diluted bacteria solution

Fig. 11-42 Process of antibacterial finishing

The antibacterial treatment was conducted to the nano-fiber with blending ratio of cellulose/PAN of 25/75, the nano-fiber membrane fabricated in this experiment exhibits excellent spinnability, thermal stability and hydrophily.

Reaction between nano-fiber and cuprammonium: the nano-fiber was impregnated into the cuprammonium and then taken out after being stirred gently for specified time by using a glass bar; after that, the solution was dried at a low temperature, in this case, the regenerated cellulose/nano composite fiber of Cu^{2+} complexation was fabricated.

Two reaction conditions are designed for the experiment as follows:

Blending ratio of nano-fiber membrane was set to be 25/75 and the surface density was set to be 40 g/m^2, the reaction time is 30 min when the concentrations of cuprammonium were set to be 0.005, 0.01, 0.015, 0.02 and 0.025 moL/L at 25 ℃, finally, the solution was dried at 50 ℃ with bath ratio of 1 : 20.

Blending ratio of nano-fiber membrane was set to be 25/75 and the surface density was set to be 40 g/m², the reaction time is 0, 15, 30, 45 and 60 min when the concentration of cuprammonium was set to be 0.01 mol/L at 25 ℃, finally, the solution was dried at 50 ℃ with bath ratio of 1 : 20.

11.5.5 Analysis of Experimental Results

11.5.5.1 Variation of Mechanical Properties after Cuprammonium Treatment with Different Concentrations and Time

As can be seen from Fig. 11-43, the average breaking strength of nano-fiber membrane treated by cuprammonium with concentration of 0.005 mol/L is similar with untreated nano-fiber membrane. To guarantee perfect antibacterial effect, the cuprammonium with concentration of 0.01 mol/L is selected, which is because, the breaking strength decreases remarkably when the concentration of cuprammonium is set to be 0.015 mol/L. From Fig. 11-44, it can be clearly observed that no obvious variation of average breaking strength of nano-fiber membrane could be found when the concentration of cuprammonium is set to be 0.01 mol/L and the time is set to be 15 or 30 min. The 30 min treatment time is selected, which is because, the average breaking strength of nano-fiber membrane decreases significantly when the treatment time is greater than 40 min, in this case, both excellent mechanical properties and antibacterial properties could be achieved.

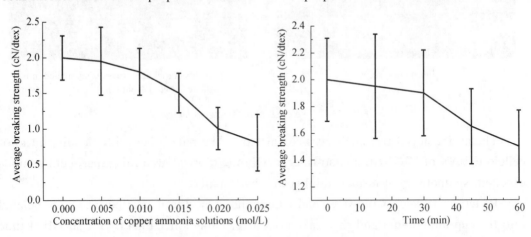

Fig. 11-43 Average breaking strength of nanofibers after cuprammonia treatment with different concentration

Fig. 11-44 Average breaking strength of nanofibers with different treatment time

11.5.5.2 Test and Analysis of Contact Angle

Variation of dynamic contact angle of electrospun nano-fiber membrane before and after cuprammonium treatment is illustrated in Fig. 11-45. It can be found that the droplet is

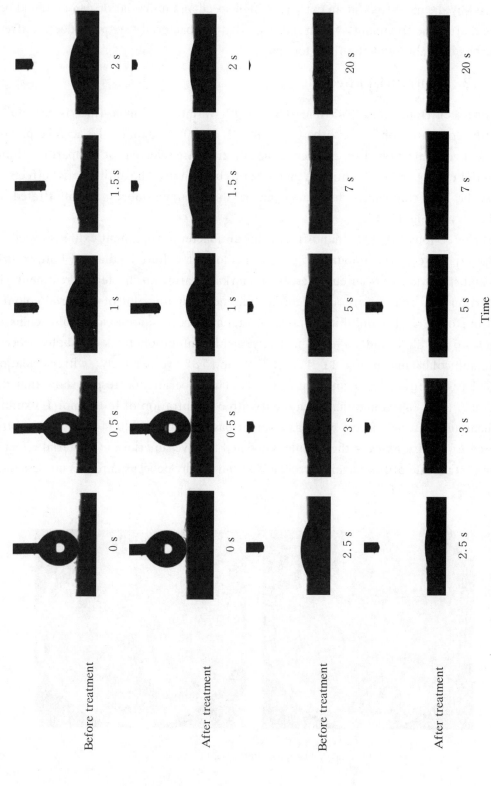

Fig. 11–45　Changes of dynamic contact angle before and after cuprammonia treatment

absorbed by the nano-fiber at an extremely high speed and no variation can be found before and after the treatment, which indicates that antibacterial treatment doesn't affect the surface wettability of the fiber too much.

11.5.5.3 Antibacterial Effect

It can be seen from Fig. 11-43 that the strength variation of nano-fiber is acceptable when the concentration is set to be 0.005 mol/L and 0.01 mol/L. To achieve precise treatment concentration, two different concentrations are selected for comparison of antibacterial effects and the treatment time is set to be 30 min. The antibacterial effects of colibacillus and staphylococcus aureus under these circumstances are illustrated in Fig. 11-46 and Fig. 11-47.

It can be found that the nano-fiber after antibacterial treatment exhibits excellent antibacterial effect, the amounts of bacterial colony in culture medium of both colibacillus and staphylococcus aureus decrease remarkably after antibacterial treatment, in comparison with that before antibacterial treatment (quite a lot of bacterial colonies could be found). The inhibition rates of staphylococcus aureus and colibacillus in Fig. 11-46 are 82% and 75% respectively; while inhibition rates of staphylococcus aureus and colibacillus in Fig. 11-47 are 72% and 78% respectively, with calculation on the basis of Eq. (11-1) and Eq. (11-2). The experimental results show that the nano-fiber after cuprammonium treatment with concentration of 0.01 mol/L exhibits excellent antibacterial effect, with assessment based on GB/T 20944.3—2008 *Antibacterial Effect Assessment*: the sample is deemed to have superior antibacterial effect if the inhibition rate of fiber against colibacillus and staphylococcus aureus is not less than 70%.

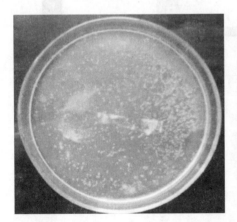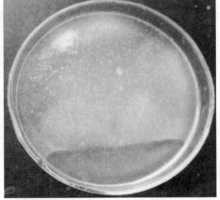

(a) Antibacterial effect against S. aureus

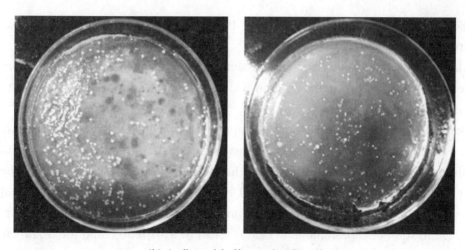

(b) Antibacterial effect against E. coli

Fig. 11-46　Antimicrobial effect against staphylococcus aureus

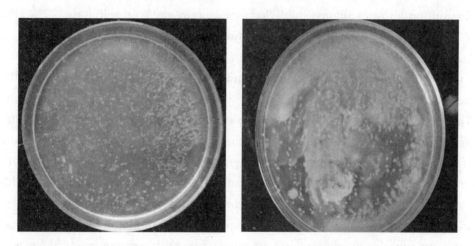

(a) Antibacterial effect against S. aureus

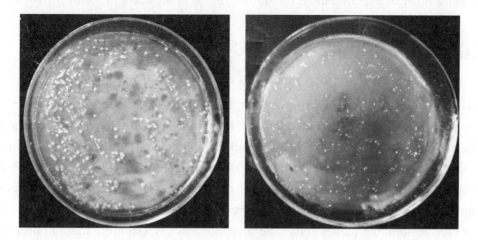

(b) Antibacterial effect against E. coli

Fig. 11-47　Antibacterial effect of solution with concentration of 0.01 mol/L

Following conclusions can be drawn based on the investigation on antibacterial properties.

The nano-fiber membrane after antibacterial treatment by cuprammonium exhibits excellent antibacterial properties, which widens the scope of applications of Cu^{2+} in the field of inorganic antibacterial agent.

Both the concentration and time of cuprammonium have direct effects on the mechanical strength of the two-component nano-fiber, the strength decreases gradually with the increasing of concentration and the strength decreases significantly when the concentration is set to be greater than 0.01 mol/L; the strength also decreases with the increasing of treatment time whereas not as obvious as concentration will cause. The best concentration of cuprammonium can be set to be 0.01 mol/L and the treatment time can be set to be 30 min, in this case, perfect mechanical properties and antibacterial properties of the membrane can be guaranteed.

The inhibition rates against staphylococcus aureus and colibacillus are 72% and 78% respectively when the concentration of cuprammonium is set to be 0.005 mol/L; while inhibition rates against staphylococcus aureus and colibacillus are 82% and 75% respectively when the concentration of cuprammonium is set to be 0.01 mol/L, where the best antibacterial effect is provided.

There are still certain disadvantages to be improved and optimized: superior antibacterial effect could be achieved based on the optimization of antibacterial treatment of nano-fiber, such as applying antibacterial agent to spinning solution. However, the selection of solvent and applying method of antibacterial agent need to be reconsidered, because the LiCl/DMAC solvent system is considered as the non-aqueous solvent system and a large amount of moisture could be found in the cuprammonium; the antibacterial fiber presents superior hygroscopicity, in comparison with common medical gauze, however, more medical experiments are required for its application in medical field.

11.6 Conclusions

In this part, the LiCl/DMAC solvent system is proposed to dissolve the PAN and natural cellulose with high polymerization degree. The activation and dissolution mechanism of cellulose are investigated and the regenerated cellulose membrane is fabricated, in addition, many experiments have attempted for the development of preparation of cellulose/PAN components fiber at nano-scale by electro-spinning. The antibacterial treatment by cuprammonium is conducted to the nano-fiber with best properties, after the optimization of electro-spinning parameters. The morphology, crystallinity, thermal properties, contact angle and antibacterial properties of regenerated cellulose membrane and its blended nano-fiber are characterized, based on SEM, XRD, TG, DSC, FTIR contact angle meter as well as GB/T 20944.3—2008 antibacterial test standards. Therefore, the following conclusions can be drawn.

In this research, to obtain the optimum activation technology and prepare the organic solution series with different mass fractions of cellulose, the LiCl/DMAC direct solvent system is proposed to conduct the activation of natural cotton fiber. To achieve the complete dissolution of natural cellulose, the moisture content in this system should be controlled strictly. The lower the temperature is, the better the dissolution effect will be, because the high-temperature activation is prone to cause the oxidative degradation of cellulose. However, further investigations on activation mechanism and optimization of activation process for natural cotton fiber are required.

The higher the mass fraction of cellulose is, the greater the viscosity of cellulose solution will be, because the mass fraction of cellulose has effects on the mechanical properties of regenerated cellulose membrane; The regenerated cellulose membrane exhibits excellent transparency, and the TG and DSC test results show that regenerated cellulose membrane presents low initial decomposition temperature and poor thermal stability when compared with natural cotton fiber; The FTIR results show that the positions of absorption peaks of cellulose before/after regeneration are almost the same, which indicates that the components before/after dissolution are the same, that is to say, no derivatization reaction occurs during the cellulose dissolution in the LiCl/DMAC solvent system, which is classified into the direct dissolution. The composition analysis of coagulating bath has great effects on swelling properties of the membrane when methanol bath, ethanol bath and water bath are selected. For these three above-mentioned coagulating bathes, the cellulose membrane exhibits higher porosity (i.e., Pr), and the porosity can even reach 60% when methanol bath and ethanol bath are selected; the high porosity proves the high swelling ratio of cellulose in the water or alcohol, and it indicates the strong hydrophily of cellulose membrane, and the best preparation technology are designed: high-speed film formation conducted by KW-4A Spin Coater, water bath selected as the coagulating bath, mass fraction of cellulose selected as 3.5%.

The experimental results of effects of single-factor such as mass fraction, electro-spinning distance, electro-spinning voltage and blending ratio show that the diameter of the fiber is uniform with low dispersion degree and electro-spinning efficiency is high when specified parameters is settled down (blending ratio = 25/75, mass fraction = 8%, electro-spinning distance = 15 cm, electro-spinning voltage = 20 kV).

PAN or natural cellulose with high polymerization degree could be dissolved in the LiCl/DMAC solvent system directly, with no derivative being generated. The surface of electrospun cellulose/PAN composite nano-fiber with diameter of 200~400 nm becomes rougher and the diameter dispersion degree increases remarkably with the increasing of cellulose component.

It can be found that no variation occurs to component of cellulose and PAN in the cellulose/PAN blending system, no compatibility in crystalline region and certain molecular-level compatibility in amorphous region could be found, with low degree of compatibility. The two-component blended nano-fiber exhibits excellent thermal properties and the ideal thermal stability could be achieved when the blending ratio is set to be 25/75. T_{max} of cellulose/PAN blending system is 295 ℃, which is 10 ℃ higher than T_{max} of pure cellulose (blending ratio = 100/0, T_{max} = 285 ℃).

Both the concentration and time of cuprammonium have direct effects on two-component nano-fiber, the strength decreases gradually with the increasing of concentration and the strength decreases significantly when the concentration is set to be greater than 0.01 mol/L; the strength also decreases with the increasing of treatment time whereas not as obvious as concentration will cause. The best concentration of cuprammonium can be set to be 0.01 mol/L and the treatment time can be set to be 30 min, in this case, excellent mechanical properties and antibacterial properties of the membrane can be guaranteed. The nano-fiber membrane after antibacterial treatment by cuprammonium exhibits excellent antibacterial properties, which widens the scope of application of Cu^{2+} in the field of inorganic antibacterial agent.

In this part, the LiCl/DMAC solvent system is proposed to dissolve the PAN and natural cellulose with high polymerization degree. The activation mechanism and dissolution mechanism of cellulose are investigated and the regenerated cellulose membrane is fabricated, in addition, many experiments have attempted for the development of preparation of cellulose/PAN two-component blended fiber at nano-scale by electro-spinning. There are still certain disadvantages to be improved and optimized for the development of antibacterial material made of regenerated cellulose and following aspects could be focused for future investigation:

To achieve the complete dissolution of natural cellulose, the moisture content in this system should be controlled strictly. The lower the temperature is, the better the dissolution effect will be, because the high-temperature activation is prone to cause the oxidative degradation of cellulose. However, further investigations on activation mechanism and optimization of activation process for natural cotton fiber are required.

During preparation of two-component blended nano-fiber, with the increasing of mass fraction of cellulose, the spinnability of blended spinning solution becomes poorer and the spinning efficiency decreases obviously when compared with pure PAN, in addition, the fiber adhesion becomes severer and the spinneret is prone to be blocked. It is difficult for the blended nano-fiber to be stored for a long time, a little solvent which will corrode the aluminum foil paper and destroy the nano-fiber still fails to be volatilized completely although the vacuum drying has been performed. The further investigation on electro-spinning technology of blending of cellulose and other component in LiCl/DMAC solvent system is required.

Only effects of electro-spinning voltage, electro-spinning distance and mass fraction on fiber morphology have been investigated for electro-spinning technology, whose parameters also include flow speed of spinning solution, diameter of spinneret, movement state and environment of collector, all these parameters also have effects on fiber morphology and properties of the fiber, so they could be investigated respectively, thus establishing complete process system of preparation of electrospun cellulose/PAN blended fiber.

The antibacterial treatment of nano-fiber has widened the scope of application of Cu^{2+} serving as antibacterial agent, superior antibacterial effect could be achieved based on the optimization of antibacterial treatment of nano-fiber, such as applying antibacterial agent to spinning solution. However, the selection of solvent and applying method of antibacterial agent need to be reconsidered, because the LiCl/DMAC solvent system is considered as the non-aqueous solvent system and a large amount of moisture could be found in the cuprammonium.

The antibacterial fiber presents superior hygroscopicity, in comparison with common medical gauze, however, more medical experiments are required for its application in medical field.

References

[1] Yang G, Zhang L. Regenerated cellulose microporous membranes by mixing cellulose cuoxam with a water soluble polymer. Journal of Membrane Science, 1996, 114(2): 149-155.

[2] Zhang Y P. Preparation and Formation Mechanism of Cellulose Membranes Prepared from NMMO Solution. 2002, Donghua University.

[3] Kalia S, Kaith B S, Vashistha S. Handbook of Bioplastics and Biocomposites Engineering Applications. John Wiley & Sons, 2011.

[4] Yang Z G. Process and Principle of Membrane Science and Technology. East China University of Science and Technology Press, 2009.

[5] Lin S. Preparation of Antibacterial Cellulose Membranes and Their Application in The Depth Treatment of Water. Fujian Agriculture and Forestry University, 2013.

[6] Lu P, Hsieh Y L. Preparation and properties of cellulose nanocrystals: rods, spheres and network. Carbohydrate Polymers, 2010, 82(2): 329-336.

[7] Ye D Y. Advances in cellulose chemistry. Journal of Chemical Industry and Engineering, 2006, 57(8): 1782-1791.

[8] Wang N, Ding E Y, Cheng R S. Nano microcrystalline cellulose surface modification research. Acta Polymerica Sinica, 2006, 1(8): 982-987.

[9] Wang C. Organic Nano Functional Materials: High Voltage Electrostatic Spinning Technology and Nano Fiber. Science Press, 2011.

[10] Wan H J. The current research status of cellulose electrostatic spinning. Modern Textile Technology, 2009, 17(5): 57-60.

[11] Zhang L. The Preparation of Inorganic/Organic Hybrid Antibacterial Membrane and Characterization. Tianjin Polytechnic University, 2011.

[12] Fong H, Chun I, Reneker D H. Beaded nanofibers formed during electro-spinning. Polymer, 1999, 40(16): 4585-4592.

[13] Wang J P. The latest progress of antibacterial fiber. Technical Textiles, 1998, 16(11): 6-10.

[14] Wang J P. The latest progress of antibacterial fiber. Knitting Industries, 2000, (5): 27-29.

[15] Qin Y M. Copper base performance and application of antibacterial fiber. Journal of Textile Research, 2009, (12): 134-136.

[16] Shang C J. Antibacterial and Anti Mite Textile Finishing. China Textile Press, 2009.

[17] Zhu P. Function Fibers and Textiles. China Textile Press, 2006.

[18] Wang J P. Antibacterial fiber with antibacterial agent system. Synthetic Fiber, 2003, 32(2): 10-14.

[19] Reneker D H, Chun I. Nanometre diameter fibres of polymer, produced by electro-spinning. Nanotechnology, 1996, 7(3): 0957-4484.

[20] Wang J P. Anti-bacterial fiber and anti-bacterial agent system(Ⅱ). Synthetic Fiber in China, 2003, 32(3): 5-9.

[21] Chen W X, Shen Z Q, Liu G F, et al. Coordination structure and and antibacterial activity of copper(Ⅱ) complex fibers. Acta Polymerica Sinica, 1998, 1(4):431-437.

[22] Qin Y M. Properties and applications of Cupron copper containing antimicrobial fibers. Journal of Textile Research, 2009, 30(12): 3.

[23] Chen W X, Shen Z Q, Liu G F, et al. Coordination structure and higher order structure of Cu(Ⅱ)-silk fibro in protein(Bombyx Mori) complexes. Chemical Journal in Chinese Universities, 2000, 21(2):309-310.

[24] Yu Z C, Chen W X, Lin R G. Study on the preparation, structure and the properties of the deodorizing and antibacterial cellulose fiber. Journal of Functional Polymers, 2002, 15(4): 5.

[25] Mamat H. Preparation and stability of cellulose LiCl/DMAc solution. Journal of Textile Research, 2010, 31(8): 6-11.

[26] Chen W X. Copper (Ⅱ) chelating ligand structure and antimicrobial properties of the fiber. Acta Polymerica Sinica, 1998, 1(4): 431-437.

[27] Wang J, Ge J, Xu h. The preparation of copper Anti-bacterial cotton fiber and its antibacterial properties. Journal of Environment and Health, 2007, 24(2): 4.

[28] Qin Z Y, Chen Y Y. Research on antibacterial activity and reaction conditions of Cu (Ⅱ) complex bamboo pulp fabrics. Journal of Beijing Institute of Clothing Technology(Natural Science Edition), 2011, 31(2).

[29] Qin Z Y. The preparation and UV-blocking performance of Cu (Ⅱ) complex bamboo pulp fabrics. Journal of Suzhou University of Science and Technology (Natural Science Edition), 2011.

[30] Gao C P. Textile antibacterial performance test methods and standards. Textile Dyeing and Finishing Journal, 2007, 29(2): 38-42.

[31] Zhao X W. Testing standards for antibacterial activity of textiles. Textile Printing, 2013, (15): 36-39.

[32] Zhao X, Zhan Y Z. Performance testing methods of antibacterial textiles. Shanghai MAO Hemp Technology, 2009, 1: 31-36.

[33] Yang S H. Chemistry of Plant Fiber. China Light Industry Press, 2001.

[34] Rong C. Research of The New Type of Cellulose Solvent System. Qingdao University, 2011.

[35] Ahn Y, Hu D H, Hong J H, et al. Effect of co-solvent on the spinnability and properties of electrospun cellulose nanofiber. Carbohydrate Polymers, 2012, 89(2): 340-345.

[36] Li L, Zhao S, Hu H Q. Development in solvent system of cellulose. Journal of Cellulose Science and Technology, 2009, 17(2): 69-75.

[37] Yan L, Gao Z. Dissolving of cellulose in PEG/NaOH aqueous solution. Cellulose, 2008, 15(6): 789-796.

[38] Hao Y P. Properties and Structure of Biaxial Cellulose Films from NMMO Solution. Donghua University, 2005.

[39] Yin Y K. Activating or solublizing process of cellulose. Journal of Cellulose Science and Technology, 2004, 12(2): 54-63.

[40] Mccormick C L, Callais P A. Solution studies of cellulose in lithium chloride and N, N-dimethylacetamide. Macromolecules, 1985, 27(2): 91-92.

[41] Mccormick C L, Callais P A. Derivatization of cellulose in lithium chloride and dimethylacetamide solutions. Polymer, 1987, 28(13): 2317-2323.

[42] Wang Y L, Cheng B W, Zhao J S. Cellulose activation. Journal of Tianjin Polytechnic University, 2002, 21(2): 83-86.

[43] Raus V, et al. Activation of cellulose by dioxane for dissolution in N, N-dimethylacetamide / LiCl. Cellulose, 2012, 19(6): 1893-1906.

[44] Potthast A, et al. The cellulose solvent system N, N-dimethylacetamide/lithium chloride revisited: The effect of water on physicochemical properties and chemical stability. Cellulose, 2002, 9(1): 41-53.

[45] Jerosch J, Lavédrine B, Cherton J C. Study of the stability of cellulose-holocellulose solutions in N, N-dimethylacetamide-lithium chloride by size exclusion chromatography. Journal of Chromatography A, 2001, 97(1-2): 31-38.

[46] Zhou X D. Effect of cellulose activation on its dissolvability. Advanced Textile Technology, 2008, 16(5): 4-7.

[47] Wang X L, Fang G Z. Influence on structure and oxidation reactivity of microcrystalline cellulose by different activation methods. Chemistry and Industry of Forest Products, 2007, 27(3): 67-71.

[48] Luo H, et al. Achieving shape memory: Reversible behaviors of cellulose-PU blends in wet dry cycles. Journal of Applied Polymer Science, 2012, 125(1): 657-665.

[49] Zhu Y, et al. Rapidly switchable water-sensitive shape memory cellulose/elastomer nano-composites. Soft Matter, 2012, 8(8): 2509-2517.

[50] Luo H. Study on Stimulus-responsive Cellulose-based Polymeric Materials. The Hong Kong Polytechnic University, 2012.

[51] Chanzy H, Maia E, Perez S. Cellulose organic solvents III. The structure of the N-methylmorpholine N-oxide-trans-1, 2-cyclohexanediol complex. Acta Crystallographica Section B: Structural Crystallography and Crystal Chemistry, 1982, 38(3): 852-855.

[52] Gu G X, Shen Y Y. Dissolution methods of cellulose in NMMO solvent and quality evaluation of the solution. Journal of Donghua University(JCR Science Edition), 2001, 27(5): 127-131.

[53] Liu H R. The development in cellulose solvent systems. Materials Review, 2011, 25(7): 135-139.

[54] Jaeger R. Electro-spinning of Ultra-thin Polymer Fibers. Wiley Online Library, 1998.

[55] Ma Z, Kotaki M, Ramakrishna S. Electrospun cellulose nanofiber as affinity membrane. Journal of Membrane Science, 2005, 265(1): 115-123.

[56] Uppal R, Ramaswamy G N. Cellulose submicron fibers. Journal of Engineered Fabrics & Fibers, 2011, 6(4):39-45.

[57] Dupont A L. Cellulose in lithium chloride/dimethylacetamide, optimisation of a dissolution method using paper substrates and stability of the solutions. Polymer, 2003, 44(15): 4117-4126.

[58] Kim C W, et al. Structural studies of electrospun cellulose nanofibers. Polymer, 2006, 47(14):

5097-5107.

［59］Wan H J. Study on Preparation and Properties of Electrospun Fibers Film from Naturally Colored Cotton. Zhejiang Sci-Tech University, 2010.

［60］Mamat H, et al. Preparation of cellulosic film by LiCl/DMAc process. Journal of Textile Research, 2011, 32(4): 33-38.

［61］Li Z. Research and application on cellulose /LiCl/DMAC solution system. Polymer Bulletin, 2010, (10): 53-59.

［62］Liu B B. Bacterial Cellulose Dissolving and Electrostatic Spinning Technology Research. Nanjing University of Science and Technology, 2012.

［63］Wei J L. High Flux Composite Ultra/nanofiltration Membrane Based on Polyacrylonitrile(PAN) Nanofibrous Substrates and Cellulose Acetate (CA) Hydrophilic Coatings. Donghua University, 2010.

［64］Qin X H, et al. Effect of LiCl on electro-spinning of PAN polymer solution: Theoretical analysis and experimental verification. Polymer, 2004, 45(18): 6409-6413.

［65］Gu S Y, J Ren, Wu Q L. Preparation and structures of electrospun PAN nanofibers as a precursor of carbon nanofibers. Synthetic Metals, 2005, 155(1): 157-161.

［66］He C, Pang F, Wang Q. Properties of cellulose/PAN blend membrane. Journal of Applied Polymer Science, 2002, 83(14): 3105-3111.

［67］Zhang R L. The Preparation and Research of Fiber Structure and Performance of Blend Spinning Polyacrylonitrile Fiber/cellulose/ionic Liquid Solution. Tianjin Polytechnic University, 2011.

［68］Xu G M. Nanometer Fiber Wikipedia Three Cellulose Acetate (CTA) Composite Membrane research. Donghua University, 2012.

［69］Zhang H, et al. Crosslinking of electrospun polyacrylonitrile/hydroxyethyl cellulose composite nanofibers. Materials Letters, 2009, 63(13): 1199-1202.

［70］Pan M L. Reed Cellulose Membrane Preparation and Its Performance Study. Tianjin University of Technology, 2013.

［71］Legnani C, et al. Bacterial cellulose membrane as flexible substrate for organic light emitting devices. Thin Solid Films, 2008, 517(3): 1016-1020.

［72］Jie X M. Preparation and separation performance of cellulose membrane from direct dissolution. Journal of Chemical Industry and Engineering (China), 2006, 57(8): 1756-1762.

［73］Levine M, Burroughs R H. Factors influencing the role of additives during the regeneration of cellulose from viscose solutions. Journal of Applied Polymer Science, 1959, 2(5): 192-197.

［74］Liu S, Zhang Li. Effects of polymer concentration and coagulation temperature on the properties of regenerated cellulose films prepared from LiOH/urea solution. Cellulose, 2009, 16(2): 189-198.

［75］Yang G, Zhang L, Feng H. Role of polyethylene glycol in formation and structure of regenerated cellulose microporous membrane. Journal of Membrane Science, 1999, 161(1-2): 31-40.

［76］Duan J J. Preparation and Characterization of Natural Macromolecule New Material. Xiamen University, 2012.

［77］Fink H P, et al. Structure formation of regenerated cellulose materials from NMMO-solutions. Progress in Polymer Science, 2001, 26(9): 1473-1524.

［78］Fink H P, P Weigel, Bohn A. Supermolecular structure and orientation of blown cellulosic films. Journal of Macromolecular Science—Physics, 1999, 38(5-6): 603-613.

［79］Xu M. Inorganic Nanomaterials Improve Cellulose Film Performance Study. Tianjin University of

Science and Technology, 2010.

[80] Zhang Y J. Preparation and Performance Study of The Farm Cellulose Film. Jiangnan University, 2012.

[81] Potthast A, et al. Degradation of cellulosic materials by heating in DMAc/LiCl. Tetrahedron Letters, 2002, 43(43): 7757-7759.

[82] Saxena S D, Gupta K S. Kinetics and mechanism of the oxidation of dimethylformamide by aquothallium (III) in perchloric acid solutions. Journal of Inorganic and Nuclear Chemistry, 1977, 39(2): 329-331.

[83] Zhu Y P. Study on crystallinity and degree of orientation of naturally colored cottons. Journal of Donghua University (Natural Science), 2009, 35(6): 626-631.

[84] Ma B M. Regenerated bamboo cellulose film prepared with Ionic liquid as solvent. Journal of Donghua University (Natural Science), 2011, 36(6): 604-607.

[85] Zhang W, Chen C J, Fu S S. Preparation and properties of cellulose/NMMO solution and membrane. China Synthetic Fiber Industry, 2011, 1: 10.

[86] Tong X T. Cellulose film preparation and properties research. Polymer Bulletin, 2013(10): 151-155.

[87] Wu J. Preparation, Characterization and Application on A Novel Separation Membrane-α-cellulose Membrane. Dalian Institute of Chemical Physics, Chinese Academy of Sciences, 2002.

[88] Dawsey T R, Mccormick C L. The lithium chloride/dimethylacetamide solvent for cellulose: A literature review. Journal of Macromolecular Science-Reviews in Macromolecular Chemistry and Physics, 1990, 30(3-4): 405-440.

[89] Rosenau T, et al. Hydrolytic processes and condensation reactions in the cellulose solvent system N, N-dimethylacetamide/lithium chloride. Part 1. Holzforschung, 2001, 55(6): 661-666.

[90] Potthast A. Hydrolytic processes and condensation reactions in the cellulose solvent system dimethylacetamide/lithium chloride. Part 2: Degradation of cellulose. Polymer, 2003, 44(1): 7-17.

[91] Chen W J. The preparation and characterization of electrospun polysulfonamide nano-scale fibers. Technical Textiles, 2013, 31(7): 7-15.

[92] Deitzel J M, Kleinmeyer J, Harris D, et al. The effect of processing variables on the morphology of electrospun nanofibers and textiles. Polymer, 2001, 42(1):261-272.

[93] Chen Z M. Preparation and characterization of PAN/CA composite nanofibers membrane. New Chemical Materials, 2011, 39(3): 76-78.

[94] Xie J J. Preparation and Evaluation of Antibacterial Bacterial Cellulose Membranes Impregnated Silver Nanoparticles. Donghua university, 2012.

[95] Liu J. Application of biological bandage and dressing in wound healing. Journal of Clinical Rehabilitative Tissue Engineering Research, 2011, 15(3): 535-538.

[96] Xu X L. Study on The Preparation and Environmental Impact of New Wound Dressing-gelatin-based Antibacterial Nanofiber Hydrogels. Donghua University, 2009.

[97] Dai H L. Preparation of Superabsorbent Medical Cotton Cloth and Performance Research. Donghua University, 2014.

[98] Liang Y Y, Wu Z. Study on the antibacterial ability of chelating fiber with silver element. Journal of Beijing Institute of Clothing Technology, 2007, 27(4): 19-24.

[99] Zhang N. The preparation of silver antibacterial fiber and its performance measurement. Journal of Beijing Institute of Clothing Technology, 2008,28(4):52-57.

[100] Zhang X Y. Cellulose in Copper Ammonia Solution and The Dissolution of Copper Ethylenediamine Solution and Regeneration. Qingdao University, 2012.